PROCESS IMPROVEMENT IN THE ELECTRONICS INDUSTRY

WILEY SERIES IN SYSTEMS ENGINEERING

Andrew P. Sage

PROCESS IMPROVEMENT IN THE ELECTRONICS INDUSTRY

YEFIM FASSER
DONALD BRETTNER
Advanced Micro Devices, Inc.
Sunnyvale, California

A Wiley-Interscience Publication
JOHN WILEY & SONS, INC.
New York / Chichester / Brisbane / Toronto / Singapore

In recognition of the importance of preserving what has been
written, it is a policy of John Wiley & Sons, Inc., to have books
of enduring value published in the United Stated printed on
acid-free paper, and we exert our best efforts to that end.

Library of Congress Cataloging in Publication Data:
Fasser, Yefim, 1931–
    Process improvement in the electronics industry/Yefim Fasser,
Donald Brettner.

        p. cm.—(Wiley series in systems engineering)
    "A Wiley-Interscience publication."
    Includes bibliographical references.
    ISBN 0-471-53638-5
    1. Electronics industries. 2. Manufacturing processes. 3. Process
    control. I. Brettner, Donald. II. Title. III. Series.

TK7836.F37 1992
621.381—dc20   91-23312
                                                          CIP

Printed and bound in the United States of America by Braun-Brumfield, Inc.

10 9 8 7 6 5 4 3 2 1

To Ida, Elina, and Natasha
Y.F.

To Amy and Nicholas
D.B.

Contents

Preface

This book describes a systematic approach to continuous process improvement through process capability studies, on-line and off-line design of experiments, and statistical process control. This approach has been introduced in the factories of Advanced Micro Devices, Inc. (AMD) and is proving to work successfully.

The development of the principles of continuous process improvement is mainly based on Shewhart, Deming, Juran, and Taguchi's philosophies, and on the application of the theories and methodology developed by some famous authors, such as G. E. Box, A. J. Duncan, E. L. Grant, A. V. Feigenbaum, and others. To develop a systematic approach, AMD also used the experience of Western Electric, IBM, General Motors, Motorola, and other companies that have achieved success by applying statistical principles in their factories.

The book is illustrated with a large number of actual case histories and examples from AMD factories. The authors will have a great sense of satisfaction if the information contained in this book can be used by other companies that are in the process of introducing statistical principles for quality improvement.

Acknowledgments

In the process of preparing the manuscript for this book, a number of people assisted by submitting examples from the plants and giving suggestions. We would like to express our appreciation to C. L. Choo, Tina Chow, Melissa Lee, Debra Polmanteer, Penny Ong, Indra Tukimin, Chris Wang, Eric Wee, Gabriel Wong, C. H. Yew, and to all the people who participated in the implementation of the continuous process improvement system.

Introduction

If you had the opportunity to compare the records of two manufacturing meetings conducted in the same place but at different periods of time, let's say an interval of 5 to 10 years, you would see a big change in the vocabulary used at these meetings. In previous years, the most frequently used words and terms were: make, inspect, ship, yield, return, retest, rescreen, LTPD, AQL, MRB, etc.

This reflected a philosophy of "make-inspect-ship." And if the product were returned, it was rescreened and shipped again—to the same or to another customer, in the hope that this time the AQL would work on the supplier's side. There was little, if any, attention paid to the process where the product was being produced. The main focus of attention was the finished product.

Today, at the same type of meeting, you would hear words (and terms) that were never heard before, for example, process control, Cpk, cycle time, lead time, ppm, precision, stability, sigma, SPC, spread, process average, process characterization, specification design, customer study, feedback, etc. This change in the vocabulary reflects the change in the philosophy of running a business. Now the attention is more concentrated on the process and product design, on process capability studies and control, on total quality improvement. This means not only making a high-quality product, but also shipping it out at the right time, having the shortest manufacturing cycle time, and other components related to total customer satisfaction. All this should be done in such a way that the customer will be motivated to buy the product for a reasonable price and the producer will also gain a reasonable return on investment.

This can be achieved most effectively when we have an optimal process for running a business. Today, this means not just having a process that is capable of manufacturing products to the customers' specification, but a process with a much tighter spread than the allowable customer specification spread. Today's target is to have processes whose variation consumes 50% or less of the customer specification range.

This type of process makes such techniques as AQL and LTPD obsolete. Even SPC will be needed less when the process spread is reduced to an absolute minimum. All the objectives mentioned above dictate the need for the development and the introduction of a system that will accommodate all the needs for continuous process improvement in all areas of an organization. This book is an attempt to share some of the experience of continuous process improvement accumulated in the Advanced Micro Devices factories.

PART I

A Systematic Approach to Process Improvement

Continuous process improvement requires a systematic approach to managing all activities related to this subject. In this part of the book we will first introduce AMD's general direction in process improvement based on Dr. Deming's 14 points and the concept of the cybernetic system.

In a separate chapter the reader will find a description of the organizational structure and what has been done to prepare our statistical leaders to fulfill the needs of the statistical staff.

AMD has developed a system of process improvement that consists of 10 major subsystems (blocks). Together these embrace all elements of continuous process improvement. The reader will find a short description of every block. Special attention is given to the "educational" block, because without educating the employees all the other blocks cannot function. A description is given of different forms of education, which includes the establishment of a special educational institution, the Sigma College. This institution is formed to give the employees continuous education. The reader may be interested in the introduction of executive audits, which can control the process improvement activities, not only by formal written reports, but by meeting with people on a regular basis. Although every organization is different and has its own specific needs, the information in this part of the book may be helpful for those who would like to introduce a systematic approach of process improvement.

Chapter **1**

General Directions for Process Improvement

1.1 DR. DEMING'S 14 POINTS FOR CONTINUOUS PROCESS IMPROVEMENT

The popularity of Dr. Deming's 14 points[1] (see Table 1.1) is very high and it continues to grow because in the last decade more and more companies have started to work on continuous process improvement, which reflects Deming's philosophy. If considering all the business activities of an organization as a macroprocess, the 14 points can be applied as a direction for continuous process improvement.

In this book the reader will become familiar with some techniques that allow us to materialize the concepts contained in the 14 points. For example, reducing the variation around a calculated target that is based on the customer's requirements is a direction for continuous improvement reflected in the 14 points. Improving the manufacturing process by design of experiments as a way of building quality into the product and eliminating the dependence on inspection reflects Point 3 of the 14 points where it says, "Cease dependence on inspection to acquire quality."

Going through the chapters of this book, the reader will find the answer of how to introduce Deming's 14 points, because the book has been written based on Deming's philosophy, which is reflected in his 14 points.

TABLE 1.1 Condensation of Deming's 14 Points for Management

1. Create constancy of purpose toward improvement of product and service, with the aim to become competitive and to stay in business, and to provide jobs.
2. Adopt the new philosophy. We are in a new economic age. Western management must awaken to the challenge, must learn their responsibilities, and take on leadership for change.
3. Cease dependence on inspection to achieve quality. Eliminate the need for inspection on a mass basis by building quality into the product in the first place.
4. End the practice of awarding business on the basis of price tag. Instead, minimize total cost. Move toward single supplier for any one item, on a long-term relationship of loyalty and trust.
5. Improve constantly and forever the system of production and service, to improve quality and productivity, and thus constantly decrease costs.
6. Institute training on the job.
7. Institute leadership. The aim of supervision should be to help people and machines and gadgets to do a better job. Supervision of management is in need of overhaul, as well as supervision of production workers.
8. Drive out fear, so that everyone may work effectively for the company.
9. Break down barriers between departments. People in research, design, sales, and production must work as a team, to foresee problems of production and in use that may be encountered with the product or service.
10. Eliminate slogans, exhortations, and targets for the workforce asking for zero defects and new levels of productivity. Such exhortations only create adversarial relationships, as the bulk of the causes of low quality and low productivity belong to the system and thus lie beyond the power of the workforce.
11. a. Eliminate work standards (quotas) on the factory floor. Substitute leadership.
    b. Eliminate management by objective. Eliminate management by numbers, numerical goals. Substitute leadership.
12. a. Remove barriers that rob the hourly worker of his/her right to pride to workmanship. The responsibility of supervisors must be changed from sheer numbers to quality.
    b. Remove barriers that rob people in management and in engineering of their right to pride of workmanship. This means, *inter alia*, abolishment of the annual or merit rating and of management by objective.
13. Institute a vigorous program of education and self-improvement.
14. Put everybody in the company to work to accomplish the transformation. The transformation is everybody's job.

1.2 SHEWHART'S CYCLE AS A REFLECTION OF DEMING'S PHILOSOPHY

Put everybody in the company to work to
accomplish the transformation. The trans-
formation is everybody's job.
—Point 14 from Deming's 14 points

Describing how to introduce Point 14, Dr. Deming recommends using Shewhart's cycle[2] as a procedure. The reader is probably familiar with the PDCA cycle, which means "plan, do, check, act" (see Figure 1.1). This is called *Shewhart's cycle*, which was first introduced in Japan by Dr. Deming in the early 1950s. The Japanese call it "Deming's cycle."

This short abbreviation "PDCA" contains a philosophy that, in our opinion, should be accepted as the rule of action by any small or large organization, group of people, or individuals. Let's try to interpret the PDCA cycle in relation to continuous process improvement.

1.2.1 Step 1—Plan

At this stage, we need to determine what we want to achieve by changing the process, who are the members of the team involved in the project, when we are planning to start and finish the project, what we need to accomplish the objective, what data are already available and what needs to be collected, in what format we want the data to be collected, what statistical tools will be used to organize and interpret the results, etc. It sounds like an ordinary procedure. But think how many times we "jump" into the same activities

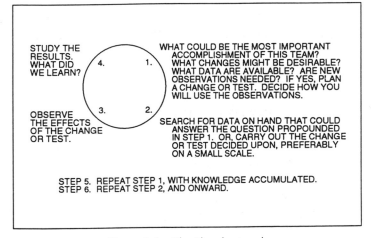

Figure 1.1   The Shewhart cycle.

without having a clear plan as to why and how we want to do it, and what we really expect from the effort.

1.2.2 Step 2—Do

Now we perform the activities according to the plan. If the project of process improvement involves the introduction of new ideas and/or principles, it is important to conduct the experiment on a small scale. How many times in our experience have we taken a "half baked" idea and tried to implement it on a broad scale, only to find that the idea would not work or that we were not prepared for it. The result is lots of frustration and loss of confidence. Only by trying the idea on a small scale, polishing it, and then introducing it on a larger scale, can we achieve progress with a small amount of investment.

1.2.3 Step 3—Observe the Effect (Check)

When the project is completed, it is important to confirm the results. For example, with a design of experiment it doesn't matter how properly we perform it, it still needs to be checked by a confirmation run.

1.2.4 Step 4—Study the Results (Act)

Any time we achieve results (good or bad) we need to study them and act accordingly. This means that if the results are positive, we need to find out exactly how we achieved this improvement so we can repeat the results and enjoy the benefit from the improvement. If the results are negative, we also need to know why, so we will never repeat the same mistakes again. This is called learning by your mistakes. Finally, we need to plan to introduce the good results on a larger scale and check these results. As you can see, we start the cycle again.

Through the years, Shewhart's cycle has been further developed and introduced by Dr. Deming in Japan, the United States, and other countries. It reflects Deming's philosophy of continuous process improvement, which is described in his 14 points.

1.3 CONTROL AS A WAY OF ACHIEVING THE GOAL

What is the commonality between an operator, technician, engineer, designer, manager, and the president of a corporation? The operator works at the workplace to fulfill an operation, the designer works on a project to develop a new product, the president manages people and directs them to achieve goals and targets. All these people perform different functions on different levels, but they have one important commonality: Using words from cybernetics, we can say that all of them are "control devices" whose actions are directed toward achieving specific objectives. Essentially, they organize an object and bring it closer to perfection.

For example, the operator's aim is to keep the process in a state of statistical control and produce parts to the specification. For the president, the aim is to keep the corporation in business, creating jobs for people and making profits and a return on investment for the corporation and its shareholders. We would have no difficulty describing the objectives of control devices such as an engineer, designer, accountant, secretary, etc.

But to succeed in control it is not enough just to know the objective. We also need to know how it can be achieved. We have to be able to influence the object under control in such a way that our plans can be fulfilled. For example, the operator needs to know how to influence the manufacturing process when the product is outside of the specification limits; the president needs to know how to act when the customer order rates are falling, etc. And most of the time this is more difficult to do than defining the objectives.

This brings us to a consideration of one of the most fundamental concepts of modern cybernetics, namely the *control algorithm*. A control algorithm is a method of achieving the stated goal or objective—a rule for action. The control algorithm for the operator, for example, is to stop (or adjust) the process when a point falls outside the control limits. For some automated operations, the control operation is very simple. The process can be corrected just by pushing a button. Even for a pilot who controls a complicated aircraft, the control algorithm is sometimes very easy. The pilot just pushes one of the multiple buttons on the control panel. In the worst case, the pilot pushes the "eject" button to leave the aircraft in an emergency situation. But how about the control algorithm of an executive? What button should the executive push when he/she is losing the competitive edge?

So far we have described the commonality of different "control devices" and the differences in the complexity of the control algorithms. Despite those differences we can consider a general control system independently of the specific peculiarities of controlling a particular object.

Figure 1.2 shows the block diagram for a general control system. Here the interaction of the object with the control device is indicated by two arrows, A and B, which represent the communication channels between the object and the control device. The "object" can be a simple process operation or an overall manufacturing process; it can be a corporation or even an entire country. The "control device" can be an operator, a manager, a president, or any other device whose aim is to control. The control device receives information about the object via Channel B and then acts upon the object via Channel A, and this *controls* it.

We would like to emphasize here the importance of Channel B. This is the feedback from the object under control. Without this element the control system doesn't work. How can you control anything if you don't know what the object needs? Can we say that we have constant feedback (Channel B) from our object of control? The reader should not have difficulty answering

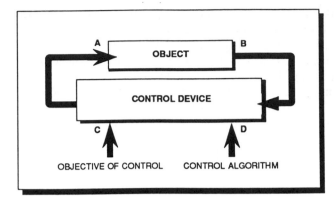

Figure 1.2 The block diagram for a general control system.

this question, having in mind a particular example. Speaking in general, we should say that SPC is an example of how a process (object) can give feedback to the control device (operator, engineer, supervisor).

Channel A has the same level of importance. You can have constant feedback (through Channel B) from the object, but if you don't know how or what to do, you cannot do anything to the object when it asks for it, and again control is impossible. Do we always react to the signals that are coming through Channel B?

The picture of a modern system of control would not be complete if we did not introduce two more elements to the system. To control effectively, we need to know *what* to do with the information received and *how* to use it to control the object. For this purpose, we introduce the following inputs to the control device: objective (or aim) of the control process (see C on Figure 1.2) and the algorithm (or method) of control (see D on Figure 1.2). This information (data) has to be fed into the control device beforehand. So, if the control system is to impose the required order on the object, it must contain two essential elements:

1. The objective of control.
2. The control algorithm showing how the objective is to be achieved.

In simpler terms, we need to know (1) where we want to go and (2) how to get there. We described a control system that is varied for any controlled object. The purpose of describing this general system is to propose a way of taking a second look at our control activities from a cybernetic point of view. In addition, if we find that the control system we are involved with is short of just one of the four elements reflected in Figure 1.2, this system will not work.

In this book the reader will find some techniques to help develop an objective of control, determine the control algorithms, and establish feedback with the object of control.

REFERENCES

1. W. Edwards Deming, "Out of the Crisis" (1982), Cambridge University Press, pp. 23–24.
2. W. Edwards Deming, "Out of the Crisis" (1982), Cambridge University Press, p. 88.

Chapter **2**

The Total Continuous Process Improvement (TCPI) Organizational Structure

---

Create a structure in top management that
will push every day on the above 13 points.
—Dr. Deming[1]

2.1 TOTAL CONTINUOUS PROCESS IMPROVEMENT (TCPI)

In the last 10 years SPC has again become a modern approach to process improvement. But in parallel with SPC, a number of other approaches have been applied for process improvement; e.g., total cycle time (TCT), just in time (JIT), total quality control (TQC), customer/supplier relations (CSR), etc. These directions are related to the process improvement program, and most of them are based on the application of statistical principles. So, to consolidate all efforts under one umbrella, some of the AMD divisions combined all activities related to process improvement under one organizational structure called *total continuous process improvement* (TCPI). Please note that the term "total" is to emphasize that process improvement relates to all manufacturing and nonmanufacturing organizations of the company, including service and management.

2.2 AN EXAMPLE OF TCPI ORGANIZATIONAL STRUCTURE

Figure 2.1 is an example of a TCPI structure at AMD's Manufacturing Services Division (MSD). As one can see from this structure, the MSD receives general directions from the corporate TCPI Steering Committee that are further developed by the MSD-TCPI board and materialized by the division plants (factories). The vice-president of the division is the chairman of the TCPI board and he conducts meetings on a regular basis. The agenda of the meetings and the minutes are distributed to all the plants so they can use them as information to conduct their local TCPI committees. The organizational structure also includes the Methodological Center which is

10

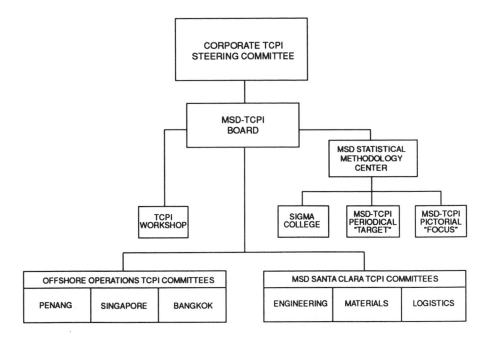

The **MSD-TCPI BOARD** includes members from all major areas and departments of the division, and meets according to the planned schedule to:

- Develop new directions.
- Revise the status report.
- Approve programs, policies, and standards.
- Discuss projects.
- Determine and approve awards.
- Perform periodic TCPI audits.

Figure 2.1   The MSD TCPI organizational structure.

responsible for developing and delivering educational programs by conducting classes, seminars, conferences, and symposiums, and also through the Sigma College (see Section 4.3). The Methodological Center is also responsible for developing short- and long-term plans related to the SPC activities and conducting continuous consultation for the project teams.

The TCPI workshop is also an element of the divisional structure. Its function is to review individual programs and prepare recommendations to the board. The members of the workshop are experts in different fields, and their recommendations to the TCPI divisional board are a result of a collective experience.

Figure 2.2 is an example of the TCPI structure at the plant operations levels. The chairman of the board is the plant director who conducts all activities related to process improvement. The plant has a full-time coordinator who is the statistical expert and internal consultant. Every operation has

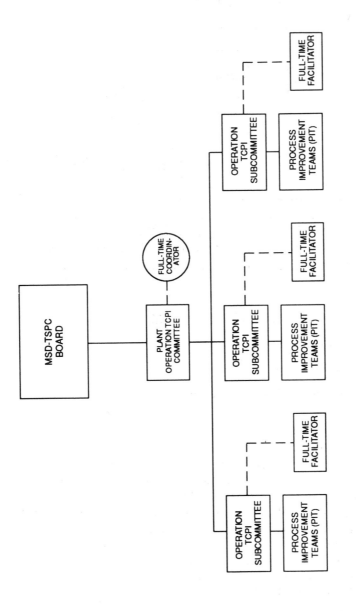

NOTES: 1. The managing director is the chairman of the TCPI committee .
       2. The operating manager is the leader of the TCPI subcommittee.
       3. The operation TCPI committee includes members from all major
          areas and departments of the plant.

Figure 2.2 The organizational structure of the MSD operations TCPI committee.

its own subcommittee, which is conducted by the operations manager. The full-time facilitator is responsible for the educational program and consultation on behalf of the factory personnel. He has a dotted-line responsibility to the plant coordinator.

One of the most important elements of the operations structure is the process improvement team (PIT). These teams include people who not only participate in the decision-making process and planning, but also materialize those decisions and plans.

All functions of the organizational structure are described in the divisional TCPI specification, which formalizes the authority and responsibilities of every person involved in process improvement activities.

2.3   DEFINING AND LOCATING AN "INDUSTRIAL STATISTICIAN"

What efforts have you made to discover
people with knowledge of statistical theory
right in your company, and given them a
chance to move into statistical work under
competent leadership?
—Dr. Deming[2]

When the management of a company decides to introduce statistical principles in its operations, the first question that usually comes up is, "Where do we find a good statistician?" We already know from management practices that when we have a problem with a machine we look in a catalog to find a new type of equipment or instrument, when we have a technological (or organizational) problem we might call a consulting firm, and when we need a qualified specialist we advertise in the newspaper or refer to an employment agency. This way of fulfilling manufacturing needs has become a management practice. But the position of a statistician, even after industry has for many years been applying statistical principles in a large number of organizations, still remains uncertain. What are the requirements for a statistician? What knowledge should the statistician have? Who is preparing statisticians for a particular field of application? How will the statistician fit in the organizational structure? All these questions came up at AMD when it was decided to introduce statistical principles in the factory. The first management step was, as usual, a request to the personnel department to search for a statistician; which meant advertisements in newspapers and magazines, calls to employment agencies, etc. There were some results from these activities, but now after getting some experience from defining the SPC organizational structure and staffing it with the right people, we can make some conclusions that may be interesting to the reader. Today, after 10 or more years of receiving the demands for statistical knowledge, a lot of companies have already developed their staff. So, if you are in luck, you can find a good statistician through the newspaper or agency. But what is a good statistician? A statistician-practitioner, to succeed in his/her position should, in our opinion, have knowledge and skills in approximately the following proportion:

Engineering knowledge in the field for which he/she was hired to work—40%.

Knowledge in applied statistics—30%.

Organizational and teaching skills—15%.

Teamwork and leadership skills—15%.

Where do you find a person with this combination of knowledge and skills? There is no definite answer to this question. You can hire a new graduate who has "pure" statistical education and help that person to develop all the other skills by on-the-job training. You also can select a person who has all the skills except statistical knowledge and find a way to help that person to fill the gap by obtaining a statistical education.

The analysis of seven years' work to build up an SPC staff at AMD has demonstrated that it is much easier, faster, and more efficient to prepare a statistician who already has all the skills except statistical, than to have a statistician who needs knowledge in engineering, management, organization, and other skills. In one of AMD's divisions, which was in the process of seeking the right candidate for a statistical position, a questionnaire was distributed. It was found that every third engineer or manager had received in the past some kind of statistical education by attending different courses more than twice. It was also found that a large number of people who had Masters degrees in statistics never used the knowledge in a real application and could be quite incapable of helping fellow employees. Based on the questionnaire, it became possible to select a group of candidates whose knowledge was close to the structure mentioned earlier and who were interested in a statistical position. Today AMD has a body of leaders who have sufficient knowledge and skills to conduct all activities relating to continuous process improvement.

2.4   GETTING THE COMMITMENT AND SUPPORT FROM MANAGEMENT

2.4.1   What Comes First?

To succeed in the introduction of statistical principles in an organization, you need management's commitment. But how can we expect commitment from a manager who doesn't know what he/she is committing to? When the manager stands in front of his/her subordinates and says, "Let's do it! I am committed!," but the employees realize the manager doesn't know what he/she is committing to, this commitment doesn't help a lot. So what comes first, the educational activities that require management's commitment, or educating management to obtain its commitment to support the same educational activities? It's like the story of the chicken and the egg; which came first?

A lot of companies start their statistical program not because they are committed to SPC, but because the customer requested it. Or, the reason for developing such programs was that the company leader saw articles on SPC in trade publications and, if IBM or Motorola did this, it makes sense to "give it a try." Usually it doesn't matter what strikes the spark to start the "fire" of SPC. The important thing here is that the company leader decided to "do it."

In this case, it is very important to find at least one person to be responsible for developing the SPC program. If the company finds the right candidate and supports that person, the company will succeed in introducing statistical principles.

2.4.2   60 Minutes SPC

Some companies started SPC activities by giving an intensive course to all their employees right away. From our experience, when you try to "sell" an idea to a large number of people at the same time, a very small percentage of them (if any) will "buy" the idea right away. At AMD we used the "snowball" approach, which means that we prepared a small group of enthusiastic people who volunteered to participate in the program. You can always find this kind of employee in any company; and it doesn't have to be the best or most knowledgeable ones. Sometimes it can be employees who are dissatisfied with what they are doing, so they might see SPC as a way to broaden their routine jobs. Or, it can be employees who feel that they are not skilled enough, and think that SPC can make them more valuable to the company. Whatever the reason, these people want to participate in forming the "nucleus" of the "snowball."

So a group of 25 employees were selected and given 20 hours of introductory courses in SPC. Then they developed small projects using real examples related to their jobs based on what they had learned (with permission from their supervisors, who agreed to "give it a try").

At that time an SPC workshop was established where these people could meet on a regular basis to discuss their first steps in SPC. When some good, impressive projects were completed, there came a need to "sell" them to the management. For this purpose, it was decided to conduct short sessions on a weekly basis by inviting different levels of management to hear the presentations of their people. At the same time, the nucleus of participants was growing from day to day. This is how we established an ongoing presentation and discussion group for SPC accomplishments called, "60 minutes SPC." Later, there came the time when the president and all his direct staff were invited to the "60 minutes SPC." That particular "60 minutes SPC" was longer than usual, but it was a real demonstration of the SPC potential. When in the meeting, the president said, "I am not surprised, I am delighted," we knew that SPC in AMD would work... and it does work. This is a short story of how the nucleus of the SPC snowball was formed.

Later a two-day SPC seminar for all AMD executives was held where the executives received an introductory course that included an overview of statistical process control (SPC), design of experiments (DOE), and some methods for applying statistical principles. In the program there was also an outline of how the AMD-SPC organizational structures would be developed, and what their role should be in this program. It is understandable that two days is a very limited time for such a program. So, it was decided that this should be an overview, just to give a feeling of what the program was all about. By discussing the SPC program with the executives, we found that most of them were already familiar with statistical principles, but this was a good forum for refreshing their knowledge. Still, the main intention of this two-day seminar was not so much to educate the executives as to illustrate the SPC philosophy, describe the importance of introducing it in the plants, and receive their commitment and support. And this was accomplished.

REFERENCES

1. W. Edwards Deming, "Quality, Productivity, and Competitive Position" (1982), Cambridge University Press, p. 17.
2. W. Edwards Deming, "Quality, Productivity, and Competitive Position" (1982), Cambridge University Press, p. 98.

Chapter 3

The General Block System of Total Continuous Process Improvement (TCPI)

Improve constantly and forever the system of production and service.
—Point 5 from Deming's 14 points[1]

Improvement should not be a one-time effort, it should be continuous. According to Dr. Deming, this means "continual reduction of waste and continual improvement of quality in every activity: procurement, transportation, engineering, methods, maintenance, locations of activities, instruments and measures, sales, methods of distribution, accounting, payroll, service to customers."[2] This means that continuous process improvement cannot be limited only to the known factors. Everyone and every department in the organization must be involved in the process of continuous improvement. To organize the process of continuous improvement, we need to apply a systematic approach that will allow us to organize the available resources and utilize the best available methods for continuous process improvement. Figure 3.1 shows the general block system of total continuous process improvement (TCPI). Below is a detailed description of each subsystem.

3.1 BLOCK 1—TOTAL CONTINUOUS EDUCATION (TCE)

Introduction of any system first requires education and training of the fundamentals related to the tactics and strategy on which the system to be introduced is based. The educational block (or subsystem) is designed in such a way that almost anyone who is on the company's payroll will receive on a continuous basis the necessary knowledge he/she needs to fulfill his/her job successfully.

Block 1 includes all elements related to the organizational structure of training and education: developing educational forms and programs, planning and control, continuous consultation, motivation and stimulation, etc. In

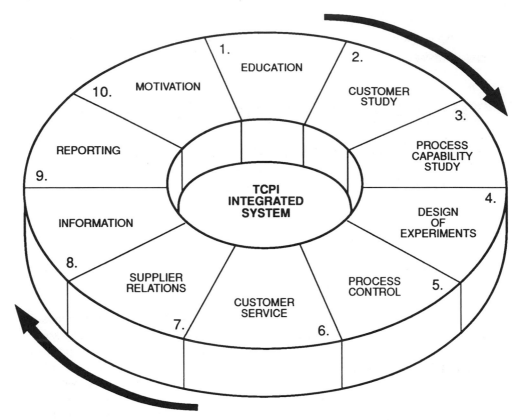

Figure 3.1 The architecture of the total continuous process improvement (TCPI) integrated block system.

a separate section we will describe other elements of the educational subsystem that we think may be interesting to the reader.

3.2 BLOCK 2—CUSTOMER STUDY

This block includes the following major elements:

1. Studying customers' requirements and translating those requirements into appropriate company requirements at every stage: design, materials, manufacturing, testing, support services, marketing, and sales.
2. Searching for the industry's best practices and comparing them with the company's practices in all aspects: planning, design, manufacturing, finished product, etc.

3. Developing long-term objectives and short-term targets based on studying customer requirements and the industry's best practices.
4. Developing the company strategy and tactics to achieve the short- and long-term objectives and targets.
5. Planning and scheduling of improvement activities, and then feedback analysis of accomplishments.

All these activities are based on a continuous process of benchmarking and application of the quality function deployment (QFD) concept as a systematic approach of customer study.

3.3 BLOCK 3—PROCESS CAPABILITY STUDY

Having translated customer requirements into operational needs, it is important to determine the capability of the process to see what should be done to make the company capable of meeting the set targets and objectives. Note here that "process capability study" by definition is not just a measurement activity. It is a way of improving the process, by bringing it into a state of statistical control, and then assessing its capability.

Long before introducing control charts on the manufacturing floor, it is important to perform capability studies to see if the process needs to be brought into control. And only when the process is stable does it make sense to apply charts for continuous control and transfer the ownership of the process to the operators. AMD's experience shows that placing control charts in the operations prematurely when the process is not in control confuses the operators and can discredit the SPC approach.

Block 3 includes the following major elements:

1. Determining the critical parameters and priorities where a process capability study should be performed.
2. Planning and performing manufacturing process capability studies.
3. Planning and performing instrument capability studies.
4. Planning and performing inspector capability studies.
5. Revising and developing operational specification requirements based on the capability studies and customer requirements.
6. Planning improvement for processes that are in a state of statistical control, but are not meeting the company's requirements (by applying the design of experiments methodology).

Process capability studies are performed on a continuous basis because the process elements (such as material, people, machines, methods, product, environment) are also continually changing.

3.4   BLOCK 4—DESIGN OF EXPERIMENTS

As we mentioned earlier, the process capability study is an important function because by bringing the process into control, we usually reduce the process variability and increase the process output. However, this type of improvement has limits that are related to the process potential. In other words, when a process is brought into a state of control the best you can expect from it is performance on its potential level. But what if this level is not good enough? Here we need to employ the design of experiment methodology to change the system of causes. This is the function of Block 4, which includes the following major elements:

1. On-line design of experiments, which includes evolutionary operations (EVOP), simplex-EVOP, Plex, and other methods.
2. Off-line design of experiments, which includes classical design of experiments and Taguchi methods. A combination of on- and off-line experiments allows us to have a greater opportunity to perform experiments by involving the manufacturing staff, and also to reduce the cycle time of research and development activities.
3. Metrology design of experiments, which are related with test and inspection systems to determine their impact on the product results.
4. Comparative experiments, which involve performing simple experiments when buying new equipment, qualifying suppliers, and other comparisons in the decision-making process.

All these activities allow us to work on continuous reduction of the process variability, determining methods for yield improvement and process optimization.

3.5   BLOCK 5—PROCESS CONTROL

This subsystem is the second largest block (after the block of continuous education) that involves a broad variety of people. Those involved are: the operators who control the process by using control charts, engineers and technicians who develop control charts and take corrective actions on the process if the operator needs help, and supervisors and managers who use the control chart information for decision making. Control chart techniques are also applied in the nonmanufacturing areas to control their processes. This block includes the following major elements:

1. Determining where and what type of control should be applied.
2. Introducing changes in the control chart parameters on a continuous basis (changing the control limits, process average, etc.).

3. Taking actions based on the control chart signals.
4. Making decisions about discontinuing or establishing new process charts.

The functions in this block are controlled by the process improvement teams (PIT) and control chart committees (C^3).

3.6 BLOCK 6—CUSTOMER SERVICE

This block includes activities related with the just-in-time approach (JIT), total cycle time (TCT), and customer complaints and returns (CCR). This block includes the following elements and activities:

1. Cycle time study
2. Line balancing
3. Optimal lot size determination
4. Inventory control
5. Return analysis and corrective actions
6. Downtime analysis

Experience shows that from the time this new block has been introduced into the system, a lot of progress is achieved in customer service.

3.7 BLOCK 7—SUPPLIER RELATIONS

It is impossible to achieve continuous process improvement if the suppliers will not work together with you to accomplish mutually beneficial goals. This block includes all activities related with improving the supplier's performance as an important element of the process. This subsystem includes the following elements:

1. Supplier selection and approval
2. Material analysis and approval
3. Supplier auditing
4. Technical and educational assistance
5. Direct-to-stock activities
6. Specification activities
7. On-time delivery control
8. Supplier comparison experiments

The general direction is to work with a minimum number of suppliers, provide them with long-term commitments, and strive for direct-to-stock performance.

3.8   BLOCK 8—INFORMATION

AMD manufacturing plants are located in different countries. To share the experience and results from process improvement, this subsystem includes the following elements and activities:

1. Periodic publication of *Target* and *Focus* magazines, which reflect the experience and results from process improvements in all locations.
2. Periodically conducting international conferences and symposiums where presentations on process improvement are made.

3.9   BLOCK 9—REPORTING

AMD developed a special form of reporting the process improvement activities, which includes:

1. Capability status report
2. Educational status report

The capability status report tracks how process Cpk's of less than 1 are eliminated, when Cpk's of 1.33 begin to dominate, and finally, how the percentage of Cpk's 1.5 or higher increases in the overall process capability.

3.10   BLOCK 10—MOTIVATION

To motivate employees for process improvement, a number of different incentives are developed and introduced; for example:

1. Awards for the "best projects of the year."
2. Awards for the "best conference presentation."
3. Awards for the "engineer of the year."
4. Awards for the "operator of the year."

The experience of having a separate activity that develops and introduces new forms of motivation and stimulation has demonstrated to us that this accelerates progress. It is important here to note that almost all the activities mentioned in the previous blocks also work as motivators. For example, the control chart motivates operators to do a better job because their improvement is visible. The presentation made by an engineer at a conference is a motivating element because of the visibility and recognition. An article published in one of the magazines is also a motivating factor. In other words, the process improvement activities by themselves are excellent elements of motivation. People get bored from doing a routine job day to day without a change. So all the systems of continuous process improvement are excellent motivators to make people work hard and learn from each other while satisfying the customer and thereby surviving in a very competitive business.

REFERENCES

1. W. Edwards Deming, "Out of the Crisis" (1982), Cambridge University Press, p. 23.
2. W. Edwards Deming, "Quality, Productivity, and Competitive Position" (1982), Cambridge University Press, pp. 30–31.

Chapter **4**

The Educational Process

Institute a vigorous program of education
and self-improvement.
  —Point 13 from Deming's 14 points[1]

Describing Point 13 in his 14 points, Dr. Deming said, "Education in simple but powerful statistical techniques is required of all people in management, all engineers and scientists, inspectors, quality control managers, management in the service organizations of the company, such as accounting, payroll, purchasing, safety, legal department, consumer service, consumer research. Engineers and scientists need rudiments of experimental design."[2]

In other words, Deming suggests that statistical education should be available to everyone in the organization, from the operators to the president. But how can this be done and who should do it? How do you provide an SPC education to a large number of people who have different functions, responsibilities, and backgrounds? And how can you make the "boring" subject of SPC interesting? Should it be mandatory or voluntary?

All these questions came up when we started the SPC program at AMD. And while there are a large number of organizations that are already familiar with the rudiments of statistical theory and applications, there are still some organizations that are in the stage of introducing statistical principles. So, for their benefit, we will describe AMD's experience of educating its employees in statistical principles.

4.1 LEVELS OF EDUCATION

Adults cannot be forced to learn. You can make them go to a class (if it is during work time), but you cannot control what they do there. At the same time, adults will demand an opportunity to obtain an education if this education will make the job easier and if they will feel more secure by obtaining new knowledge. What management can do is develop an atmosphere where people will look for an opportunity to learn. This kind of atmosphere can be developed as follows.

First of all, the educational program should be tailored as close as possible to the individual professional needs. Second, the opportunity to apply the new knowledge should be available. And third, the efforts of applying the new knowledge should be recognized in such a way that it is easy to see the difference between those who try hard to improve themselves and their jobs, and those who think they know enough already and believe no change in their performance is necessary.

For example, in one of the operators' interviews, which are held at AMD on a continuous basis as feedback to improve operator performance, a question was asked, "What is your opinion about teaching the operators machine setup and small repairs and maintenance?" All the operators

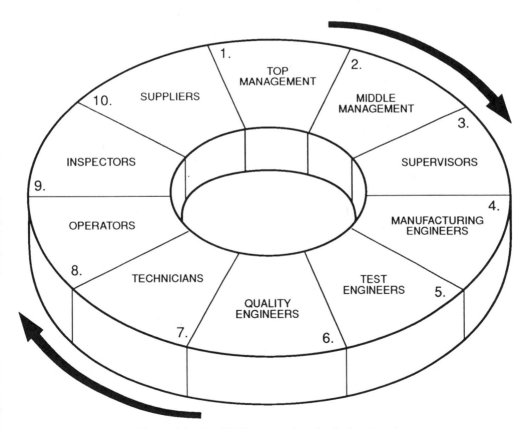

Figure 4.1 The TCPI system: Levels of education.

responded positively, but there were a number who commented, "I am in favor of this if I will get the opportunity to apply this knowledge." The operators were not asking for extra money, they were asking for "the opportunity to apply the knowledge," otherwise there is little motivation to learn. How many times have your subordinates attended different types of courses and come back to the plant and forgotten everything? The only reminder of the experience is the colorful binder received at the seminar, which now sits gathering dust on the bookshelf. This is why at AMD, after giving an overall introduction of the statistical principles, separate courses were developed for different levels. This permitted the opportunity to tailor the educational program to the individuals' needs. If an engineer needs to know more about design of experiments, a manager needs to know more about statistics related to the decision making process, or if an operator needs to know more about his role in process control, the technician needs to know techniques in statistical troubleshooting, etc. Figure 4.1 demonstrates the different levels of statistical education programs.

4.2 USING DIFFERENT FORMS OF EDUCATION

Developing the educational program at AMD, we were looking for different forms to make the education interesting, motivational, effective, and continuous. A number of different forms (see Figure 4.2) were chosen and developed. Together they make "boring" statistics more interesting. Let's briefly describe all the 10 forms reflected in Figure 4.2.

Classroom Education
This traditional form of education can be enhanced by inviting the employees to participate in the teaching process by bringing examples from the workplace and presenting these examples to the class. After the first day of instruction, the students were requested to do some homework and to bring the results to the attention of the class. What is important here are the mistakes that were made by the students in the first days of education. There is probably no better way to teach people than by demonstrating real mistakes.

A second powerful tool for making the classroom activity effective and interesting is simulating real situations directly in the class. For example, in giving a course of evolutionary operations (EVOP), the class is divided into three to four groups and asked to perform an EVOP study by simulating a real process. For this purpose, we would recommend using the EVOP game developed by Box.[3] While describing the process capability study for attribute and variable data, it is very effective to use the Quincunx* to demonstrate distribution inference, and to use the sampling bowls* to explain sampling

---

*Both the Quinconx and sampling bowls can be purchased from the Quantum Company, P.O. Box 769, Clifton Park, New York 12065, (518) 877-5236.

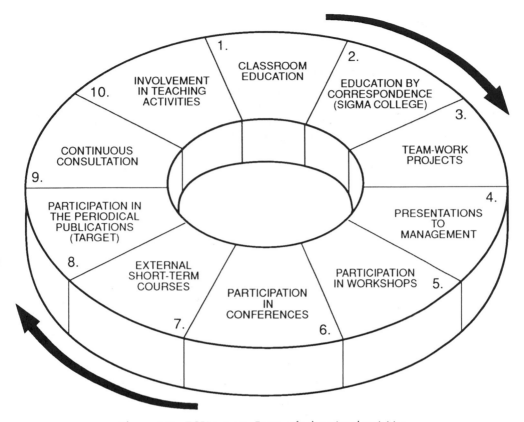

Figure 4.2   TCPI system: Forms of educational activities.

errors and process control principles. In delivering statistical knowledge to the employees, we would recommend not using the professor-student system. We would suggest making an atmosphere that promotes sharing knowledge.

Education by Correspondence
Taking into consideration that people cannot always get together, teaching by correspondence is a great form of education (see Section 4.3).

Teamwork Projects
Experience shows that after receiving a 20- to 30-hour course of education, students may feel slightly confused. And when they go back to their work-place, there is not always the opportunity right away to start some projects to practice the new tools they learned. So after two to three months, they forget what they learned. This is why an effective SPC educational program requires that after attending a course, employees have to make at least one project by themselves or in participation with a team. Completing the project is a part

of the course. Students receive a certification of accomplishment after the project is submitted to the SPC Methodological Center.

Presentations to the Management
Sometimes engineers receive an SPC course but afterward, the way in which they perform their everyday work doesn't change a bit. However, if the manager requires the engineers to use a t-test or F-test for different kinds of comparisons, or a control chart and a histogram to present the process capability, the engineers are motivated to go back to the handouts they received at the SPC course and try to utilize the knowledge. So AMD's SPC specification is developed in such a way that the knowledge received in the class will never be forgotten because it is used in everyday work. In other words, it is important to generate a demand for the knowledge that is received, and this demand generates an interest for more knowledge.

Participation in the Workshops
The workshop is a place where people can "polish" their statistical tools. People come here, for example, before submitting an analysis that is based on statistical principles, or when there are disagreements between departments and the disagreements can be resolved by using statistical principles. The workshop's procedure reminds one a little bit of a courtroom. The chairman (the "judge"), a knowledgeable person in the field, gives the parties the opportunity to speak up. The other participants (the "jury") can ask questions and describe their opinions. The statistical consultant (the "lawyer") gives his/her opinions to the chairman who makes the final decision. Usually the chairman's decision is accepted by all the participants. The workshop is really an excellent form of sharing knowledge where everybody wins.

Participation in Conferences
AMD regularly organizes internal SPC conferences where people come together to share their results in process improvement. In 1991 the AMD Manufacturing Services Division, for example, held its sixth TCPI conference where 32 SPC projects on different topics were presented. The process of preparing and delivering presentations, listening to how other people present, meeting with the management, receiving an award for a good job, and listening to the audience applauding you, all makes for a system of educational growth, motivation, recognition, and a sense of accomplishment.

External Short-Term Courses
At AMD most of the educational activities are performed internally. However, sometimes there is a need to participate in external seminars; to be able to get special knowledge, meet new people, discuss similar problems, and make new friends. Again, the AMD SPC system discourages the participant of an outside seminar to come back, put the binder on the shelf, and consider

the "mission accomplished." The system encourages the person to make a report on what can be applied from the knowledge he/she received.

Participation in the Internal Periodicals
AMD publishes two periodicals on a regular basis. *Target* is a periodical where people present SPC project results and share new ideas related with process improvement. *Focus* is a pictorial that reflects all SPC activities by using the camera. This is a form which emphasizes people's faces and their activities. Involving people in a publication makes them work on their knowledge and vocabulary, since one needs to know the results of a project very well to be able to put it in a two-page written format.

Continuous Consultation
While developing new work habits and applying new tools, people should be able to get assistance whenever they get stuck. This consultation is available to AMD's employees by using personal contact, telephone conversations, postal and electronic mail systems, and other forms of communication, with SPC facilitators and other specialists who can help clarify project issues.

Involvement in Teaching Activities
After two to three years of working together to introduce statistical principles, a large team of engineers and managers were given a level of education good enough to make them SPC instructors. These people are now involved in educating and trying new SPC techniques and are also involved in the continuous development of new courses. They are the faculty of the MSD-SPC Sigma College.

The application of the 10 forms of education and training mentioned above allows AMD to have a continuous system of education that fulfills the needs of continuous process improvement.

4.3 SIGMA COLLEGE

When an organization is concentrated in one location or in at least one country, it is much easier to organize a continuous educational program and use the best resources available. But when an organization is located in a number of different countries, it is more difficult to do this. For example, AMD's manufacturing services division (MSD) headquartered in Santa Clara, California, has manufacturing facilities in Singapore, Malaysia, and Thailand. Because of this we decided to establish an institution called the MSD Sigma College. This institution was established in 1990, and now technicians, engineers, managers, and supervisors are the students of the Sigma College. The college provides an ongoing education in statistical methodology to employees who require the knowledge for continuous improvement. The objective is to give the same opportunity for all the employees to receive the education necessary to work successfully in the organization.

Form of Education

The educational program of the Sigma College is a two-year program and was developed based on MSD's specific needs. The program is conducted mainly by correspondence with consultation or tutorial sessions held regularly by the faculty instructors.

The approved applicants receive educational material and assignments for homework on a continuous basis. The completed assignments are then submitted to the local Sigma College faculty for review and appraisal. The students receive scores for each homework assignment they submit ranging from 0 to 10, the passing score being 5. After completing and passing 24 homework assignments (on the average of one per month), the student is required to sit for a written and oral examination to graduate. The graduates will receive a diploma, "Certified Practitioner of Statistical Technology" from management. The program was developed in such a way that the graduates can use the education obtained at the Sigma College to apply for a "Certified Quality Engineer" certificate from the American Society for Quality Control (ASQC).

Sigma College Board

The board (see Figure 4.3) has the function of providing the policies and directives of the Sigma College. These are set by the chairman with the consensus of all the vice-chairmen, who are then responsible for ensuring compliance to the policies and for the progress and development within their respective branches.

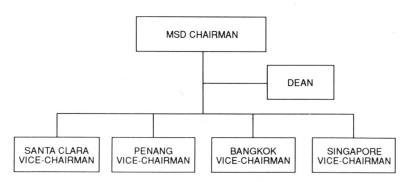

Figure 4.3 The Sigma College board.

The Dean
Based on the policies set forth by the college board, the dean plans a training strategy and program outline in which a body of knowledge is developed with a list of recommended reading materials. The dean also approves the selection of instructors recommended by the branch vice-chairman.

The Assistant Deans
Having obtained the program outline, the assistant deans are responsible for assisting the dean in compiling general training packages that would have widespread applications. They also develop special application packages for specific functions when the need arises. In order not to "reinvent the wheel," commercial educational packages may be evaluated for compatible applications. The assistant deans are also responsible for the training of the faculty instructors and the upkeep of the standards of the faculty.

The Administrators
The administrators are responsible for the registration of Sigma College students, the maintenance of the student flies that contain records of their academic achievements in the Sigma College, and the distribution of the educational materials and assignments to the students.

The Faculty Instructors
The branch faculty (see Figure 4.4) consists of a group of voluntary teachers who are selected from engineers and supervisors by the local TCPI Committee and approved by the MSD Sigma College board using the following

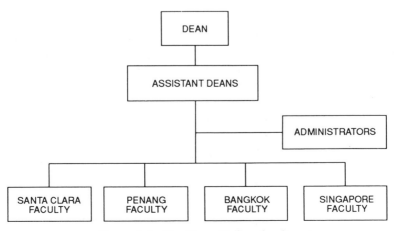

Figure 4.4 The Sigma College faculty.

criteria:

1. A college degree as a minimum.
2. Completion of all the TCPI education programs.
3. Ability to teach and conduct training.
4. Active application of statistical techniques at work.

Who Can Apply to the Sigma College
To apply to the Sigma College, one needs to be an educable employee of AMD. Completion of basic statistics, statistical process control, and statistical design of experiments courses (which are available at AMD) is preferable, and prior participation in any of the MSD international conferences/symposiums is an advantage.

The initial phase of this form of employee self-improvement shows that this is a very effective form of education. We can see this by the enthusiasm of the students, by their participation in the projects and their display of teamwork, as well as their presentations to the International SPC conferences, and their individual performance improvement.

4.4 THE *TARGET* PERIODICAL

Publication of the results from introducing statistical principles and sharing the experiences of how these results have been achieved are very important components in the TCPI system. It is difficult to calculate the return on investment from such a publication. It gives us the opportunity to share knowledge and disseminate experience through all the plants. It also prevents us from "reinventing the wheel" and gives recognition to people for their achievements, makes people think, and teaches them how to write reports. Putting all these pieces together makes the magazine an important component of the overall system of process improvement.

In January 1988, the AMD Manufacturing Services Division established a periodical publication called *Target*. The name of this periodical reflects AMD's direction, which at that time was to improve product quality and customer service by continually reducing the variation around the target. Below is an article extracted from our first *Target* publication that illustrates the beginning of a new way of thinking, a new philosophy, and a breakthrough in process improvement to achieve quality levels that are now leading us toward "six sigma quality."

"Looking to the Historical Chart," by B. C. Foo, AMD Plastic Production Manager

I picked this chart [see Figure 4.5] because if reflects the positive changes that have occurred since we started to apply statistical methods in our department.

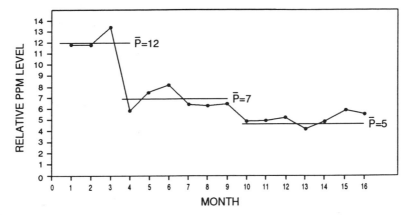

Figure 4.5 Monthly epoxy die attach monitoring chart.

The real beginning of SPC activity in our department was initiated by the first SPC class in 1986.

Then, in trying to apply what I learned in the class to manufacturing, I found that I needed the help and support of my subordinates. So, I started to conduct formal and informal classes. From day to day the number of charts increased, but the results were not visible right away. This was because we didn't know how to react to the charts. Looking back through time, I would say that learning how to take action on a control chart is the most important component in SPC application.

Epoxy die attach was a process that needed a lot of improvement, so we started there. Control charts were established on all workstations. The SPC team started to concentrate its efforts based on indications from the control charts. It became very interesting to see how our efforts impacted the ppm level.

The chart used in this article is a monthly cumulative chart for reporting purposes only. The chart shows that after four months of applying a simple p chart on each machine and by making weekly Pareto analyses to see where the major problems to attack were, the quality level improved 42%. We can also see that starting from month 10 there was a strong tendency toward a new process average, which improved 29% from the previously achieved level. This statement of improvement can be made with 99.7% confidence and 5% accuracy because of the large quantity inspected.

The most important thing is that our own confidence has been increased by seeing the results. We feel that by resolving the problems related to defect mode 1, the ppm level will drop (conservatively speaking) another 25%.

Our statistical experiments show the superiority of material B over material A related to defect mode 1. Material B allowed us to improve the machine capability. The Pareto analysis made from the data of May–September 1987

**Pareto Analysis of
Die Attach Monitoring**

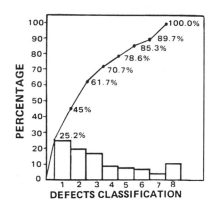

**Pareto Analysis
Breakdown of D/A Defects
May - Sept 1987**

Defect Mode	%	Cumulated %
1	25.2%	25.2%
2	19.8%	45.0%
3	16.7%	61.7%
4	9.0%	70.7%
5	7.9%	78.6%
6	6.8%	85.3%
7	4.4%	89.7%
8	10.3%	100.0%

Figure 4.6 Pareto Analysis of Die Attach Monitoring.

[see Figure 4.6] shows that if we take care of the three major defect modes, the ppm level can be lowered 61.7%. This is only one example of how we can create savings from statistical knowledge.

REFERENCES

1. W. Edwards Deming, "Out of the Crisis" (1982), Cambridge University Press, p. 24.
2. W. Edwards Deming, "Quality, Productivity, and Competitive Position" (1982), Cambridge University Press, p. 47.
3. G. E. P. Box and N. R. Draper, "Evolutionary Operation" (1969), John Wiley & Sons, Inc., pp. 141–147.

Chapter 5

The Executive Process Improvement Audit

There is a classic story written by Gogol, called "Revisor." The story is about people who live quietly in a small town, and suddenly they hear a rumor that the Revisor is coming to their town to conduct an audit. All the town was in a state of commotion. Even those who had nothing to be worried about were in a panic. A lot of talk and preparations were made to be sure that the Revisor would get the appropriate attention. Finally, the Revisor left the town very pleased and happy. In the morning of the next day, when the people of the town could draw a sigh of relief, they found that the Revisor was an impostor and that the real Revisor was on his way to the town.

In our day, events like this cannot happen and the reaction to an audit is different. However, in some cases the expectation of an audit generates an unpleasant state of anxiety. Here we will describe an audit system that is awaited with pleasure. This is a system that motivates employees to do the best they can to become recognized as a team and as individuals who contributed to the team's success. It is important to note that the desire to be recognized is a normal human trait, and if management makes the effort to fulfill this natural desire, the return on investment will be very high.

Process improvement involves almost all the people of an enterprise. The top management of an organization receives hundreds of pages of different kinds of reports where the figures reflect people's achievements. Under almost every column and graph are people's results (good or bad). But who did this? Who contributed the most? Who needs help to do even better? What is the real cause of improvement? Who deserves recognition? These types of questions are continuous, but executives will never get all the answers by reading reports and having meetings mainly with the employees who report directly to them. The only way to have the full picture of what is going on in an operation is by regularly conducting executive audits. This is an excellent form of meeting with people and taking oral reports from different levels in the organizational structure.

The AMD Manufacturing Services Division started to apply executive audits three years ago. The system of the executive process improvement audit (EPIA) includes a number of audits that are performed on different levels as preparation to the audit of the senior management. And, finally, the ultimate goal of the audit is to be prepared for a customer audit. Below is the description of the executive audit.

5.1   THE STEPS OF THE EXECUTIVE PROCESS IMPROVEMENT AUDIT

To have continuous and frequent feedback from the process, the executive audit has a number of preparation levels that involve a large number of participants. Here we will give a short description of all audit levels.

1. The Operation Manager's Audit

On this level the audit is performed monthly and is based on a program that includes all customer, corporate, and divisional requirements related to process improvement. The program is almost the same for all levels of the audit. The only difference between them is that the lower the level of the audit, the more details are included in the program.

To perform the audit, the manager forms a working team, which consists of representatives from every department, including operators and inspectors. The results of the audit are analyzed by the Process Improvement Subcommittee and a report is approved and submitted to the operations staff. No later than a week after the audit is performed, a meeting with all the workers is held where the results are discussed, and new actions to correct the discrepancies found by the audit are agreed upon.

2. The Managing Director's Audit

This audit is performed quarterly by the managing director who is responsible for all operations. The procedure for this audit is almost the same as previously described for the manager's audit. The only difference here is that the director is using the information received from the previous monthly audits conducted by the manager. The manager randomly selects some areas of interest for personal auditing, and the rest of the time he/she spends meeting with employees at their workstations, observing the rhythm of work, looking at the control chart results, asking and answering questions, and looking at what is going on. The manager includes the results from this audit in the quarterly staff meetings. In addition, the manager also publishes a short report about information from this audit, which is delivered to all the people of the plant through their direct supervisors.

3. The Vice-President's Audit

This audit is conducted twice per year, usually just before the International TCPI Conference. This allows, the vice-president to deliver the results to a large group of people who participate in the international conferences, and at the operator's symposium which is conducted immediately following the international conference. The vice-president's audit is performed by an executive team that also includes the local management. The results of all plant audits conducted by the vice-president and by the other levels are analyzed and published in a divisional report that is delivered to the plant for information and action. The division report includes actions and suggestions related with continuous process improvement that are based on the information obtained from all the plants. This helps prevent the same mistakes occurring at different plants, and distributes good ideas and initiative throughout the division.

5.2   THE CLOSED-LOOP INFORMATION SYSTEM

If you want employees to improve, tell them how they are doing. Without giving your subordinates constant feedback on how they are doing, how will they know if they need to improve? This auditing system is one way to tell people how well they work on process improvement.

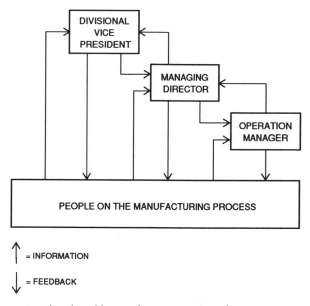

Figure 5.1   The closed loop information system from a process audit.

As one can see from Figure 5.1, the auditors from any level visiting the manufacturing process obtain information from the people who work on the process. As feedback, the results, opinions, and impressions are delivered directly to the people and/or to their direct supervisors by the summarized audit results.

The experience of introducing this type of system shows that when the customer comes to conduct an audit, the plants are better prepared and receive less criticism. It is understandable that it doesn't matter how well the plant is prepared or how many audits were conducted before the customer came, there will always be some criticisms. (What kind of auditor would you be if you didn't find anything negative?) But experience shows that these observations are significantly less. The customer's constructive recommendations are included in the system as feedback on self-improvement.

5.3 CONCLUSION

It is very difficult to overestimate the effectiveness of such a hierarchical procedure as auditing. It is also difficult to describe the effect on employee morale of meeting on a continuous basis with the leaders of different levels, where employees can express their feelings, give recommendations, and ask for advice. This form of auditing provides continuous feedback to management on how its decisions, plans, and targets are being accomplished.

There is one more element in this auditing work. The opportunity of learning from each other. Sitting in the executive conference room and looking at a control chart projected on the wall is one thing, and "touching" the chart directly at the operation and listening to the operators explanation is absolutely another thing. There is no other way that an executive can meet with his/her people, especially when we are talking about a large company. To get benefits from an audit, it is most important not to make it just "another form of formal management." There should be a human aspect involved here, which means not only looking at the figures on the chart, but trying to get information that is not reflected in the reports; information gotten from talking to employees who can't always come to top management, but have a lot to say. An example of a checklist used in the executive audit is given in Appendix A. This is just an example and should be tailored to every organization to reflect the goals of the audit.

PART II

Process Control and Capability Studies

The first major portion of this part is dedicated to the concept, methodology, application, and interpretation of classical and modern statistical tools for process control. The construction of control charts for variables and attributes is described by using a step-by-step procedure and illustrated with actual examples. Examples of interpretation are also given.

The second major portion of this part is the test and inspection capability study. The methodology introduced in this portion can help us to determine whether the test equipment is adequate and which inspectors need to be retrained.

The third major portion of this part is dedicated to the methodology of process capability studies for variable and attribute data, and calculating capability indexes. Special attention is given to the concept of working in a low ppm environment and to Motorola's six sigma quality philosophy. A special section is also dedicated to the problem of how to control and calculate capability indexes when the process has an abnormal distribution, and how to design engineering specification limits by statistical methods.

All together, this part covers a relatively broad variety of topics that will demonstrate how to bring the process into a state of statistical control and how to monitor processes in a low ppm environment. In this part the reader will find the answer to the question, "What can and should be done to improve the quality before expending efforts to change the system?"

Chapter **6**

Some Important Probability Distributions

Some of the most important categories of probability distributions that are frequently applied in the procedures for process improvement are:

1. Normal distribution
2. Binomial distribution
3. Poisson distribution
4. Geometric distribution

We will describe these distributions in this chapter, and will present some examples of their application. However, the major applications of the distributions mentioned above can be found in later chapters of this book.

6.1 THE NORMAL PROBABILITY DISTRIBUTION (OFTEN CALLED THE "NORMAL CURVE")

6.1.1 Properties of the Normal Distribution

The normal probability distribution is one of the most useful and important distributions in statistical analysis. Even if no distribution of empirical data can be exactly normally distributed, many engineering characteristics can be approximated by the normal distribution function. The normal distribution has two parameters, μ and σ. Equation (6.1) defines a family of normal

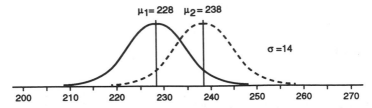

Figure 6.1 Two normal distributions with different averages (μ), but with the same deviation (σ).

curves:

$$f(x) = \frac{1}{\sigma\sqrt{2\pi}} \exp\left[-\frac{1}{2}\left(\frac{x-\mu}{\sigma}\right)^2\right] \qquad -\infty < x < \infty \qquad (6.1)$$

where σ is the standard deviation of the population, μ is the mean of the population, π is a mathematical constant equal to 3.1416, and x is a continuous variable.

When values are specified for μ and σ, a particular normal curve is identified. For example, let's assume $\mu = 228$ and $\sigma = 14$. This identifies the normal distribution on the left-hand side in Figure 6.1. One can see that the normal curve is symmetrical with a central peak, and the mean (μ) is located at the center of the distribution. The distribution in Figure 6.1 has a bell-shaped curve where the mean, median, and mode are all equal. We can also observe that the normal curve is highest in the center and approaches the horizontal axis in each tail of the distribution, stretching out infinitely in each direction and approaching, but never touching, the horizontal axis.

6.1.2   How Is the Normal Curve Affected if μ and σ Change?

If the μ is changed, say from 228 to 238, and the σ remains the same ($\sigma = 14$), the only change we will observe is that the distribution will shift to a higher position on the x axis (see Figure 6.1, dotted line). Now, if the σ changes, let's say from 14 to 7, but the μ remains the same, the distribution will change as shown in Figure 6.2. As one can see, with a larger σ the distribution has a greater spread. By continuous reduction of the standard deviation, we will achieve a greater concentration of the distribution around the mean. The smaller the σ, the higher the central portion of the curve and the closer the curve is to the x axis in the tails. As illustrated later in this book, the main effort related with process improvement will be to reduce the standard deviation (σ) and achieve greater uniformity of the process or product by reducing the variability around the target.

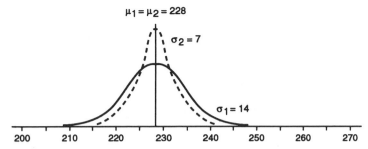

Figure 6.2 Two normal distribution with the same average (μ) but with different standard deviations (σ).

6.1.3 Standardization

The most important use of the normal distribution is related to the area under the curve. The total area under the curve represents the total of the relative frequencies for the distribution, which is 1.00. Since the curve is symmetrical about the mean, the area under either side of the mean is 0.5. Formulation of statistical inferences is generally based on the proportion of the area under the normal curve for specified intervals along the horizontal axis. It can be calculated by integration of the function between the two points, which is quite difficult. Integration of one particular normal distribution, that is, the standard normal distribution, has been completely calculated, and the results are set up in a table called the "Z table" (see Appendix C). The standard normal distribution is a normal distribution with zero mean ($\mu = 0$) and unit standard deviation ($\sigma = 1$). The formula for the standard normal distribution can be derived by setting $\mu = 0$ and $\sigma = 1$ in (6.1), and we will have

$$f(x) = \frac{1}{\sqrt{2\pi}} \exp\left[-\frac{x^2}{2}\right] \qquad (6.2)$$

The standard normal distribution is plotted in Figure 6.3. As one can see from Figure 6.3, almost all (99.72%) of the distribution is within three standard deviations of the mean. Since using the Z table is much easier than doing the integration, most problems involving the normal distribution are solved by converting the given normal distribution into the relevant points on the standard normal distribution and then using the table. Any normal distribution can be transformed into the standard normal by using the following equation:

$$z = \frac{x - \mu}{\sigma} \qquad (6.3)$$

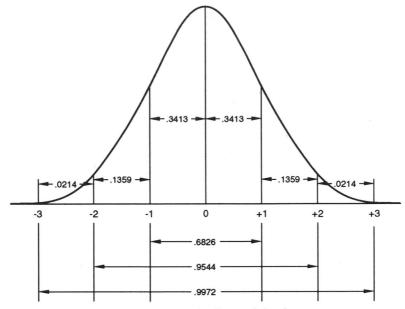

Figure 6.3   The standard normal distribution.

where z is the transformation value of x, μ is the mean of the x values, and σ is the standard deviation of the x values.

We will use the following example to illustrate the conversion and use of the standard normal distribution.

Example   Suppose the average tin plating thickness for 100 electronic parts is 500 μ'' (microinches) and the standard deviation is 20 μ''. Assuming that the plating process is in a state of statistical control and the tin thicknesses are normally distributed, what is the probability of the parts with a tin thickness (x): (a) more than 550 μ'', (b) less than 460 μ'', or (c) between 500 and 520 μ''.

Solution
  a. More than 550 μ'':   The mean of 500 μ'' is equal to the mean 0 in the standardized normal distribution. The standard unit for 550 μ'' is determined as follows:

$$z = \frac{x - \mu}{\sigma} = \frac{550 - 500}{20} = 2.5$$

The shaded area in Figure 6.4 represents the probability of the parts with tin thickness more than 550 μ''.

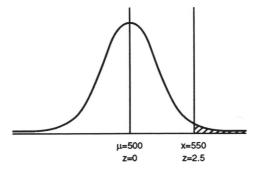

μ=500   x=550
z=0   z=2.5

Figure 6.4   Area under the normal curve for tin thickness (x) more than 550 μ''.

From the normal distribution table in Appendix C, we find that

$$P_{(Z \geq 2.5)} = 6.210E - 03 = 0.006210 = 0.621\%$$

This means the probability of an electronic part having a tin thickness more than 550 μ'' is only 0.621%.

b. Less than 460 μ'':   Converting 460 μ'' to standard units yields

$$z = \frac{x - \mu}{\sigma} = \frac{460 - 500}{20} = -2.0$$

The minus sign indicates that the point in interest is to the left of the mean (see Figure 6.5), but when we use the table in Appendix C to find the area under the curve, the minus sign is not considered. Therefore,

$$P_{(Z \leq -2.0)} = P_{(Z \geq 2.0)} = 2.275E - 02 = 0.02275 = 2.275\%$$

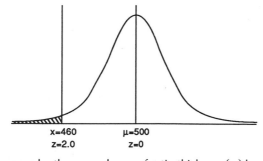

x=460   μ=500
z=2.0   z=0

Figure 6.5   Area under the normal curve for tin thickness (x) less than 460 μ''.

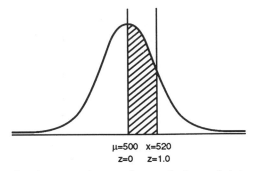

Figure 6.6 Area under the normal curve for tin thickness (x) between 500 μ'' and 520 μ''.

Thus, we can make a conclusion that there will be around 2.275% of the parts with a tin thickness less than 460 μ''.

  c. Between 500 to 520 μ'':   Converting 520 μ'' to standard units, we have

$$z = \frac{x - \mu}{\sigma} = \frac{520 - 500}{20} = 1.0$$

From Appendix C, we find

$$P_{(Z \geq 1.0)} = 1.587E - 01 = 0.1587 = 15.87\%$$

But this is the probability of z larger than 1.0 (i.e., $x \geq 520$ μ''). Since we want to know the probability between $z = 0$ ($x = 500$ μ'') and $z = 1.0$ ($x = 520$ μ'') (as shown in Figure 6.6), we use the overall area of the half-distribution, which is $0.5 - P_{(Z \geq 1.0)}$, then we will get the value for $P_{(0 < Z < 1.0)}$, as follows:

$$P_{(0 < Z < 1.0)} = 0.5 - 0.1587 = 0.3413 = 34.13\%$$

Thus, we can make a conclusion that the probability of a part having the tin plating thickness of between 500 μ'' and 520 μ'' is 34.13%.

6.2 THE BINOMIAL DISTRIBUTION

In any manufacturing environment there are many applied problems in which we are interested in the probability of an event occurring "x times in n trials." For example, we may be interested in the probability of finding one nonconforming unit in an inspection sample of 1000 units, or we may be interested in the probability of returned lots from 10,000 lots shipped to a customer, etc. To borrow terms used in games of chance, we say that in each of these examples we are interested in the probability of getting x successes and $n - x$ failures in n trials.

In our situation, this means finding a nonconforming unit or having a returned lot (x) is a "success," the number of conforming units or lots accepted $(n - x)$ are "failures." For a manufacturing practitioner, this terminology is a little bit confusing, but this is the language used in the theory of probability.

In the problems that we will study in this section, it will always be assumed that the number of trials is fixed, that the probability of a success is the same for each trial, and that the trials are all independent. To solve problems that meet the conditions mentioned above, we use the binomial distribution formula which is described below:

The probability of getting x successes in n independent trials is

$$f(x) = \binom{n}{x} p^x (1 - p)^{n - x} \qquad \text{for } x = 1, 2, \ldots, n \qquad (6.4)$$

where p is the constant probability of a success for each trial. The mean of the binomial distribution will be $\mu = p$ and the standard deviation will be

$$\sigma = \sqrt{\frac{p(1 - p)}{n}}$$

It is customary to say here that the number of successes in n trials is a random variable having the *binomial probability distribution* or simply the binomial distribution. Technically, the use of the binomial distribution to generate the probabilities of the various outcomes of an experiment is justified only if the experiment is characterized as a Bernoulli process (i.e., a series of Bernoulli trials). There are two conditions which need to be met for a Bernoulli process:

1. The outcome on any one trial must fall into one of two mutually exclusive and collectively exhaustive categories.
2. The outcome of any given trial of the experiment must be independent of the outcomes on the other trials. When trials are independent, it means that the probability of success remains constant from trial to trial.

Below is an example of applying the binomial distribution.

Example   Assuming that the average nonconforming rate (p) for a process is 0.02, what is the probability that in a random sample of 50 parts there will be 3 nonconforming parts?

Solution

Substituting $x = 3$, $n = 50$, $p = 0.02$ into Equation (6.4), we get

$$f(3) = \binom{50}{3}(0.02)^3(1-0.02)^{50-3}$$

$$= \frac{50!}{3!(50-3)!}(0.02)^3(0.98)^{47}$$

$$= 0.061$$

6.3   THE POISSON DISTRIBUTION

The Poisson distribution takes its name from Simèon Poisson, a French mathematician (1781–1840). It is a discrete probability distribution that is very similar to the binomial probability distribution described earlier in this chapter. The Poisson distribution is usually applied to discover the behavior pattern of a process in which some event of interest occurs at varying random intervals over a continuum of time, length, or space. Such a process is called a Poisson process, which we will describe in this section with some typical examples.

6.3.1   The Poisson Process

Consider, for example, a horizontal line as representing the continuous passage of time, and dots on the line as representing the occurrence of some given event such as machines arriving for repair from the manufacturing floor to the maintenance department (see Figure 6.7). Or, the horizontal line can be considered as representing a coil of aluminum wire that is used in the wire bonding machine to connect the electronic device to the lead frame. Dots on the line indicate the location of defects on the wire (see Figure 6.8).

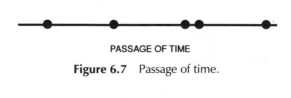

PASSAGE OF TIME

Figure 6.7   Passage of time.

FEET OF ELECTRICAL WIRE

Figure 6.8   Feet of electrical wire.

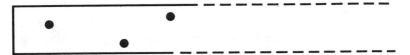

Figure 6.9   Length of a roll of metal used to produce lead frames.

Another example of a Poisson process can be a rectangular configuration representing a roll of metal used to produce lead frames for a molding process. The dots within the configuration indicate the location of the defects or blemishes (see Figure 6.9).

Theoretically, each continuum described above can be subdivided into subsections so that, at the most, one event (as depicted by a dot) could occur in each subdivision (see Figure 6.10). Provided that the underlying "experiment" meets the requirements of a Bernoulli process, the subdivisions can be treated as a sequence of Bernoulli trials. The Poisson distribution can then be used to generate the probability distribution, indicating the relative frequency one could expect to observe the occurrence of the event over repeated observations. Each segment of the process observed will ordinarily be comprised of a large number of these small subdivisions.

For example, in the case of the coil of electrical wire, the concern might be with the number of defects (blemishes) per 1000 feet. Each subdivision would be one foot in length; thus, each segment of interest would contain 1000 subdivisions. In order for the subdivisions to qualify as a series of Bernoulli

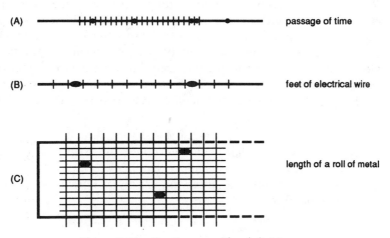

Figure 6.10   Continuum with subdivisions.

trials, the following constraints must be met:

1. The subdivisions must be sufficiently small that the probability of two or more events occurring in any one subdivision is so unlikely that it may be considered to be zero.
2. The occurrence or lack of occurrence of the event in question in one subdivision must have no effect on the probability of occurrence of the event in any other subdivision.

In the discussion above, the reader might notice the similarity to a binomial experiment. However, there is a fundamental difference here. In a binomial experiment we are concerned with the number of successes out of a specified number of determined trials. Each trial and its outcome can be observed. For example, when inspecting a sample of 1000 units we are concerned with the number of successes (or nonconformities) out of a specified number of 1000 units. In this case, we know the total number of opportunities to be nonconforming units. This number is obviously 1000 units.

In the case of the Poisson distribution, the number of successes is again the subject of concern, but it becomes virtually impossible to determine the total number of trials. For example, when we inspect a wafer we can find five scratches, but we cannot count the number of "non-scratches." For a Poisson distribution, the total number of trials (or inspected units) is the total number of opportunities for the event to occur. The difficulty is in observing occasions when the event did not occur. For example, how can we determine the total number of opportunities for a defect to occur in some specified length of wire? Fortunately, it is not necessary to know the number of opportunities for a success (or reject) to have occurred in order to utilize the Poisson distribution. The only information we need is the expected or mean number of successes for the segment of concern for the process under investigation. This information can be derived from the historical data of the process and is designated by the symbol λt, where λ refers to the expected number of successes over some specified segment of the process and t refers to the multiple of the segment on which λ is based. For example, if λ refers to the expected number of defects per 200 feet of aluminum wire and you are interested in the distribution of the number of defects per 2000-foot interval, then

$$\lambda t = 10\lambda$$

In the special case where λ has been determined for exactly the segment of interest to the experimenter, $t = 1$ and λt simplifies to λ.

In general, the probability of observing exactly x Poisson successes over the interval t is equal to

$$P(X) = \frac{(\lambda t)^x e^{-\lambda t}}{x!} \tag{6.5}$$

where x is a random variable representing the number of occurrences (or successes) and is restricted to a nonnegative integer, λt is the expected mean number of occurrences of the event in question over the interval t, and e is the base of the natural logarithm or 2.71828.

Both the mean and the variance of the Poisson distribution always equal the same value, namely λt. Thus, the Poisson distribution is totally characterized by the single parameter λt, or its mean.

6.3.2 Examples of Applying the Poisson Distribution

Example 1   Records show that the average rate of the arrival of machines to be repaired at the maintenance department is 0.8 arrivals per hour. Use the Poisson distribution to find the probability that: (a) exactly two machines will arrive in one hour, (b) five or less machines will arrive in one hour, and (c) five or more machines will arrive in one hour.

Solution
  a. Since $\lambda = 0.8$, $t = 1$,

$$\lambda t = 1 \times 0.8 = 0.8$$

Substituting $\lambda t = 0.8$ and $x = 2$ into the Poisson distribution formula

$$P(x) = \frac{(\lambda t)^x e^{-\lambda t}}{x!}$$

we have

$$P(2) = \frac{(0.8)^2 e^{-0.8}}{2!}$$

From Appendix L, we find that

$$e^{-0.8} = 0.449$$

Substituting this value into the formula for $P(2)$, we get

$$P(2) = \frac{(0.8)^2 (0.449)}{2 \times 1} = 0.1438 = 14.38\%$$

Conclusion: The probability that exactly two machines will arrive in one hour is 14.38%.

  b. To find the probability of five or less machines arriving in one hour, we can use the Poisson distribution formula to derive the probabilities of exactly 0, 1, 2, 3, 4, and 5 machines arriving as described in (a), and then add them

TABLE 6.1   The Cumulative Probability of Occurrence of Machine Arrival in One-Hour Intervals ($\lambda t = 0.8$)

Number of Arrivals in One-Hour Intervals	Probability of Occurrence
0	0.449
1	0.808
2	0.952
3	0.990
4	0.998
5	0.999
6	0.999

together to find the answer. An easier way is to use the Poisson distribution table (see Appendix J), which will give the same answer. Table 6.1 is reproduced from Appendix J and shows the cumulative probability of occurrence for one to six machines. Using this table, we can find that the probability of five or less machines arriving in one hour is

$$P(x \leq 5) = 0.999$$

c. The probability of five or more machines arriving in one hour will be

$$P(x > 5) = 1 - 0.999 = 0.001 = 0.1\%$$

Based on this information, the maintenance department can schedule its activities and evaluate its workforce utilization.

Example 2   In the previous example, λ has been determined for exactly the segment of interest to the analyst, i.e., $t = 1$, and λt simplifies to λ. The purpose of this example is to show the method of calculations when λ is based on one interval, but the analyst wants to use more than one interval. In other words, when $t > 1$.

In this example, using the same information as in Example 1, we want to find the probability of more than six machines arriving in a two-hour interval. In this case, since $\lambda = 0.8$, $t = 2$, we have

$$\lambda t = 2 \times 0.8 = 1.6$$

Again, we have two ways to obtain an answer to the question. One is to use the Poisson distribution formula to find, separately, the probability of $0, 1, 2, 3, 4, 5, 6$ machines arriving in an interval of two hours, adding the results and

TABLE 6.2   The Cumulative Probability of the Occurrence of Machine Arrival in Two-Hour Intervals ($\lambda t = 1.6$)

Number of Arrivals in Two-Hour Intervals	Probability of Occurrence
0	0.201
1	0.524
2	0.783
3	0.921
4	0.976
5	0.993
6	0.998
7	0.999
8	0.999

subtracting them from 1 as follows:

$$P(x > 6) = 1 - [P(0) + P(1) + P(2) + P(3) + P(4) + P(5) + P(6)]$$

The second method, which is much easier, is to use Table 6.2 and find the answer directly. Table 6.2 is reproduced from Appendix J. Using this table, we find that the probability of six or less machines arriving in two hours is 0.998. So, the probability of more than six machines arriving in two hours is

$$P(x > 6) = 1 - 0.998 = 0.002$$

Conclusion: The probability that there will be more than six machines in a two-hour interval is only 0.002.

6.4   THE GEOMETRIC DISTRIBUTION

If we know the average proportion nonconforming rate (p) for the process, how many units should we inspect from this process to find the first nonconforming unit? To answer this question, the geometric distribution can be used. If x is the trial on which the first nonconforming unit occurs and p is the probability of a unit being nonconforming, the probability that the first nonconforming unit will occur on trial x is

$$P_{(x)} = p(1-p)^{x-1} \tag{6.6}$$

where $0 < p < 1$ and $x = 1, 2, 3, 4, \ldots$. Figure 6.11 depicts this type of distribution. Equation (6.6) expresses the fact that we have one nonconforming unit

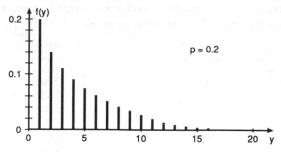

Figure 6.11   Geometric probability function.

followed by $(x - 1)$ conforming units. The cumulative distribution function is

$$P_{(x \le n)} = \sum_{x=1}^{n} p(1-p)^{x-1} = 1 - (1-p)^{n} \qquad (6.7)$$

This is the probability that the first nonconforming unit will be found somewhere among the first n items observed. It is important to note that the geometric distribution has the property of being memoryless. That is, if the outcome under consideration has not occurred in the first n trials, the probability it will not occur during the next n trials is the same as the probability it will not occur in the first n trials. For example, if we had already inspected 1000 units and no nonconforming units were found, the probability that no nonconforming units would be found continuing the inspection is the same as when we inspected the first 1000 units.

The mean of the geometric distribution is

$$\mu = \frac{1}{p}$$

and the variance is

$$\frac{(1-p)}{p^2}$$

Let's give some simplified examples to become familiar with the application of the geometric distribution.

6.4.1 Examples of Applying Geometric Distribution

a. Using Equation (6.7), we can find that in repeated rolls of a balanced die, the probability that the first "6" will occur on the fifth roll is

$$\frac{1}{6}\left(\frac{5}{6}\right)^{5-1} = \frac{625}{7776} = 0.08$$

b. If we know that the average nonconforming rate (p) for the process is 0.05, then the probability that the first one of the six units coming from this process at a given time will be nonconforming is

$$P_{(x=6)} = (0.05)(1-0.05)^{6-1} = 0.039$$

c. As another example, if we know that the process average (p) is 0.0833 and we take units from the process one by one, the probability that the first nonconforming unit will occur in trial number 10 would be

$$P_{(x=10)} = (0.0833)(1-0.0833)^{9} = 0.0381$$

In Chapter 7, the reader will find a real application of the geometric distribution when describing the cumulative count control chart.

6.5   THE NORMAL DISTRIBUTION AS AN APPROXIMATION OF THE BINOMIAL DISTRIBUTION

The normal distribution gives a very good approximation of the binomial distribution when p gets close to 0.5, even when the sample size is very small (see Figure 6.12). However when p deviates from 0.5, the approximation gets worse (see Figure 6.13). But at the same time, for values of p significantly

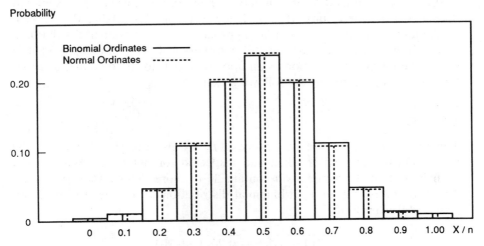

Figure 6.12 Comparison of the normal distribution with the binomial distribution for which $n=10$ and $p=0.50$. [Reprinted with permission from "Quality control and Industrial Statistics," by A. J. Duncan, Richard D. Irwin, Inc., Homewood, IL (1974), p. 94.]

Probability

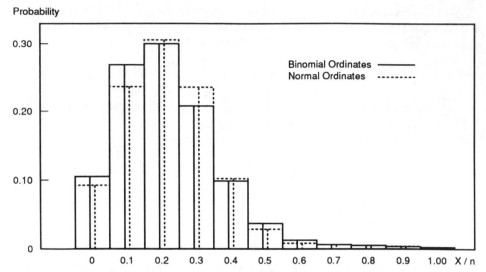

Figure 6.13 Comparison of the normal distribution with the binomial distribution for which $n=10$ and $p=0.20$. [Reprinted with permission from "Quality Control and Industrial Statistics," by A. J. Duncan, Richard D. Irwin, Inc., Homewood, IL (1974), p. 95.]

different from 0.5, the approximation of the normal distribution to the binomial distribution gets better the larger the value of n. Even if p is as low as 0.10 or as high as 0.90, if n is larger than 50, the normal approximation does not give bad results. If p is below 0.10 or above 0.90, the Poisson distribution is commonly used to approximate the binomial distribution, although the normal distribution still does very well as long as $np \geq 5$.[1] The use of the normal distribution as an approximation to the binomial distribution may be illustrated by the following example.

6.5.1 Example

The average defective rate for a process ($\bar{p}$) is 0.20. A random sample of 100 units from the process is taken and inspected for defectives. What is the probability that the sample will contain 30 or more defects (i.e., $\bar{p}=0.30$)? This probability is given by a binomial distribution with a mean of 0.20 and standard deviation of

$$\sigma = \sqrt{\frac{\bar{p}(1-\bar{p})}{n}} = \sqrt{\frac{0.20(1-0.20)}{100}} = 0.04$$

This binomial distribution can be approximated by the normal curve, the mean of which is 0.20 and the standard deviation of the mean is 0.04. The answer to the question is given approximately by the area of the normal distribution:

$$z = \frac{x - \mu}{\sigma} = \frac{0.30 - 0.20}{0.04} = 2.5$$

From the Z table in Appendix C, it is seen that the probability of getting a normal variable which lies above $z = 2.5$ is 0.0062.

6.6 THE POISSON DISTRIBUTION AS AN APPROXIMATION OF THE BINOMIAL DISTRIBUTION

For small values of p and large values of n, say for $p < 0.10$ and $np < 5$, the binomial distribution can be better approximated by the Poisson distribution than by the normal distribution. Figure 6.14 illustrates this point. When n is large, but p is small, we sometimes have a "gray area" in which the normal distribution and Poisson distribution approximations work almost as well. However, the Poisson distribution is more convenient and preferable. The fact that the Poisson distribution provides a good approximation under specific conditions is quite useful, since computation involving the binomial distribution becomes very complicated. Below is an illustrative example showing the application of the Poisson approximation.

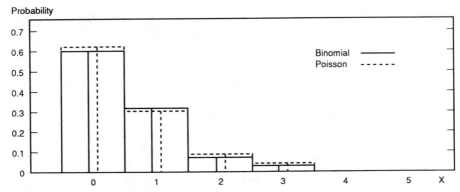

Figure 6.14 Comparison of binomial and Poisson ordinates when $p = 0.02$ and $n = 25$. [Reprinted with permission from "Quality Control and Industrial Statistics," by A. J. Duncan, Richard D. Irwin, Inc. Homewood, IL (1974), p. 96.]

6.6.1   Example

A manufacturing process has an average nonconforming rate of $p = 0.00005$. Use the Poisson approximation on the binomial distribution to find the probability that by inspecting 10,000 units, at least two nonconforming parts will be found.

Solution

The Poisson distribution formula is as follows:

$$f(x) = \frac{\lambda e^{-\lambda}}{x!} \qquad \text{where } \lambda = np$$

To find the probability of at least two nonconforming parts being found, we need to subtract from one the probabilities that zero or one nonconforming units will be found. Since $np = 10,000(0.00005) = 0.5$, and $e^{-0.5} = 0.607$ (see Appendix L), we find that

$$f(0) = \frac{(0.5)^0(0.607)}{0!} = 0.607$$

$$f(1) = \frac{(0.5)^1(0.607)}{1!} = 0.304$$

and the probability that at least two nonconforming units will be found is

$$1 - (0.607 + 0.304) = 0.089$$

REFERENCES

1. A. J. Duncan, "Quality Control and Industrial Statistics" (1974), Richard D. Irwin, Inc., p. 94.
2. A. J. Duncan, "Quality Control and Industrial Statistics" (1974), Richard D. Irwin, Inc., p. 94, Figure 4.13.
3. A. J. Duncan, "Quality Control and Industrial Statistics" (1974), Richard D. Irwin, Inc., p. 95, Figure 4.14.
4. A. J. Duncan, "Quality Control and Industrial Statistics" (1974), Richard D. Irwin, Inc., p. 96, Figure 4.15.

Chapter 7

Statistical Process Control

7.1 WHAT IS A PROCESS?

When we began to introduce statistical process control (SPC) at AMD, we faced a situation where some internal organizations and departments expressed resistance to SPC only because they didn't view their activities as a process. This is why it is very important in the beginning of this chapter to define what a process is.

In the literature one can find a broad variety of definitions of the word "process." We adopted the definition from the *AT&T Statistical Quality Control Handbook*,[1] where a process is defined as "...any set of conditions, or set of causes, which work together to produce a given result." According to this definition, the term "process" can be applied to any meaningful activity such as manufacturing, engineering, marketing, accounting, management, inspection, planning, purchasing, maintenance, or any other activity that can be viewed as a constant system of causes. This means that statistical process control can and should be applied to all elements of an enterprise. This will allow us to predict, control, and improve the quality in all areas that are related to customer satisfaction.

The AMD experience shows that before starting to teach the fundamental statistical principles, it is very important to define the processes on which employees are working to determine their responsibilities in the processes and operationally define the customer expectations from their processes. Previously, we were paying more attention to the product quality and less to other elements of the process, such as cycle time, on-time delivery, etc. But for a customer, making parts to the target or delivering those parts at the right time has the same value. It is interesting to note that when statistical principles were initially introduced in the nonmanufacturing areas, employees were required to make flowcharts of their processes. There was some slight resistance. But finally when they put on paper exactly the flow they were performing in their jobs, they discovered that a lot of things could be improved and the cycle time could be reduced. Treating the payroll, purchasing, shipping, and other activities as a manufacturing process will allow us to reduce the overall variability in the macroprocess of the overall organization. This is the only way to achieve total customer satisfaction.

7.2   THE CONCEPT OF CONTROL

7.2.1   Definition of Control

> ...for our present purpose a phenomenon
> will be said to be controlled when, through
> the use of past experience, we can predict, at
> least within limits, how the phenomenon may
> be expected to vary in the future....
> —Dr. Walter A. Shewhart[2]

If you, as a manager or executive, would review the topics of all the meetings you conducted (or participated in), you would find that most of them are related to one particular question, "Why aren't the results the way we planned?" Or, putting it another way, "How do we make things happen the way we want them to?" This is the major question related to the concept of control.

To describe the nature of control in one of our SPC classes, we used a technique based on Dr. Shewhart's illustration.[3] We asked the students to write the letter *a* 20 times, trying to make them all look alike (see examples in Figure 7.1). Analyzing the results, we came to the conclusion that although all the students were trying hard to make all the *a*'s alike, they couldn't. The result did not come as a surprise to them. The students accepted the result as an empirically established fact. But, why does this happen? What does it mean in respect to control? Why can't we do a simple thing like making all the *a*'s look alike? As an answer to these questions, the students started to name the causes that, in their opinion, introduce the variation among the *a*'s: the vibration of the desk, the smoothness of the paper, the uniformity of the pencil lead, the nervousness of the students, etc. But do we really know all the causes that introduce variations in the *a*'s? Probably not. We accept any human limitations that we don't know all the reasons for. If we knew all the reasons, we would have a better understanding of a certain part of nature.

The purpose of this class experiment is to illustrate to the students that in general we are limited in doing what we want to do; that to do what we

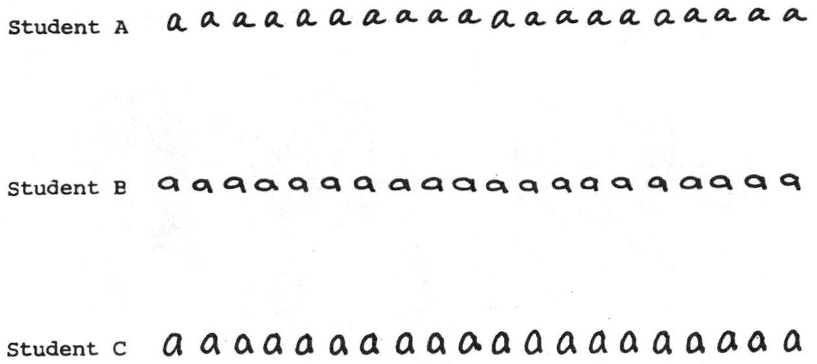

Student A   *a a a a a a a a a a a a a a a a a a a a a*

Student B   *a a a a a a a a a a a a a a a a a a a a a*

Student C   *a a a a a a a a a a a a a a a a a a a a a*

Figure 7.1   An example to demonstrate variations.

intend to do (even in such a simple test as making *a*'s look alike) requires almost infinite knowledge compared with that which we now possess.

So, if we are willing to accept the axiom that we cannot do exactly what we want to do, and cannot hope to fully understand why we cannot, we should also accept the axiom that a controlled quality will not be a constant quality. Instead a controlled quality must be a *variable quality*. This is Dr. Shewhart's first characteristic of control. Taking a second look at the example in Figure 7.1, we can see that the *a*'s of Student A are different from those of Student B. There is something about Student A's letters that make them different from Student B's letters. Neither group of *a*'s is alike. Each group varies within a certain range. But, at the same time, each group is distinguishable from the others. This distinguishable and, as it were, *constant variable within limits*, is defined by Dr. Shewhart as the second characteristic of control.

The reader may, perhaps, remember examples when someone who was presenting quality results was trying to explain the variation from month to month without accepting the fact that controlled quality must be a variable quality (see Figure 7.2 for an illustrative example). We need to get used to the fact that in all forms of production an element of chance enters. We cannot expect, for example, to receive the same process yield from day to day, or month to month. The specific problem that should concern us is the formulation of a scientific basis for prediction, taking into account the element of chance. And if we want to reduce the influence of chance, we need to accumulate more knowledge about the process, because in real life any annoying cause or phenomenon will be treated as a chance cause.

What we really need to be worried about is the excess variation of the process. How much can the quality of a product be varied and yet be controlled? The answer to this question will be given in the following sections.

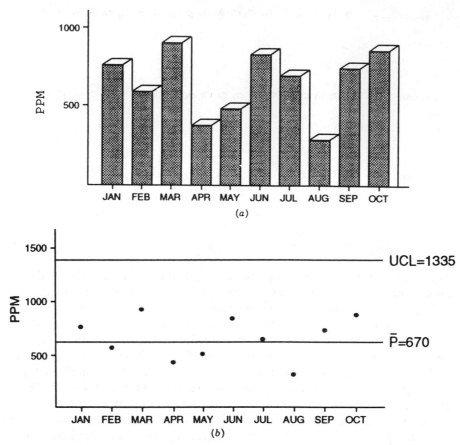

Figure 7.2   An illustrative example of the concept of control. (a) A bar chart showing monthly PPM level. This example is based on an actual foil from a manager's presentation to an executive meeting, and below are the possible reasons for explaining the different ppm results.

January—The beginning of the year is always like that.

February—The improvement is due to people getting back into a working mode.

March—The manager didn't know exactly what happened, but maybe it was because he/she was on vacation.

April—The manager came back from vacation (this is for sure).

May—'' · · · , etc.''

(b) A p-chart with moving range based on the same data. The p-chart shows that for a period of ten months, the process was stable, and the between month variation is a result of the process capability. The only way to reduce this variation is to improve the process.

7.3   CONTROL CHARTS

7.3.1   The Birth of Control Charts

In a memorandum prepared on May 16, 1924, Dr. Walter A. Shewhart of the Bell Telephone Laboratories made the first sketch of a modern "control chart."[4] This case can be considered as the first of a new method of statistical process control. Further developing his idea, in 1931 Dr. Shewhart published a book, *Economic Control of Quality of Manufactured Products*, which became the fundamental axiom of subsequent applications of statistical methods to process control.

In recent years the technical progress generated a demand for other forms of control charts. However, Dr. Shewhart's control chart still remains active and, together with other forms of process control, is successfully applied in the industry.

In this chapter we will give a brief description of the theory, the methods of construction and interpretation of control charts, and their applications and further interpretation by using real examples

7.3.2   What Is a Control Chart?

A control chart may be defined as a graphical method for evaluating whether a process is or is not in a state of statistical control.[5] It consists of a center

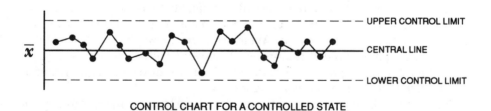

CONTROL CHART FOR A CONTROLLED STATE

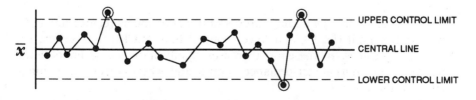

CONTROL CHART FOR A OUT-OF-CONTROL STATE

Figure 7.3   Examples of control charts.

line, a pair of control limits located above and below the center line, and measurement values (points plotted on the chart which represent the condition of the process). If all plotted values are lying within the control limits without any particular tendency, the process is considered to be in a state of statistical control. However, if some points fall outside the control limits or show a peculiar form, the process is said to be out of control (see Figure 7.3).

7.3.3   The Purpose of a Control Chart

Dr. Walter A. Shewhart suggests that the control chart may: (1) serve to define the goal or standard for a process that the management might strive to attain, (2) be used as an instrument for attaining the goal, and (3) serve as a means for judging whether the goal has been reached. It is thus an instrument to be used in specification, production, and inspection, and when so used, brings three phases of industry into an independent whole.[6]

In other words, the control chart approach can be used in all major stages of manufacturing: In the beginning when we need to establish optimal goals and standards; when we need to control the process to achieve the goal; and ultimately the control chart can be used in determining how we reach the goal, or how far we are deviating from the established goal or standards. Experience shows that applying control charts in this broad sense gives the most effective results.

7.3.4   Introduction to Control Charts

Figure 7.4 represents different records of quality performance. They are adapted from a report submitted to management. What information can we obtain from these types of charts? Which of them shows the greatest stability or control? Can we predict the process behavior? Did we achieve the goal? Do we have a stable process? These are all important questions that cannot be answered by the run charts in Figure 7.4.

Now let's look at the same patterns after transforming them into control charts (see Figure 7.5). Now the data "talk." We can see the average capability of the systems; we can see if the system is in a state of statistical control or not; and we can see where and in what period the system was disturbed, etc.

This is to show that by adding three additional elements [center line (CL), upper control limit (UCL), and lower control limit (LCL)] to a simple run chart, the usefulness of the chart increases significantly. These limits determine when an unusual performance occurs and when corrective action is needed on the process.

To improve a process, it is very important to be able to recognize when a process performs unusually and when it is performing normally. Sometimes we think that if the points on the chart vary, the underlying production

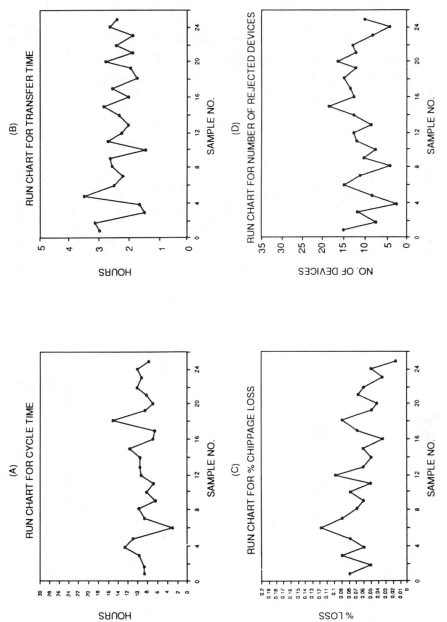

Figure 7.4 Run charts for different quality performances.

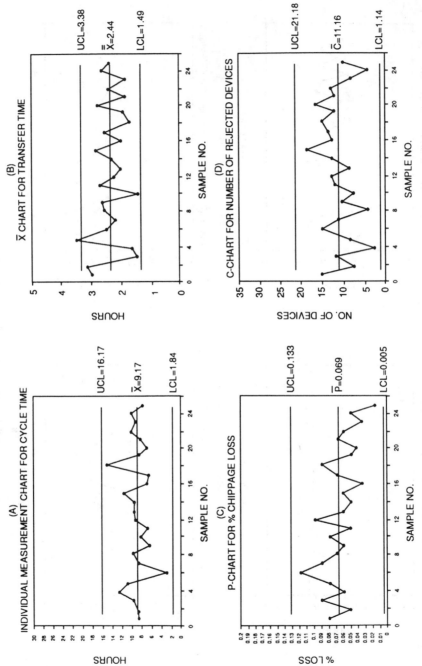

Figure 7.5 Control charts for different quality performances.

conditions (quality, cycle time, etc.) are also varying. In reality, such conditions may be as uniform as possible and the points still vary, simply by chance.

In this chapter we will demonstrate the concept and application of control charts in manufacturing areas. We will also show how their use impacts on quality improvement and cost reduction.

7.3.5 Some Important Concepts Related to Process Control

Before describing the methodology of process control, it would be useful to become familiar with some concepts, terms, and definitions related to process control.

Two Types of Data

When applying statistical principles for process control or for other purposes, we often use either the term *variable* or *attribute*, indicating the type of data available. This distinction helps us select the proper statistical technique needed for a particular purpose.

When making a record of an actual measured quality characteristic, such as the strength of a wire bond in grams, or the speed of an electronic device in nanoseconds, etc., the quality is said to be expressed by *variables*. When a record shows only the number of units conforming and the number of units failing to conform to the specification requirements, it is said to be recorded by *attributes*. In this case the measurement units can be expressed in fraction defective, percent nonconformities, parts per million, etc. For example, the quality level of a semiconductor assembly process is often stated in parts per million. More recently, the terms "conformance" and "nonconformance" have become popular. This chapter will familiarize the reader with control charts for variables and attribute data and the types of the controls that can be applied for both types of data.

Causes of Variation

After a broad study, Dr. Shewhart determined that all manufacturing processes display variation that can be divided into two major components: (1) chance causes of variation that are attributed to inherent variation (currently called *random variation*) and (2) assignable causes of variation that are attributed to intermittent variation. (Dr. Deming has termed assignable causes of variation as *special* causes and chance causes as *common* causes.)

Dr. Shewhart also concluded that assignable causes could be economically discovered and removed from the process, but that chance (random) causes could not be economically discovered and removed without basic changes in the process (system). By random causes we mean a large variety of small influences on the process lying behind a particular measurement. For example, even when measuring the same part several times and trying to keep all conditions as consistent as possible, we will still get slightly different readings. This can be caused by floor vibration, temperature changes, operator influence, and other small sources of variation that we cannot even detect. All of these sources, which have a minor effect individually, together introduce variability into the resulting measurements. There is bound to be natural variation to every process, and this can be observed if the measuring instrument can recognize the variation. The control chart has a unique ability to detect and identify causes.

Let's take a second look at Figure 7.5. As one can see, Figure 7.5(b) has one point that is outside the upper control limit (UCL) and Figure 7.5(c) shows a trend. But Figure 7.5(d), having all the points within the control limits, also has nine consecutive points above the center line. All those are evidence of unnatural patterns associated with assignable causes. Only Figure 7.5(a) has a natural pattern, which means that it represents a process that is in a state of statistical control, so the variation displayed on the chart belongs only to chance causes. This particular chart represents a process that should be left alone, because of its "natural" behavior. This is the only process of the four shown for which future values can be predicted.

However, if we are not satisfied with the capability of this process, we need to change the system (which is the responsibility of management and cannot be done just by troubleshooting on the manufacturing floor). The remaining three charts represent situations where action should be taken to find the assignable causes and bring the process back in control or to understand, as in the case of Figure 7.5(c), why the process is improving. Knowing when to act on the process and when not to act is an important economic issue. By missing the opportunity to interfere with the process, we may have losses from rejects and rework. But, interfering with the process when only common causes impact the process will increase the variability in the process. This is why control charts are so important for quality improvement.

7.3.6   Control Charts for Variables

In the electronics industry many process characteristics are measured on a continuous scale. This is why control charts for variables are used intensively. The $\overline{X}$-R chart is one of the most popular techniques when working with variable data. This chart provides information not only about the mean value of the quality characteristic but also about its variability. Another important chart to control variable data is the $\overline{X}$-S chart, which is becoming more and more popular in recent years because of the increase of computer availability. This chart has the same function as the $\overline{X}$-R chart, the main difference being that the process variability is controlled, not by the range (R) but by the standard deviation (S). Both the $\overline{X}$-R chart and $\overline{X}$-S chart require several measurements for every point on the chart. However, there are some situations when the number of measurements is limited or very expensive to produce. In this case, an *X-R* chart for individuals is used where every point on the chart represents only one measurement.

These are the three Shewhart control charts, which have a broad application in the industry when variable data is available. In addition there are other newer charts such as the cumulative sum (CUSUM) control chart, control charts with modified control limits, the moving average control chart, and other additions or modifications of Shewhart's control charts. Some of these charts will be illustrated in a separate section.

In this section we will only discuss the classical Shewhart control charts for variable data: $\overline{X}$-R, $\overline{X}$-S, and control charts for individual measurements. Listed below are step-by-step procedures to construct these charts.

7.3.6.1 An Example of Making an $\overline{X}$-R Chart

Trimming is the process of removing the link bars from device lead frames. The quality of this process is measured by using an optical comparator. Considering that the measurement process generates variable data, an $\overline{X}$-R control chart can be applied here. The procedure for making the chart by using the data from the trimming process follows.

Procedure	Example
Step 1	*Step 1*
Determine the subgroup size (n) and the frequency of sampling. For an $\overline{X}$-R chart, usually a subgroup sample size (n) of 4 or 5 is practical. To calculate the control limits, the number of subgroups (k) should be at least 20, preferably 25. The frequency of sampling depends on the production rate and other economical and technical conditions (see Section 7.6).	In this case, it was decided to randomly select a subgroup of $n = 5$ units every two hours, and the lead length was measured. A total of $k = 22$ subgroups was sampled.
Step 2	*Step 2*
Record the data on a datasheet. The datasheet should be designed so that it is easy to compute the values of the average ($\overline{X}$) and range (R) for each subgroup.	The results are recorded in a datasheet as shown in Table 7.1.

TABLE 7.1   Datasheet for the Lead Dimension Measurements

Sample Number	Readings					$\bar{X}$	R
	X_1	X_2	X_3	X_4	X_5		
1	171	172	171	173	174	172.2	3.0
2	172	171	172	175	173	172.6	4.0
3	169	170	170	174	171	170.8	5.0
4	173	174	171	174	175	173.4	4.0
5	174	173	172	172	175	173.2	3.0
6	170	171	170	171	170	170.4	1.0
7	172	170	172	169	174	171.4	5.0
8	173	172	174	170	174	172.6	4.0
9	171	168	172	171	170	170.4	4.0
10	172	170	172	173	175	172.4	5.0
11	170	170	173	171	172	171.2	3.0
12	171	174	171	173	171	172.0	3.0
13	170	172	171	174	170	171.4	4.0
14	170	170	171	172	171	170.8	2.0
15	173	172	173	171	171	172.0	2.0
16	175	171	171	171	172	172.0	4.0
17	174	175	170	173	173	173.0	5.0
18	170	170	173	172	171	171.2	3.0
19	169	170	173	172	170	170.8	4.0
20	170	172	172	175	170	171.8	5.0
21	170	170	170	171	170	170.2	1.0
22	170	171	173	173	171	171.6	3.0
Average						171.7	3.5

Procedure	Example
Step 3	*Step 3*
Compute the average ($\bar{X}$) for each subgroup by the following formula: $$\bar{X} = \frac{X_1 + X_2 + \cdots + X_n}{n}$$	The average for each subgroup is computed and recorded in the "$\bar{X}$" column of the datasheet (see Table 7.1). As an example, for subgroup 1: $$\bar{X} = \frac{171 + 172 + 171 + 173 + 174}{5}$$ $$= 172.2$$

Procedure	Example
Step 4	*Step 4*
Find the range (R) for each subgroup by the following formula: $$R = X_{\text{(largest value)}} - X_{\text{(smallest value)}}$$	The range for each subgroup is computed and recorded in the "R" column of the datasheet (see Table 7.1). As an example, for subgroup 1: $$R = 174 - 171 = 3.0$$
Step 5	*Step 5*
Calculate the overall average ($\overline{\overline{X}}$) by summing all the $\overline{X}$'s and then divide by the number of subgroups (k): $$\overline{\overline{X}} = \frac{\overline{X}_1 + X_2 + \cdots + \overline{X}_k}{k}$$	The overall average ($\overline{\overline{X}}$) is calculated as follows: $$\overline{\overline{X}} = \frac{172.2 + 172.6 + \cdots + 171.6}{22}$$ $$= 171.7$$
Step 6	*Step 6*
Compute the average value of the range ($\overline{R}$) by totaling R for all subgroups and then dividing by the number of subgroups (k): $$\overline{R} = \frac{R_1 + R_2 + \cdots + R_k}{k}$$	The average range ($\overline{R}$) is computed as follows: $$\overline{R} = \frac{3.0 + 4.0 + \cdots + 3.0}{22}$$ $$= 3.5$$
Step 7	*Step 7*
Calculate the control limits for an $\overline{X}$ chart and an R chart by the following formulas:	The control limits are calculated as follows:
$\overline{X}$ *chart:*	$\overline{X}$ *chart:*
Center line (CL) $= \overline{\overline{X}}$	CL $= 171.7$
Upper control limit (UCL)$=\overline{\overline{X}}+A_2\overline{R}$	UCL $= 171.7 + 0.577(3.5) = 173.7$
Lower control limit (LCL) $= \overline{\overline{X}}-A_2\overline{R}$	LCL $= 171.7 - 0.577(3.5) = 169.6$

TABLE 7.2   Factors for Computing Control Limits

n	A_2	A_3	B_3	B_4	D_3	D_4	E_2
2	1.880	2.659	0	3.267	0	3.267	2.660
3	1.023	1.954	0	2.568	0	2.575	1.772
4	0.729	1.628	0	2.266	0	2.282	1.457
5	0.577	1.427	0	2.089	0	2.115	1.290
6	0.483	1.287	0.030	1.970	0	1.924	1.184
7	0.419	1.182	0.118	1.882	0.076	1.924	1.109
8	0.373	1.099	0.185	1.815	0.136	1.864	1.010
9	0.337	1.032	0.239	1.761	0.184	1.816	1.050
10	0.308	0.975	0.284	1.716	0.223	1.777	0.975

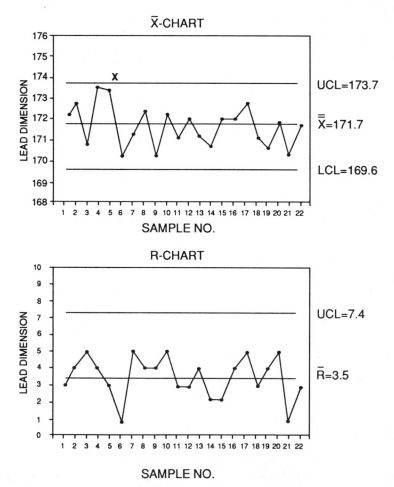

Figure 7.6   An $\bar{X}$-R chart for lead dimension measurements.

Procedure	Example
R chart: Center line (CL) = $\overline{R}$ Upper control limit (UCL) = $D_4\overline{R}$ Lower control limit (LCL) = $D_3\overline{R}$ A_2, D_4, and D_3 are the coefficients determined by the size of a subgroup (n) and are shown in Table 7.2. Note that for R chart LCL is not considered when n is smaller than 7. A more detailed table is shown in Appendix B.	*R chart:* CL = 3.5 UCL = 2.115(3.5) = 7.4 LCL = not considered
Step 8 Construct the control chart.	*Step 8* The $\overline{X}$-R chart is constructed as shown in Figure 7.6.
Step 9 Interpret the results based on the chart pattern. *Note:* See Figure 7.15 for tests for special causes.	*Step 9* It was observed that in the $\overline{X}$ chart from points 3 to 5 there were 2 out of 3 points in Zone C. This indicates that there may be assignable causes in this period. The process improvement team decided to investigate. The results showed that link bars were stuck on the trim bar in that period. Corrective actions have been taken.

7.3.6.2 An Example of Making an $\overline{X}$-S Chart

To increase the sensitivity of controlling the tin-plating process quality of an electronics component, it was decided to replace the existing $\overline{X}$-R chart with a more sensitive $\overline{X}$-S chart. The step-by-step procedure for making this chart follows.

Procedure	Example
Step 1 Determine the subgroup size and frequency of the sample. Usually for an $\overline{X}$-S chart, a sample size (n)	*Step 1* In this particular example, a sample size (n) of 15 was taken from the process in a four-hour interval.

Procedure	Example
of more than 10 is used. The frequency of the sample depends on the production rate and other economical and technical conditions (see Section 7.6). A number of at least 20 to 25 subgroups (k) is needed to calculate the control limits.	*Note:* For simplicity here, the calculations will use a sample size of $n = 5$.
Step 2	*Step 2*
Record the data on a datasheet. The datasheet should be designed so that it is easy to compute the values of average ($\bar{X}$) and sigma (S) for each subgroup.	The results were recorded on a datasheet as shown in Table 7.3.

TABLE 7.3   Datasheet for the Tin-Plating Thickness Measurements

Lot Number	Readings (microinches)					$\bar{X}$	S
	X_1	X_2	X_3	X_4	X_5		
1	638	520	629	571	556	582.8	44.7
2	652	640	556	682	573	620.6	48.1
3	571	613	566	591	540	576.2	24.6
4	636	566	575	553	636	593.2	35.6
5	695	596	554	688	627	632.0	53.9
6	661	594	530	542	493	564.0	57.0
7	622	593	599	662	626	620.4	24.4
8	566	688	661	641	662	643.6	41.6
9	609	611	628	601	625	614.8	10.2
10	526	578	568	637	592	580.2	35.9
11	644	584	574	567	550	583.8	32.1
12	596	636	646	547	631	611.2	36.3
13	638	571	669	672	602	630.4	39.0
14	511	694	632	562	493	578.4	75.3
15	575	605	654	602	646	616.4	29.5
16	667	592	670	683	692	660.8	35.6
17	625	678	669	673	622	653.4	24.6
18	652	598	699	659	719	665.4	41.7
19	659	652	668	665	615	651.8	19.2
20	561	613	643	610	664	618.2	34.9
21	546	603	524	607	559	567.8	32.4
22	645	565	679	651	673	642.6	40.9
Average						614.0	37.1

Procedure	Example
Step 3	*Step 3*

Step 3

Compute the average ($\overline{X}$) for each subgroup by the following formula:

$$\overline{X} = \frac{X_1 + X_2 + \cdots + X_n}{n}$$

Step 3

The average for each subgroup was computed and recorded in the "$\overline{X}$" column on the datasheet (see Table 7.3). As an example, for subgroup 1:

$$\overline{X} = \frac{638 + 520 + 629 + 571 + 556}{5}$$

$$= 582.8$$

Step 4

Find the sigma (S) for each subgroup by the following formula:

$$S = \sqrt{\frac{\Sigma(X - \overline{X})^2}{n-1}}$$

Step 4

The sigma for each subgroup was computed and recorded in the "S" column on the datasheet (see Table 7.3). As an example, for subgroup 1:

$$S = \sqrt{\frac{(638 - 582.8)^2 + (520 - 582.8)^2 + \cdots + (556 - 582.8)^2}{5-1}}$$

$$= 44.7$$

Step 5

Calculate the overall average ($\overline{\overline{X}}$) by summing all the $\overline{X}$'s and then dividing by the number of subgroups (k):

$$\overline{\overline{X}} = \frac{\overline{X}_1 + \overline{X}_2 + \cdots + \overline{X}_k}{k}$$

Step 5

The overall average ($\overline{\overline{X}}$) was calculated as follows:

$$\overline{\overline{X}} = \frac{582.8 + 620.6 + \cdots + 642.6}{22}$$

$$= 614.0$$

Procedure	Example
Step 6	*Step 6*
Compute the average value of the sigma ($\bar{S}$) by totaling S for all subgroups and then dividing by the number of subgroups (k): $$\bar{S} = \frac{S_1 + S_2 + \cdots + S_k}{k}$$	The average sigma ($\bar{S}$) was computed as follows: $$\bar{S} = \frac{44.7 + 48.1 + \cdots + 40.9}{22}$$ $$= 37.1$$
Step 7	*Step 7*
Calculate the control limits for an $\bar{X}$ chart and an S chart by the following formulas:	The control limits were calculated as follows:
$\bar{X}$ chart:	*$\bar{X}$ chart:*
Center line (CL) = $\bar{\bar{X}}$ Upper control limit (UCL)=$\bar{\bar{X}}+A_3\bar{S}$ Lower control limit (LCL)=$\bar{\bar{X}}-A_3\bar{S}$	CL = 614.0 UCL = 614.0 + 1.427(37.1) = 667.0 LCL = 614.0 − 1.427(37.1) = 561.0
S chart:	*S chart:*
Center line (CL) = $\bar{S}$ Upper control limit (UCL) = $B_4\bar{S}$ Lower control limit (LCL) = $B_3\bar{S}$	CL = 37.1 UCL = 2.089(37.1) = 77.6 LCL = not considered
A_3, B_4, and B_3 are the coefficients determined by the size of a subgroup (n) and are shown in Table 7.2. Note that for the S chart LCL is not considered when n is smaller than 6. A more detailed table is shown in Appendix B.	
Step 8	*Step 8*
Construct the control chart.	The $\bar{X}$-S chart was constructed as shown in Figure 7.7.

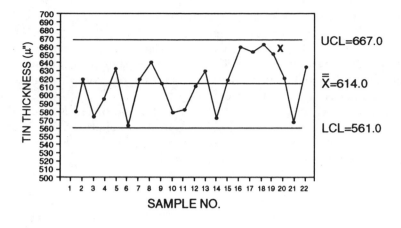

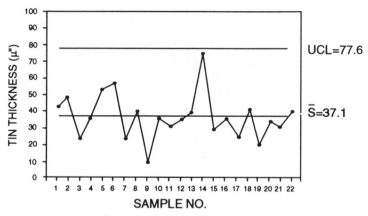

Figure 7.7 An $\bar{X}$-S chart for tin-plating thickness measurements.

Procedure	Example
Step 9	*Step 9*
Interpret the results based on the chart pattern. *Note*: See Figure 7.15 for tests for special causes.	It was observed that in the $\bar{X}$ chart from points 15 to 19 there were 4 out of 5 points in Zone B and beyond. This indicates that the process is not behaving normally. Therefore, the process improvement team decided to investigate the cause for the occurrence of high tin thickness, which would deplete the costly anodes more quickly.

7.3.6.3 An Example of Making an X-R Chart

The control chart for individual measurements (X-R chart) is useful when only one observation per lot or batch is available. And, since this type of chart is not as sensitive as an $\overline{X}$-R chart, it is recommended to use 2 sigma limits here. What makes this chart useful is that the control limits can be directly compared to the tolerance limits.

The incoming chemical developing solution for photoresist is received on a regular basis at two batches per day. To control the quality of the chemicals, the control chart for individual measurements has been installed. The step-by-step procedure for making this chart follows.

Procedure	Example
Step 1	*Step 1*
Determine the subgroup size. *Note*: Usually a minimum of 30 individuals in a subgroup of at least 2 should be collected. Meanwhile, a simple run chart can be maintained and compared with specification limits.	Since there are two batches per day, a subgroup sample size of $n = 2$ is rational.
Step 2	*Step 2*
Randomly select the sample from the lot (or batches) and record on the datasheet.	A random sample was taken and measured from each batch. A total of 30 individual measurements arranged in $k = 15$ subgroups, were collected as shown in Table 7.4.
Step 3	*Step 3*
Compute the process average as follows: $$\overline{X} = \frac{\begin{array}{c}X_{A1} + X_{B1} + X_{A2}\\ + X_{B2} + \cdots + X_{A15} + X_{B15}\end{array}}{k \times n}$$	The process average ($\overline{X}$) is computed as follows: $$\overline{X} = \frac{\begin{array}{c}0.305 + 0.304 + 0.301\\ + 0.303 + \cdots + 0.305 + 0.306\end{array}}{15 \times 2}$$ $$= 0.304$$

TABLE 7.4   The Normality Values of Photoresist Developer

Subgroup Number	Measurements		
	X_A	X_B	Range
1	0.305	0.304	0.001
2	0.301	0.303	0.002
3	0.305	0.302	0.003
4	0.306	0.304	0.002
5	0.301	0.304	0.003
6	0.305	0.304	0.001
7	0.307	0.301	0.006
8	0.305	0.304	0.001
9	0.302	0.305	0.003
10	0.304	0.305	0.001
11	0.306	0.304	0.002
12	0.306	0.305	0.001
13	0.301	0.305	0.004
14	0.302	0.304	0.002
15	0.305	0.306	0.001
Totals		9.121	0.033
Averages		0.304033	0.0022

Procedure	Example
Step 4	*Step 4*
Compute the range (R) for each subgroup as follows:	The range for each subgroup is calculated and recorded on the datasheet (see Table 7.4). For example, for subgroup 1:
$$R = x_{(largest\ value)} - x_{(smaller\ value)}$$	$$R = 0.305 - 0.304$$ $$= 0.001$$
Step 5	*Step 5*
Compute the average range ($\bar{R}$) as follows:	The average range ($\bar{R}$) is computed as follows:
$$\bar{R} = \frac{R_1 + R_2 + \cdots + R_k}{k}$$	$$\bar{R} = \frac{0.001 + 0.002 + \cdots + 0.001}{15}$$ $$= 0.0022$$

Procedure	Example
Step 6	*Step 6*
Calculate the 2 sigma control limits for the X chart as follows:	The 2 sigma control limits for the X chart are calculated as follows:
$$\text{UCL}_X = \bar{X} + \left(\tfrac{2}{3}\right) E_2 \bar{R}$$	$$\text{UCL}_X = 0.304 + \left(\tfrac{2}{3}\right)(2.66)(0.0022)$$
$$\text{LCL}_X = \bar{X} - \left(\tfrac{2}{3}\right) E_2 \bar{R}$$	$$= 0.308$$
where E_2 is a coefficient determined by the size of a subgroup (n) and is shown in Table 7.2.	$$\text{LCL}_X = 0.304 - \left(\tfrac{2}{3}\right)(2.66)(0.0022)$$ $$= 0.300$$
Step 7	*Step 7*
Calculate the control limits for the R chart as follows:	The control limits for the R chart are calculated as follows:
$$\text{UCL}_R = D_4 \bar{R}$$	$$\text{UCL}_R = (3.267)(0.0022)$$
$$\text{LCL}_R = D_3 \bar{R}$$	$$= 0.007$$
where D_4 and D_3 are coefficients determined by subgroup sample size (n) and can be found in Table 7.2.	$$\text{LCL}_R = \text{not considered}$$
Step 8	*Step 8*
Construct the control charts.	The *X-R* chart was constructed as shown in Figure 7.8.
Step 9	*Step 9*
Interpret the control charts. *Note:* See Figure 7.15 for tests for special causes.	*Interpretation:* As one can see, all the points in both the X and R charts are within control limits and show a normal pattern. Therefore, we can conclude that the quality of the chemical solution is consistent from batch to batch. Also, the specification range for this parameter is 0.29 to 0.31. From the chart, we can

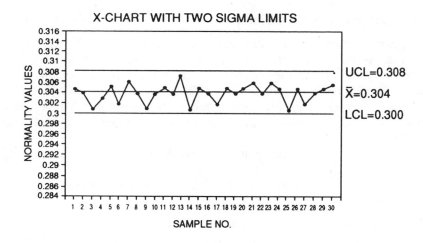

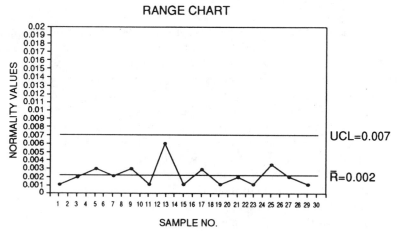

Figure 7.8 An example of an *X-R* chart.

Procedure	Example
	see that all the batches are within the specification limits. However, all the measurements are very close to the upper specification limits. Action should be taken to bring the process average to the nominal of the specification.

7.3.7 Control Charts for Attributes

Many quality characteristics can only be observed as attributes; i.e., by classifying each item inspected into one of two categories, either conforming or nonconforming to the specifications. Note that if a unit of product has at least one defect, then the unit is called "defective" or "nonconforming." (In this book the two terms are often used interchangeably.) These quality characteristics are controlled by a group of techniques called attribute control charts.

There are several types of attribute charts developed for different cases:

1. p chart, the control chart for fraction (or percent) defective.
2. np chart, the control chart for number of defective units.
3. c chart, the control chart for number of defects.
4. u chart, the control chart for number of defects per unit.

These charts have broad applications for attribute data. Even when quality characteristics can be observed by variable data, there are often situations where using attribute control charts is more appropriate. For example, in the electronics industry there are thousands of parameters and each can be controlled by an $\overline{X}$-R chart. But this would be totally impractical and uneconomical. Not all of them need to be separately controlled by variables. This is why a number of quality characteristics that can be measured on a continuous scale are transformed into attributes by applying some kind of "go" and "no-go" gauges or instruments. The attribute control chart can also be used to control a group of parameters. For example, by introducing a p chart at the end of the manufacturing process (or subprocess), we can prioritize the areas for process capability studies.

In recent years these charts have been applied in the nonmanufacturing and service areas. For example, attribute control charts are used to control the on-time delivery process, purchasing process, and other working activities that were not treated as processes before.

Besides being used for their simplicity, control charts for attributes are also popular because of their low cost in comparison to variable control charts. There are several reasons why the cost is lower. First, because the data are often already collected for other purposes. Second, because the cost of computing and charting is less. And finally, because one chart for attributes may be applied to a number of quality characteristics, whereas separate $\overline{X}$-R charts are necessary for each measured quality characteristic.

In the previous section, when describing control charts for variables, we mentioned that the sample size needed to get the same information using variable data is much less than when using attribute data. Here we are saying that control charts for attributes are more economical than control charts for variables. This may sound contradictory, but it is not. No dimension should be chosen to maintain $\overline{X}$-R charts unless there is an opportunity to save costs related to excess variability, low yields, scrap and rework, or otherwise to improve the quality of the manufacturing product. The decision about when to use attribute or variable charts is an engineering, managerial, and economic decision.

7.3.7.1 The p Chart

A p chart represents the proportion of defective units compared to the total. That is, the "proportion defective." It is the third most sensitive chart (after the $\bar{X}$-S chart and $\bar{X}$-R chart) for identifying special causes. The chart also forms the same kinds of patterns as the $\bar{X}$-R or $\bar{X}$-S chart (cycles, trends, freaks, shifts, etc.) and can easily be interpreted once the controlled variables are known.

The center line ($\bar{p}$) on the chart represents an average of the fraction defective of the process. It can be considered as the process capability if the control chart is in a state of statistical control.

The control limits on a p chart are usually ± 3 sigma limits. However, the distance of these limits from the center line depends on the sample size used to monitor the process. If the sample size is constant, the limits are also constant. By decreasing the sample size the limits become wider, or by increasing the sample size the limits become narrower. The p chart can be used in both cases; whether the sample size is constant or not. However, when we can use constant sample sizes, it is more convenient to apply an np chart (described in Section 7.3.7.3).

Although the p chart still remains one of the most versatile and widely used techniques, its application has become less and less effective and more difficult to interpret when the process approaches a low ppm quality level. The reason is because a p chart is based on the theory that values of the subgroup p_s which have a binomial distribution can be approximately described by a normal distribution with the mean and standard deviation estimated by p and $[(p(1-p)/n_s]^{1/2}$, respectively.

It is important to remember here that the normal curve gives an excellent approximation of the binomial distribution when p is close to 0.5. The more p deviates from 0.5 (50%), the worse the approximation gets. At the same time, for values of p significantly different from 0.5, the approximation of the normal distribution to the binomial distribution gets better when the value of n increases.[7] In practice, a normal approximation of the binomial is considered legitimate as long as $np \geq 5$. This means that if applying a p chart for a process with, let's say, a 500 ppm quality level, the subgroup sample size should be at least 10,000 units, so np will be equal to 5. The necessity of large sample sizes make the p chart difficult to use as a tool for real-time process control.

There are other techniques that have been developed as substitutes for a p chart when working in a low ppm environment, but the p chart still remains one of the most important tools. It can be successfully used to control processes with 10,000 ppm and above; for reporting processes where the combined sample sizes are very large; and in a number of supporting and nonmanufacturing areas. The procedure for constructing a *p* chart is as follows.

7.3.7.2 An Example of Making a p chart

In a hypothetical manufacturing process, an inspection is used to control the process. Because of the difference in the subgroup sample size, let's assume a *p* chart is appropriate to use to control the quality level at this stage of the process. An example of this type of chart follows.

Procedure	Example
Step 1	*Step 1*
Determine the subgroup sample size by using the following formula: $$n = \frac{2.3}{\bar{p}}$$ where $\bar{p}$ is the estimated average proportion defective and 2.3 is a constant that will allow us to have at least one defective part 90% of the time.	From previous information, we assume that the average proportion defective is $$\bar{p} = 0.03$$ Thus, the minimum sample size for each subgroup is $$n = \frac{2.3}{\bar{p}} = \frac{2.3}{0.03} = 77$$
Step 2	*Step 2*
The frequency of sampling depends on a number of economic and technical factors. In general, the larger the sample and shorter the time interval, the better centered the sampling process will be.	In this particular case, the frequency of sampling was dictated by the production rate, because a sample was specified to be taken from every lot produced.
Step 3	*Step 3*
Take random samples from the process and inspect them according to the specifications to find the number of defective units. *Note:* To construct a control chart, the number of subgroups (k) should be at least 20 to 25.	Twenty-five samples are taken and inspected. The results are shown on a datasheet in Table 7.5.

TABLE 7.6 Datasheet for the Results of an Incoming Inspection (Samples of Equal Size)

Lot Number	Sample Size (n)	Number of Rejects (np)	Fraction Defective (p)
1	120	4	0.033
2	120	3	0.025
3	120	7	0.058
4	120	3	0.025
5	120	4	0.033
6	120	1	0.008
7	120	5	0.042
8	120	2	0.017
9	120	4	0.033
10	120	3	0.025
11	120	1	0.008
12	120	2	0.017
13	120	1	0.008
14	120	3	0.025
15	120	1	0.008
16	120	2	0.017
17	120	3	0.025
18	120	1	0.008
19	120	5	0.042
20	120	4	0.033
21	120	6	0.050
22	120	3	0.025
23	120	4	0.033
24	120	1	0.008
25	120	5	0.042
	$\sum n = 3000$	$\sum np = 78$	

Procedure	Example
Step 4	*Step 4*
Compute the fraction defective (p) for each subgroup: $$p = \frac{\text{number of defective units}}{\text{number of units in a subgroup}}$$ $$= \frac{np}{n}$$	As an example, for subgroup 1, $$p = \frac{5}{150} = 0.033 = 3.3\%$$ The fraction defectives for all the subgroups are recorded on the datasheet in Table 7.5 (see column 4).

Procedure	Example

Step 5

Find the average fraction defective ($\bar{p}$) which will be the center line for the control chart:

$$\bar{p} = \frac{\text{total number of defective units}}{\text{total number of units inspected}}$$

$$= \frac{\Sigma np}{\Sigma n}$$

Step 5

The average fraction defective is calculated as follows:

$$\bar{p} = \frac{120}{3750} = 0.032$$

Step 6

Considering that the control limits will change depending on the size of the subgroup (n), try to avoid having a large variation between the size of samples. As a rule of thumb, if the subgroup size (n) varies no more than 25% to 30% from the average sample size ($\bar{n}$), the limits can be calculated using $\bar{n}$ for all of the chart. If the variation is large, separate limits for every sample should be computed:

$$\bar{n} = \frac{\text{total number inspected in all samples}}{\text{number of samples in the series } (k)}$$

$$= \frac{\Sigma n}{k}$$

Calculate the upper control limit (UCL) and the lower control limit (LCL):

$$\text{UCL} = \bar{p} + 3\sqrt{\frac{\bar{p}(1-\bar{p})}{n}}$$

$$\text{LCL} = \bar{p} - 3\sqrt{\frac{\bar{p}(1-\bar{p})}{n}}$$

Step 6

Determine the $\bar{n}$:

$$\bar{n} = \frac{\Sigma n}{k} = \frac{3750}{25} = 150$$

Comparing $\bar{n}$ to all the samples, you can see that subgroup 6 has a sample size of 80 which deviates 47% from $\bar{n}$:

$$\frac{150-80}{150} = 0.47 = 47\%$$

Also subgroup 16 has a sample size of 250 which deviates 67% from $\bar{n}$:

$$\frac{250-150}{150} = 0.67 = 67\%$$

The rest deviate in a range of 15% to 25%, so we calculate an overall limit for all the samples except for samples 6 and 16, which will have their own limits. For overall samples:

$$\text{UCL} = 0.032 + 3\sqrt{\frac{0.032(1-0.032)}{150}}$$

$$= 0.075$$

Procedure	Example
Note: When the LCL is negative, we assume that there is no lower limit.	$LCL = 0.032 - 3\sqrt{\dfrac{0.032(1-0.032)}{150}}$ $= -0.011$ For sample 6: $UCL = 0.032 + 3\sqrt{\dfrac{0.032(1-0.032)}{80}}$ $= 0.091$ $LCL = 0.032 - 3\sqrt{\dfrac{0.032(1-0.032)}{80}}$ $= -0.027$ For sample 16: $UCL = 0.032 + 3\sqrt{\dfrac{0.032(1-0.032)}{250}}$ $= 0.065$ $LCL = 0.032 - 3\sqrt{\dfrac{0.032(1-0.032)}{250}}$ $= -0.003$
Step 7 Construct the control chart.	*Step 7* The *p* chart is constructed as shown in Figure 7.9.
Step 8 Make an interpretation based on the control chart pattern. *Note:* See Figure 7.15 for tests for special causes.	*Step 8* *Interpretation:* As one can see from Figure 7.9, point 16 is out of control, which indicates that the process needs to be investigated.

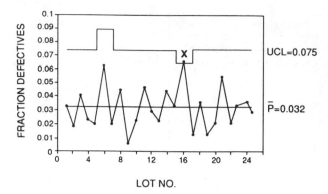

Figure 7.9   A p-chart with unequal sample sizes for a manufacturing process inspec tion.

7.3.7.3 The np Chart
There are situations in manufacturing when it is possible to keep the subgroup sample size constant; e.g., when we take constant sample sizes from the process or from incoming lots to control a given level of quality. In such cases the *np* chart may be used where the actual number of defective units is plotted.

7.3.7.4 An Example of Making an np Chart
Let's assume that from every incoming lot of a certain item, a constant sample will be selected and inspected. In this case, an *np* chart will be appropriate to control the lot quality. A step-by-step procedure to construct this chart follows.

Procedure	Example
Step 1	*Step 1*
Determine the minimum number of units in a sample (n) such that the sample will contain at least one defective unit most of the time. The formula for n is $$n = \frac{2.3}{\bar{p}}$$ where $\bar{p}$ is the estimated average proportion defective and 2.3 is a constant that will allow us to catch at least one defective part 90% of the time.	Based on the available data, the average proportion defective is estimated as $$\bar{p} = 0.03$$ Thus, the minimum sample size for each subgroup is $$n = \frac{2.3}{\bar{p}} = \frac{2.3}{0.03} = 77$$ In this example, it was decided to use a sample size of $n = 120$.

Procedure	Example
Step 2 Take *n* units from each lot and inspect them according to the specification to determine the number of defective units. *Note:* To construct a control chart, the number of subgroups (k) should be at least 20 to 25.	*Step 2* Twenty-five subgroup samples are taken and inspected. The results are shown on a datasheet in Table 7.6.

TABLE 7.6   Datasheet for the Results of an Incoming Inspection (Samples of Equal Size)

Lot Number	Sample Size (n)	Number of Rejects (np)	Fraction Defective (p)
1	120	4	0.033
2	120	3	0.025
3	120	7	0.058
4	120	3	0.025
5	120	4	0.033
6	120	1	0.008
7	120	5	0.042
8	120	2	0.017
9	120	4	0.033
10	120	3	0.025
11	120	1	0.008
12	120	2	0.017
13	120	1	0.008
14	120	3	0.025
15	120	1	0.008
16	120	2	0.017
17	120	3	0.025
18	120	1	0.008
19	120	5	0.042
20	120	4	0.033
21	120	6	0.050
22	120	3	0.025
23	120	4	0.033
24	120	1	0.008
25	120	5	0.042
	$\sum n = 3000$	$\sum np = 78$	

Procedure	Example
Step 3 Calculate the process average ($\bar{p}$) by dividing the total number of defective units in all subgroups by the total number of units inspected. $$\bar{p} = \frac{\Sigma np}{k \times n}$$	*Step 3* $$\bar{p} = \frac{78}{25 \times 120} = 0.026$$
Step 4 Calculate the control limits and center line (CL): $$CL = n\bar{p}$$ Upper control limit (UCL): $$UCL = n\bar{p} + 3\sqrt{n\bar{p}(1-\bar{p})}$$ Lower control limit (LCL): $$LCL = n\bar{p} - 3\sqrt{n\bar{p}(1-\bar{p})}$$	*Step 4* $$CL = 120 \times 0.026$$ $$= 3.12$$ $$UCL = 3.12 + 3\sqrt{3.12(1-0.026)}$$ $$= 8.35$$ $$LCL = 3.12 - 3\sqrt{3.12(1-0.026)}$$ $$= -2.11$$

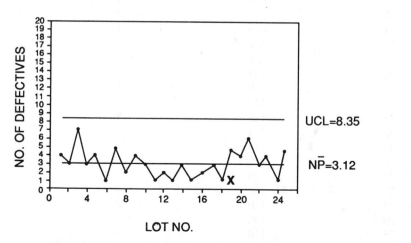

Figure 7.10   An *np*-chart for an incoming inspection.

Procedure	Example
Note: The lower control limit is not considered when its value is negative.	The lower control limit became negative, so its value is not considered.
Step 5	*Step 5*
Construct the control chart.	The *np* chart is constructed as shown in Figure 7.10.
Step 6	*Step 6*
Make an interpretation based on the control chart pattern.	*Interpretation:* Although in Figure 7.10 no points are outside the control limits, from point 10 to 18 there are nine consecutive points below the average. This indicates that a change in the process occurred. An investigation should be conducted to see what the reason was for this change.
Note: See Figure 7.15 for tests for special causes.	

7.3.8 Control Charts for a Number of Defects

Here we will consider a situation in which each sample unit may have several defects. For example, a silicon wafer is inspected for minor blemishes on the surface. Theoretically, there is an infinite number of places where a blemish can occur, but the probability of such a blemish occurring in any given place is very small (virtually zero). Still, there may occur an average of c blemishes per wafer.

In general, we mean a situation where n is large and p is small and $np = c$ is a finite number. In such cases, the application of the c chart (which is based on the Poisson distribution) would be appropriate. The variable plotted on the c chart would be the number of defects in each sample unit, and the sample size must remain constant from sample to sample. That is, the opportunity for a defect to occur must remain constant. In our example this means that the wafer size must remain constant. This does not mean that a constant unit should be one wafer. It can be any number of wafers as long as the type, size, and the number of wafers are kept constant.

If the area of opportunity varies from sample to sample, which may happen in the practical application of the c chart, it is possible to standardize the unit size and use a u chart, where $u = c/k$ and k is the number of inspected units in each sample. Below are the procedures for the two types of charts: c chart and u chart.

7.3.8.1 An Example of Making a c chart

Let's assume there is a production test where we are interested in the number of defects in each unit produced. If the sample size taken from each lot is constant, we can use a *c* chart. A step-by-step procedure to construct this chart follows.

Procedure	Example
Step 1	*Step 1*
Take samples from the process with a constant number of units and inspect them to determine the number of defects according to the specification.	A total of 25 samples with 16 units each are taken from the process and inspected. The results are shown in Table 7.7.

TABLE 7.7 Datasheet for the Results of the Production Test (Samples of Equal Size)

Lot Number	Sample Size (n)	Number of Defects (c)	Defects per Unit (u)
1	16	8	0.50
2	16	6	0.38
3	16	7	0.44
4	16	1	0.06
5	16	6	0.38
6	16	4	0.25
7	16	7	0.44
8	16	6	0.38
9	16	7	0.44
10	16	4	0.25
11	16	5	0.31
12	16	2	0.13
13	16	5	0.31
14	16	4	0.25
15	16	8	0.50
16	16	7	0.44
17	16	4	0.25
18	16	10	0.63
19	16	6	0.38
20	16	3	0.19
21	16	4	0.25
22	16	7	0.44
23	16	6	0.38
24	16	2	0.13
25	16	7	0.44
	$\sum n = 400$	$\sum c = 136$	

Procedure	Example
Note: The number of units for a sample should be large to contain at least two to five defects most of the time.	
Step 2	*Step 2*
Calculate the average number of defects in each sample ($\bar{c}$), which will be the center line for the control chart: $$c = \frac{\text{total number of defects}}{\text{total number of samples}}$$ $$= \frac{\Sigma c}{k}$$	The average number of defects in each sample ($\bar{c}$) is $$\bar{c} = \frac{\Sigma c}{k}$$ $$= \frac{136}{25}$$ $$= 5.44$$
Step 3	*Step 3*
Compute the upper control limit (UCL) and lower control limit (LCL): $$\text{UCL} = \bar{c} + 3\sqrt{\bar{c}}$$ $$\text{LCL} = \bar{c} - 3\sqrt{\bar{c}}$$	The control limits are calculated as follows: $$\text{UCL} = 5.44 + 3\sqrt{5.44}$$ $$= 12.43$$

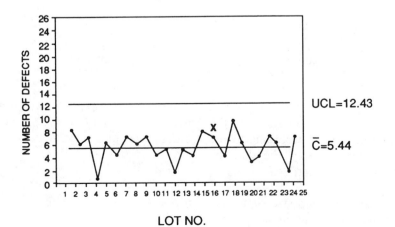

Figure 7.11   A *C*-chart for a production test.

Procedure	Example
Note: When the value of the lower control limit is negative, it means there is no lower control limit for this process.	$LCL = 5.44 - 3\sqrt{5.44}$ $= -1.56$
Step 4 Construct the control chart.	*Step 4* The c chart is constructed as shown in Figure 7.11.
Step 5 Make an interpretation based on the control chart pattern. *Note:* See Figure 7.15 for tests for special causes.	*Step 5* *Interpretation*: We can see in Figure 7.11 that there are no points outside the control limits. However, looking closer, one can find that from point 1 to point 16 there are more than 14 points consecutively alternating up and down. This is an indication that the process is behaving abnormally.

7.3.8.2 An Example of Making a u Chart

Let's assume another inspection is done where we are interested in the number of defects per unit. In this case, let's assume the sample size may vary from lot to lot, so a u chart is the right tool to use. A step-by-step procedure to construct this chart follows.

Procedure	Example
Step 1 Take sample units from the process and inspect them to determine the number of defects according to the specifications. Group the data by lots, type of product, etc. *Note:* The subgroup size should be large enough to contain at least two to five defects most of the time.	*Step 1* Twenty-five samples are taken and inspected. The results are shown on a datasheet in Table 7.8.

TABLE 7.8   Datasheet for the Results of an Inspection for Defects per Unit (Samples of Unequal Size)

Lot Number	Sample Size (n)	Number of Defects (c)	Defects per Unit (u)
1	16	6	0.38
2	16	4	0.25
3	16	5	0.31
4	25	1	0.04
5	25	5	0.20
6	25	8	0.32
7	16	7	0.44
8	16	6	0.38
9	36	6	0.17
10	36	4	0.11
11	36	3	0.08
12	36	2	0.06
13	16	5	0.31
14	16	6	0.38
15	25	3	0.12
16	25	7	0.28
17	25	4	0.16
18	16	8	0.50
19	16	6	0.38
20	36	3	0.08
21	36	4	0.11
22	16	5	0.31
23	16	6	0.38
24	16	5	0.31
25	16	7	0.44
	$\sum n = 574$	$\sum c = 126$	

Procedure	Example
Step 2	*Step 2*
Calculate the number of defects per unit (u):	As an example, for subgroup 1:
$$u = \frac{\text{number of defects per subgroup}}{\text{number of units per subgroup}}$$ $$= \frac{c}{n}$$	$$u = \frac{6}{16} = 0.38$$ The defects per unit for all the subgroups are recorded on the datasheet (see column 4 in Table 7.8).

Procedure	Example
Step 3	*Step 3*

Step 3

Find the average number of defects per subgroup ($\bar{u}$), which will be the center line for the control chart:

$$\bar{u} = \frac{\text{total defects for all subgroups}}{\text{total units for all subgroups}}$$

$$= \frac{\Sigma c}{\Sigma n}$$

Step 3

The average number of defects per subgroup ($\bar{u}$) is calculated as follows:

$$u = \frac{\Sigma c}{\Sigma n} = \frac{126}{574} = 0.22$$

Step 4

Calculate the upper control limit (UCL) and the lower control limit (LCL):

$$\text{UCL} = \bar{u} + 3\sqrt{\frac{\bar{u}}{n}}$$

$$\text{LCL} = \bar{u} - 3\sqrt{\frac{\bar{u}}{n}}$$

Here the limit values change according to the value of n.

Note: When the value of the lower control limit is negative, it means there is no lower control limit for this process.

Step 4

In our example, we have three different sample sizes. For $n = 16$:

$$\text{UCL} = 0.22 + 3\sqrt{\frac{0.22}{16}} = 0.57$$

$$\text{LCL} = 0.22 - 3\sqrt{\frac{0.22}{16}} = -0.13$$

For $n = 25$:

$$\text{UCL} = 0.22 + 3\sqrt{\frac{0.22}{25}} = 0.50$$

$$\text{LCL} = 0.22 - 3\sqrt{\frac{0.22}{25}} = -0.05$$

For $n = 36$:

$$\text{UCL} = 0.22 + 3\sqrt{\frac{0.22}{36}} = 0.45$$

$$\text{LCL} = 0.22 - 3\sqrt{\frac{0.22}{36}} = -0.01$$

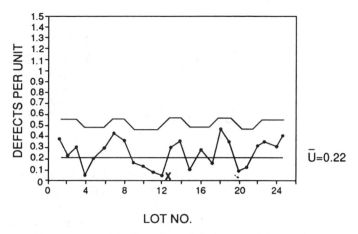

Figure 7.12 A U-chart for a defects per unit inspection.

Procedure	Example
Step 5	*Step 5*
Construct the control chart.	The u chart is constructed as shown in Figure 7.12.
Step 6	*Step 6*
Make an interpretation according to the pattern of the control chart. *Note:* See Figure 7.15 for tests for special causes.	*Interpretation*: We can see from the u chart in Figure 7.12 that all the points are within the control limits. However, from point 7 to point 12 there are six points consecutively decreasing. This is an indication that there may be a small trend in the process. An investigation should be conducted to find out the cause.

7.4   UNDERSTANDING THE CONTROL CHART SIGNALS

In one of the SPC executive audits, when the visitors came close to a workstation, the operator smilingly announced, "Everything is in control." We frequently hear people saying "in control," but what does this term really mean? In short, "control" is when we can predict what will happen (see Section 7.2). For instance, when the operator said that everything was "in

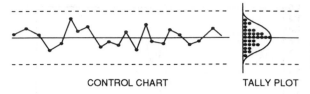

CONTROL CHART   TALLY PLOT

Figure 7.13   A control chart with a normal pattern.

control" he meant that he could predict the process behavior and, if necessary, take corrective actions to make the process behave the way he wants it to. The tool he uses to predict the process is a control chart.

In applying control charts as a statistical method for prediction, it is very important to know how to read and interpret the pattern of the chart. The experience of introducing SPC in the factory showed that it is much easier to learn how to design and introduce a control chart than it is to learn how to extract the necessary information and use it for prediction and prevention. This section contains some techniques that will allow you to understand what the chart is "telling" you.

7.4.1   The Process Is "In Control"

When the special causes have been eliminated from the process, the control chart will have a natural pattern as in Figure 7.13. The points on a natural pattern will be distributed as follows:

1. About two-thirds of them will fall near the center line.
2. A few points will fall close to the control limits.
3. The points will be located back and forth across the center line.
4. The number of points will be balanced on both sides of the center line.
5. There will be no points beyond the control limits.
6. If we tally and plot all the points on one side of the chart, they will form a symmetrical distribution.

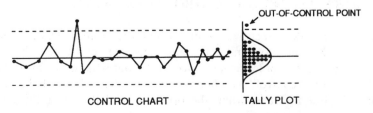

OUT-OF-CONTROL POINT

CONTROL CHART   TALLY PLOT

Figure 7.14   A control chart with points outside the control limits.

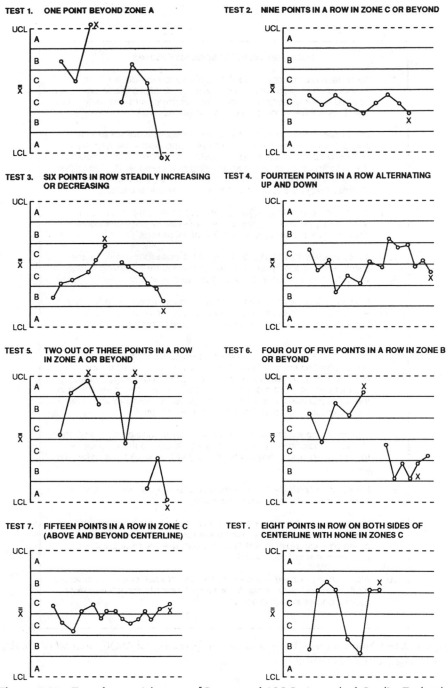

Figure 7.15 Tests for special causes. [Courtesy of ASQC, *Journal of Quality Technology,* October (1984).]

Notes on Tests for Special Causes

1. These tests are applicable to $\bar{X}$ charts and to individuals (X) charts. A normal distribution is assumed. Tests 1, 2, 5, and 6 are to be applied to the upper and lower halves of the chart separately. Tests 3, 4, 7, and 8 are to be applied to the whole chart.

2. The upper control limit and the lower control limit are set at three sigma above the centerline and three sigma below the centerline. For the purpose of applying the tests, the control chart is equally divided into six zones, each zone being one sigma wide. The upper half of the chart is referred to as A (outer third), B (middle third) and C (inner third). The lower half is taken as the mirror image.

3. When a process is in a state of statistical control, the chance of (incorrectly) getting a signal for the presence of a special cause is less than five in a thousand for each of these tests.

4. It is suggested that Tests 1, 2, 3, and 4 be applied routinely by the person plotting the chart. The overall probability of getting a false signal from one or more of these is about one in a hundred.

5. It is suggested that the first four tests be augmented by Tests 5 and 6 when it becomes economically desirable to have earlier warning. This will raise the probability of a false signal to about two in a hundred.

6. Tests 7 and 8 are diagnostic tests for stratification. They are very useful in setting up a control chart. These tests show when the observations in a subgroup have been taken from two (or more) sources with different means. Test 7 reacts when the observations in the subgroup always come from both sources. Test 8 reacts when the subgroups are taken from one source at a time.

7. Whenever the existence of a special cause is signaled by a test, this should be indicated by placing a cross just above the last point if that point lies above the centerline, or just below it if it lies below the centerline.

8. Points can contribute to more than one test. However, no point is ever marked with more than one cross.

9. The presence of a cross indicates that the process is not in statistical control. It means that the point is the last one of a sequence of points (a single point in Test 1) that is very unlikely to occur if the process is in statistical control.

10. Although this can be taken as a basic set of tests, analysts should be alert to any patterns of points that might indicate the influences of special causes in their process.

Figure 7.16 Notes on tests for special causes. [Courtesy of ASQC, *Journal of Quality Technology*, October (1984).]

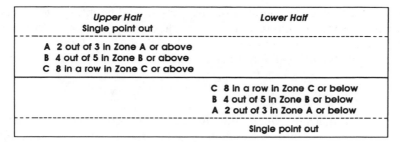

Upper Half Single point out	Lower Half
A 2 out of 3 in Zone A or above B 4 out of 5 in Zone B or above C 8 in a row in Zone C or above	
	C 8 in a row in Zone C or below B 4 out of 5 in Zone B or below A 2 out of 3 in Zone A or below
	Single point out

Figure 7.17 Summary of tests of unnatural patterns. [Adapted by permission from Western Electric Company, *Statistical Quality Control Handbook* (1956), p. 27.]

7.4.2 The Process Is "Out of Control"

When a point made from a subgroup of measurements falls outside its control limits, the process is considered to be "out of control." This means that the process is disturbed by one or more assignable causes of variation (see Figure 7.14). This is the simplest and most frequently used criteria for determining an "out of control" condition. However, a process can also be considered "out of control" when all the points fall inside the control limits. This situation occurs when unnatural patterns of variation are present in the process. For example, it is "not normal" for nine or more consecutive points to fall above or below the center line. There are several other symptoms that indicate an unnatural pattern.

Lloyd S. Nelson[8] collected a set of tests which can be used to determine the existence of special causes in a process (see Figures 7.15 and 7.16). At AMD these tests are printed on the back of the control chart forms so they can be used as a reference to interpret the control chart patterns. Experience has shown that at first people have some difficulty applying and reacting to all of the tests, but after awhile they are able to read the pattern and make decisions about the process, just by a glance of the chart.

Application of the Tests

To introduce testing for unnatural patterns on the manufacturing floor, AMD adapted the use of the theory of runs from the Western Electric Company[9] as a standard operating procedure. Figure 7.17 is the complete set of tests for unnatural patterns that is included in the AMD-SPC specification.

7.5 INTERPRETING THE CONTROL CHART PATTERNS

Now that the reader is familiar with the test for unnatural patterns, let's try to interpret some of the patterns that are frequently seen in practice, starting with the $\overline{X}$-R chart. As we know, the $\overline{X}$ and R charts work together as one statistical technique and, because of this, they should be interpreted together. There is no sense in interpreting an $\overline{X}$ chart when the R chart is not in control. However, the causes affecting the $\overline{X}$ chart and the R chart are different and should be discussed separately.

7.5.1 Interpretation of the $\overline{X}$ Charts

The $\overline{X}$ chart indicates where the process is centered and represents the average of the distribution created by this process. If the center of the distribution shifts, the pattern of the $\overline{X}$ chart will also shift. Figure 7.18 is an $\overline{X}$-R chart that represents the results from a gold-plating process capability study. To observe the process behavior for the period of study, no adjust-

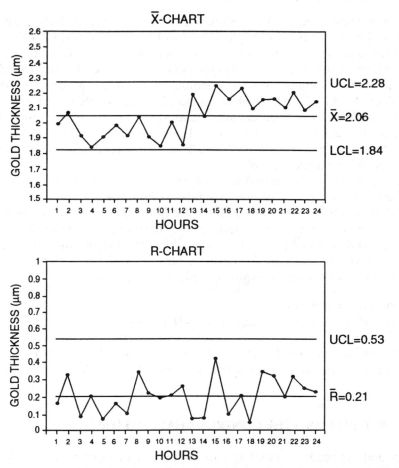

Figure 7.18 A sudden shift in process average caused by the replacement of test equipment.

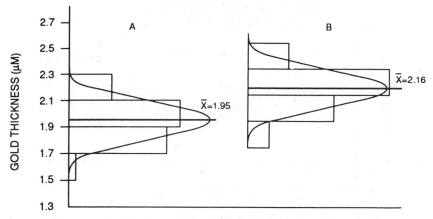

Figure 7.19   Distributions associated with a sudden shift in level.

ments or other major changes in the system were allowed, knowing that the product conforms to the specification. As one can see, a sudden change or shift in the process average occurred at the 13th hour of running the process.

Figure 7.19 shows the distributions from the control chart data. Distribution A represents the first part of the chart (before the shift) and Distribution B represents the second part of the chart (after the shift). The center of Distribution B is higher than the center of Distribution A, and this reflects a shift of the control pattern in the same direction. Looking for an explanation for the change in the process level, we found that during the capability study the measuring equipment was changed because it broke down. The new tester affected all the measurements in the same way from the time it was introduced in the process.

7.5.2   What Can Cause a Shift?

The $\bar{X}$ chart can be affected by a number of different causes that usually have one thing in common: When they enter the process they are capable of immediately affecting all of the product or the measurements. For example, changing the setting of a wire bond machine or changing the temperature in a plating bath affects all the parts being produced. What is interesting here is that this type of cause influences the process average without affecting the process spread. You can see that the R chart in Figure 7.18 is in a state of control when the $\bar{X}$ chart is shifting. A shift in the process level does not always occur as suddenly as was demonstrated in our example. It can also occur gradually, exhibiting a trend. This can be caused, for example, in a wafer sawing process when the tool gradually deteriorates or in a plating process when the solution gradually changes, etc.

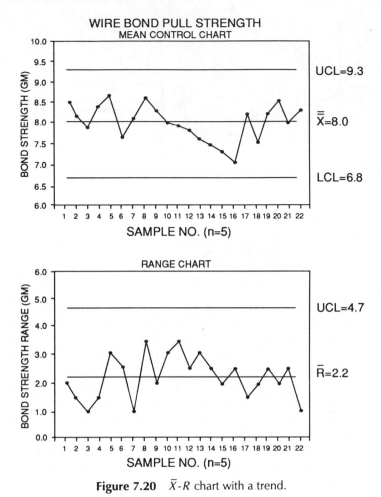

Figure 7.20 $\bar{X}$-R chart with a trend.

Figure 7.20 shows a real example of an $\bar{X}$-R chart for aluminum wire bond pull strength. In the $\bar{X}$ chart a trend is observed from point 8 to point 16. An engineer investigated the process and found that this was caused by aluminum scrap built up at the bond tool. Corrective action was taken by cleaning the bonding tool, and the problem was solved.

It is important to remember in the interpretation of an $\bar{X}$ chart that its pattern reacts to a type of cause that affects all the product at once or in the same general way. This is the most common type of cause that shows up on the $\bar{X}$ chart. So changes in machine setting, material, operator, inspection, test equipment, process parameter, environment, etc., will immediately disturb the natural pattern on the $\bar{X}$ chart. A change of the control chart pattern will not tell you where to look or what the cause was, but it will tell you *when* to look for trouble.

7.5.3   Interpreting the *R* Chart

The range (*R*) chart measures the process uniformity and reacts to changes in variation or spread. When the process is constant and all units of the product get the same treatment, the *R* chart will generate a natural pattern indicating process uniformity. If the *R* chart signals that the process is out of control, this means that some units are being treated differently than others. The uniformity of a process is reflected by the magnitude of the *R* chart data. The lower the magnitude, the better is the process.

7.5.3.1   What Changes the R Chart Pattern?

When we described the commonality of the causes capable of affecting the $\overline{X}$ chart, we emphasized that when entering the process they impact all the product at once or in the same general way. Here it is different. The causes that affect an *R* chart have another commonality; i.e., they are able to treat part of a product differently from the rest of the product. For example, an unstable tester at different times will give different readings for the same value. This special cause will influence different parts differently. The range chart will show a large variation which has nothing to do with the product. Or, a machine with a worn-out bearing does not work the same way every time so it treats the parts differently, etc.

Figure 7.21 illustrates an example when poorly trained operators introduced excess variation of the product (see Part A). After retraining the operators, the variability of the process was significantly reduced because the inspectors became more capable of recognizing the product values with better precision (see Part B). There is a large number of causes which will affect an *R* chart. Some of these causes are: (1) operator or inspector fatigue

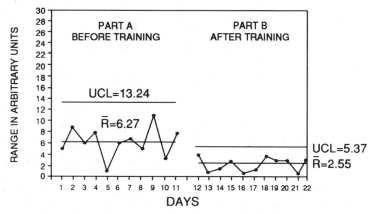

Figure 7.21 A shift in the range chart showing the results of additional operator training.

or poor training, (2) nonuniform material received from the suppliers, (3) poor maintenance of the equipment, (4) unstable test equipment, (5) mixtures of different lots, etc. For example, gold plating behaves differently when parts are "chamfered" than when they are "burnished."

The R chart is very sensitive to many types of causes. It is the best method of detecting mixtures, freaks, interactions, etc. The subject of control chart interpretation is very large and cannot be covered completely here. However, we would like to bring to the reader's attention a special type of unnatural pattern that is frequently misinterpreted as a natural "good" process. This type of unnaturalness is called "stratification." A case history related to this subject follows.

Case History 7.1: Why Is the Process so Good?

Working on a supplier process capability study (see Case History 9.3 for details) an engineer came up with the first results. Figure 7.22(a) is an R chart of glass thickness for a ceramic part. As one can see, the variation is consistent and the spread is very narrow. In the beginning, the engineer was happy, but when he tested the pattern for special causes, the chart did not pass test 7 (see Figure 7.15); all the points are clustered around the center line (in Zone C). This is a signal that a special cause is affecting the process. The range chart pattern is not considered natural and because of this the $\overline{X}$ chart cannot be analyzed and the capability index cannot be calculated. In other words, we cannot predict the process even when the range chart looks "so good." Here we are dealing with a process "sickness" and the diagnosis is *stratification*.

What Is Stratification?

Stratification is a form of stable mixture which is characterized by an artificial constancy.[10] Stratification shows up on a control chart as unnaturally small fluctuations with no points near the control limits. Stratification results when samples are consistently taken from different distributions, in such a way that one or more units in every sample will come from each of the distributions. When sampling is done this way, we ignore randomness, one of the most important requirements in making control charts. In this particular case, the supplier was "trying his best" to take a representative sample. The product came from the process in trays, so to have a representative sample that would reflect all the tray, he took samples of five units, one from each corner and one from the center of the tray. So a subgroup on the $\overline{X}$-R chart represents only one particular tray from which samples of five units are taken from constant locations. When taking a sample in this manner, the value of R will consist mostly of the difference between the highest and lowest levels that reflect two particular locations in the tray. Taking successive values of R in this way will cause them to differ very little from each other. Figure 7.22(b) is a control chart from the same process after changing the procedure of sampling. At this time, samples of 5 have been taken randomly from the

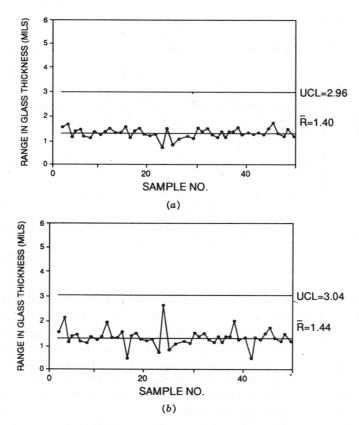

Figure 7.22 A range chart with all the points clustered around the center line versus a range chart with a natural pattern. (a) Every point represents the range from a sample of $n=5$ taken from one particular tray and from a constant location. (b) Every point represents a range from a sample of $n=5$ taken randomly from the process (unrelated to the location of the tray).

overall process so one subgroup will represent several trays. The chart is in control and the process capability can be calculated. Now it is really in "good control." Before it was artificial.

7.5.4   Interpretation of Attribute Control Charts

Attribute control charts such as p charts, np charts, c charts, and u charts have the same interpretative characteristics. Here we will use the term "p chart" to represent all other attribute control charts. As we know, a p chart can represent the combination of a number of characteristics or can represent a single characteristic. The interpretation of a p chart mainly depends on the knowledge of how many and what type of characteristics are combined together. Figure 7.23 represents a hypothetical c chart placed at the end of

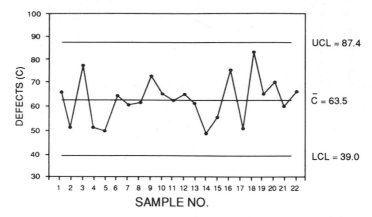

Figure 7.23   An overall hypothetical C-chart for open and short defects.

the line for open and short failure analysis. The pattern demonstrates a state of statistical control. Can we say that the process average of this chart represents the process capability? Figure 7.24 is a group of hypothetical c charts of separate defect types that are included in the overall c chart. As one can see, all the charts are out of statistical control. But together on the overall c chart they represent a "statistically balanced" process. This example demonstrates that it is very difficult to interpret an attribute control chart that represents a large group of variables.

To interpret an attribute control chart, it is more convenient to have a number of charts on various characteristics plotted closer to the origin of the process. Only after making a breakdown of the overall control chart into the component characteristics or sources of product can a real interpretation of the pattern be made.

7.5.5   Interpreting Control Charts for Individual Measurements

The control chart for individual measurements (X chart) is based on control limits calculated from the moving range (MR) (see Section 7.7.5 for control charts for individual measurements). This type of chart is mainly applied for variable data. However, in special cases it also can be used for attribute data. In general, the interpretation of the X chart is almost the same as for $\overline{X}$-R charts and p charts, but here the interpretation should be made with considerable caution. When we interpret an X chart, in the beginning we look for trends that will appear the same way as on the $\overline{X}$ chart. Second, we observe the pattern to see if the fluctuations are becoming narrower or wider. The variation on the X chart demonstrates the level of uniformity or consistency. The X chart usually does not have a moving range chart, so from

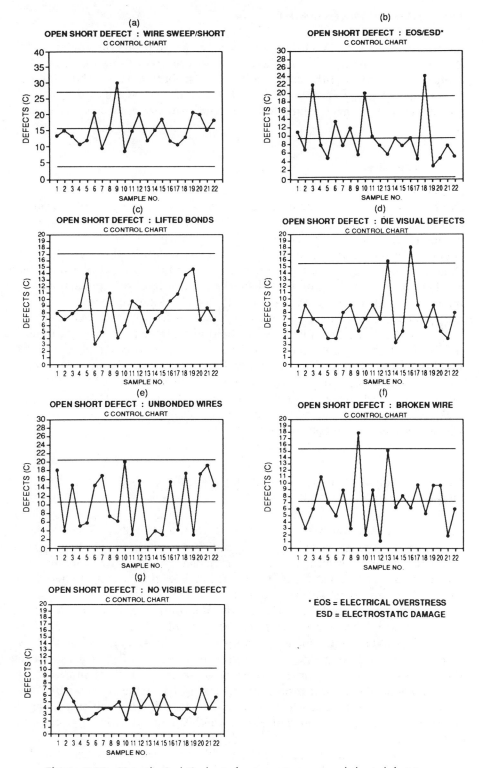

Figure 7.24 Hypothetical C-charts for separate open and short defect types.

the observation of the X chart fluctuation, we get information similar to an R chart. And, finally, we observe if the pattern stays far away from the control limits. If this is the case, we have a distribution with a short tail on one side and a long tail on the other. The X chart can be affected by any of the causes which affect either the $\overline{X}$-R chart or any control chart of attributes.

7.5.6 Summary

$\overline{X}$ charts indicate whether there is stability in the distribution's center. R charts demonstrate the magnitude of the process spread. These two charts should always be interpreted together; however, the R chart should always be interpreted first. Control charts for individual measurements are in general interpreted the same as an $\overline{X}$-R chart; however, this type of chart should be interpreted with considerably more caution. Control charts for attribute data are mainly used to show the general level of a process and to indicate overall trends. They are more difficult to interpret because usually one p chart may include too many characteristics. A breakdown of the overall p chart by its components can help in interpretation. Interpretation of control charts cannot be described in such a way to cover all situations which we meet in real life. Any control chart or results from a capability study have their own interpretation. Some of them are simple and others need more knowledge and practice. Only by applying the control charts, and making some wrong interpretations in the beginning, can we learn the wisdom of reading the control chart patterns.

7.6 SUBGROUP SAMPLE SIZE AND FREQUENCY OF SAMPLING WHEN USING CONTROL CHARTS

First let's illustrate how the sample size affects a control chart. The standard deviations of such statistics as p, $\overline{X}$, and R vary inversely with $\sqrt{n}$. The larger the subgroup sample size (n), the smaller the standard deviation and the closer 3 sigma limits will be to the center line of the control chart (see Figure 7.25). In other words, the larger the sample size, the tighter the 3 sigma control limits, and the more sensitive the control chart will be to detect shifts and drifts. However, the larger the sample size, the more time is required to obtain and plot a subgroup on the chart. This means that by increasing the subgroup size, we may lose the chance to detect a change in the process at the right time. Taking large samples at short intervals, of course, would be the best protection against shifts in the process. But this results in extra costs. So, which is better, to take large samples at longer intervals or small samples at shorter intervals? The answer will be given later in this section.

There is also a second aspect of this problem that will be addressed in this section. As we know, the better the quality, the larger is the sample size

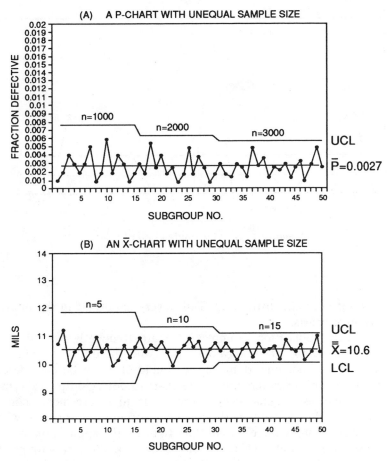

Figure 7.25   Illustration of the subgroup size effect on the control limits.

needed to detect a lack of control. In other words, if the fraction defective (p) is small, the sample size should be large enough to have some nonconforming units. Otherwise we may have a situation as demonstrated in Figure 7.26. The occurrence of just one defective item may indicate a lack of control while the process may actually be in control. This is why the attribute control chart became more and more difficult to use in a low ppm environment. The solution to this problem will be discussed later in this section.

7.6.1   The Sample Size and Frequency of Sampling when Using an $\bar{X}$-R Chart

Shewhart[11] suggested a sample size of 4 as the ideal subgroup for an $\bar{X}$-R chart. This size is small enough not to mask the changes in the process and large enough to form a nearly normal distribution of averages, even when the

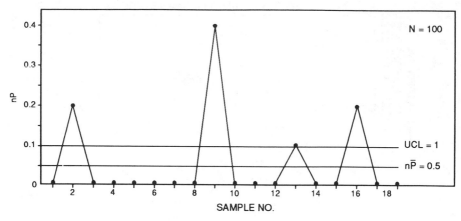

Figure 7.26 "Pyramid of Egypt" chart.

process is not normally distributed. This is very important in the interpreta-tion of control charts.

In industrial applications, samples of 5 are frequently used to make it easier to compute the process average. But, with the use of calculators and computers, this is not important and a sample of 4 can be considered the optimal. However, if it is the aim of the chart to detect shifts in the process average as small as 1 sigma, sample sizes of 15 to 20 are more economical than samples of 4 or 5.[12] But as the practical working rule suggests, when we need to use subgroup samples larger than 10 to 15, the application of the $\overline{X}$-S chart is more appropriate. There are also situations where subgroups of 2 or 3 can be justified (e.g., when the cost of measurements is very high or when the method of measuring the parameter of interest is destructive).

In general, the subgroup sample size for an $\overline{X}$-R chart may be as small as 2 or as large as 10 to 15; however, the optimal subgroup size is 4. As for the frequency of sampling, there is no general rule here; but it is better to take small samples more frequently than large samples less frequently. For exam-ple, instead of taking samples of $n = 10$ every two hours, it is better to take samples of 5 every hour. When beginning the control chart application, the frequency of taking samples can be greater, and as soon as the process is stabilized, the frequency can be reduced. Also, the frequency of taking subgroups may be expressed not only in terms of time. Sometimes it is more convenient to set a scale related to the production rate. For example, taking a sample of 4 units every 1000 units produced.

It is important to note that as soon as the subgroup sample size is determined, it should be kept as constant as possible while the process is being monitored. If for some reason the sample size must be changed, then the control limits must be recalculated.

7.6.2 The Sample Size and Frequency of Sampling when Using p or np Charts

When the quality of a process output is measured by 100% inspection, the decision about the sample size or frequency of sampling for process control is very simple. Based on the general production rate and other administrative considerations, a constant time frame is determined when samples should be taken. This can be on an hourly, bihourly, or daily basis. However, when a control chart is based on a sample that is especially selected for control purposes, the question of sample size often arises. There are several ways to determine the optimal sample size needed to monitor the process by a p or np chart. One way is to use a sample size large enough to have $np \geq 5$, so we can use the normal distribution as an approximation of the binomial distribution. For example, if the assumed process average ($\bar{p}$) is 0.0025 (2500 ppm), the subgroup sample size (n) should be at least 2000 to have $np \geq 5$ ($0.0025 \times 2000 = 5$).

Some authors[13] suggest that the sample size should be large enough to get one or more defective items per sample at least 90% of the time. For example, if the process average ($\bar{p}$) is, let's say 0.002, and if the probability of $c \geq 1$ is equal to 0.90 (i.e., $P[c < 1] = 0.10$), we would need to have $n\bar{p} = 2.3$, which means the sample size should be $n = 2.3 \div 0.002 = 1150$. Or, in other words, if the process average is $\bar{p} = 0.002$ (0.2%) and we use a subgroup sample size of $n = 1150$, the control chart (np or p) consisting of 100 points will have only 10 of the points located on the zero level. The rest of the subgroups will contain one or more nonconforming units.

7.6.3 The Sample Size in a Low PPM Environment

In the electronics industry a lot of processes run in a range of 50 to 500 ppm. In this case, to have $np \geq 5$, the sample size should be in a range of 10,000 to 100,000. This shows that Shewhart control charts for attribute data become obsolete in a low ppm environment. Because of this, the p (or np) chart is recommended to be used only when the subgroup sample size is affordable and not very large. For example, the AMD-SPC specification requires that Shewhart control charts are to be used only when the process level is 2500 ppm or higher. In this case, the minimum sample size will be 2000. However, for monitoring the process and reporting purposes when large sample sizes are available, the p chart with moving range control limits can be used; even when the process quality level is lower than 2500 ppm. In this case, the subgroup sample size is derived by using the formula:

$$\frac{2.3}{\bar{p}}$$

which will give a sample size large enough to get one or more nonconforming units in a subgroup 90% of the time. To control a process below 2500 ppm, the cumulative count control (CCC) chart is used (see Section 7.7.2) where the subgroup can be any sample size. For c and u charts the AMD-SPC specification requires that the sample size be large enough to be able to get at least one nonconformity per unit.

Frequency of sampling is more an issue of economics. When losses are high from running an uncontrolled process, the interval between samples should be small; when the inspection cost is high and time consuming, the interval can be large; but when the process has been brought into control, the frequency of inspection can be reduced.

7.6.4 Control Limits

The conventional principle of setting control limits on Shewhart control charts is ± 3 sigma. However, under certain circumstances, charts using 2 sigma or even 1.5 sigma limits are more economical. For example, 2 sigma limits can be established on a critical process where not being able to detect a problem would create a greater loss than the cost of looking for a problem when there isn't one. Sometimes it is more economical to use charts with 3.5 to 4 sigma limits if the cost of looking for problems is very high.[14] These decisions relate to economic and technical decisions that should be made by management if we want to use SPC efficiently.

The control limits should be based on at least 25 subgroups. However, when it takes a long time to collect the 25 samples, preliminary calculations can be made from 8 to 10 subgroups. After more samples are collected, modification of limits should be made. Anytime an assignable cause of variation is discovered and removed, new control limits should be calculated from 25 new subgroups.

7.6.5 The Control Chart and the Operator

Prior to introducing SPC in the plants, operators were instructed that their primary responsibility was to make parts to the specification requirements. Even now when control charts are attached to their machines and the operators personally plot the points on the charts and react to the charts signals, they still believe that their major responsibility is to make parts to the specification. In an operator survey conducted by AMD, on the question, "What is your major responsibility?" 89% responded, "Making parts to the specification." And this is understandable. But how about the control chart? How about the responsibility for keeping the process in a state of statistical control? At AMD all the operators received a specially designed SPC course where they learned how to construct, interpret, and react to a control chart. They were told about their rights, responsibilities, and benefits related to the introduction of SPC.

The AMD-SPC specification established special requirements related to the application of control charts. This document describes who should do what if the chart shows that the process is not in control. However, the workers do not react to the control chart in the same way as the specification because initially the legitimacy of the control chart has not been fully established. Maybe this is because they know that having a point outside the control limits does not necessarily mean that they have produced bad parts while having a measurement outside the specification limits definitely means they have bad parts.

There are a number of things that should be done to make the control chart legitimate. First of all, a control chart should not be transferred to an operator until the process is brought into control. Experience shows that when a control chart is in place and the operator is not capable of running the process within the control limits, the operator becomes frustrated and the control chart has a negative impact. This does not mean that an operator cannot participate in a capability study to bring the process into control. But, when the process is in a period of study, the operator should not be responsible for the control chart.

Second, when a control chart is transferred to an operator, it is very important to fully agree on who does what when the chart signals a process disturbance. The operator's responsibility should be clearly specified and he/she should be given the opportunity to fulfill his/her responsibilities. The assignments of responsibility may vary from plant to plant; however, the AMD-SPC specification gives the outline for these responsibilities, which can be altered according to the specifics of organizational structure and operation.

Figure 7.27 is an example of a matrix showing the responsibility of different types of personnel relative to the application of the control charts. Based on the divided responsibilities, the process improvement team (PIT) conducts a process capability study to determine if the process is in a state of statistical control or if it is necessary to bring the process into control. At the same time, the team decides the type of control chart to be used, the parameter to be controlled, and the specification requirements to be applied. The engineer calculates the required sample size and frequency of sampling, the preliminary control limits, and the center line. After the control chart parameters are established, the design of the control chart is approved by the supervisor or manager who is responsible for this particular process. The operator monitors the process quality by plotting points on the chart and reviews the results on a continuous basis. If a signal of process disturbance occurs, the operator takes corrective actions according to the procedure and, if necessary, stops the process and asks for help. The quality inspector also

	Activities	PIT	Operator	Quality Inspector	Supervisor	Engineer
1	Process capability studies to bring the process into control.	⊗	X	X		X
2	Construct the control chart.	⊗	X		X	⊗
3	Transfer the control chart to the operator.	⊗	X	X	X	X
4	Plotting the chart.		⊗			
5	Review the results.		⊗	X	X	X
6	Act on the results.		⊗			X
7	Reporting to the Control Chart Committee.	⊗	X		X	X
8	Analyzing the results for further improvement.	X	X	X	X	⊗

⊗ = Takes the main responsibility.

X = Shares the responsibility.

Figure 7.27   Division of responsibilities relative to the application of control charts.

reviews the control chart on a periodic basis and helps the operator maintain control. According to an approved schedule, the process improvement team (PIT) reports the results of the process behavior for a particular period to the control chart committee. This committee decides about changing the control limits, subgroup sample size, frequency of sampling, and other issues related to control chart design or process improvement. The data manually plotted on the chart are also entered into the computer to accumulate historical data for further investigation. The engineer uses these data for process performance evaluations that may give additional information of what should be done to bring the process to the 6 sigma quality level.

7.7   SOME NEW AND FORGOTTEN OLDER STATISTICAL TECHNIQUES

The application of Shewhart control charts made it possible to improve quality and reduce the reject rate to a low ppm quality level. This generated a demand for new types of charts to control such "close to perfection" processes. An example of such a technique is the cumulative count control chart (CCC chart) which is based on a geometric distribution. This chart can be used to control processes on any ppm quality level.

Continuous process improvement by reducing the variability around the target brought back to life an old technique called pre-control, which fits very well in the low ppm environment. This technique can be used when working with a zero defects program. Another interesting technique is using multi-vari charts. They illustrate the variation within parts, between parts, and from measurement time to time, which allows us to concentrate our efforts in the most critical areas.

The simple technique of the control chart for individuals, with control limits that have been calculated based on the moving range, has become popular again since it can be used not only in the manufacturing process, but also in the supporting and nonmanufacturing areas. It can be applied to attribute and variable data.

When the process spread is significantly reduced and the Cp becomes 2 or larger, there are sometimes situations when the control limits for the $\bar{X}$ Chart can be relaxed, then the control chart with modified control limits is the right tool to apply. This tool is based on the assumption of a normal distribution.

As the expression says, "New is the forgotten old." The demand for different types of statistical techniques has made people take a second look at the old, forgotten tools and give them a new life. In addition to this, a series of new statistical developments has been made in recent years. The statistical "toolbox" has become larger. Now it is just a matter of making people familiar with these techniques and methods. In this section we will limit ourselves by describing only a small group of the statistical techniques that have been mentioned above.

7.7.1 Controlling a Process in a Low PPM Environment

Previously we mentioned that the p chart is based on the binomial distribution approximated by the normal distribution, and that this approximation works well as long as $np \geq 5$. This creates a problem when we are trying to apply the conventional p chart in a low ppm environment. Let's illustrate this by using an example.

Suppose we want to control a process with a quality level of 200 ppm. If we want the p chart to be effective, we need to satisfy the requirements related to the normal approximation. This means we need a subgroup sample size large enough to have $np \geq 5$. In our case we would need $n = 25,000$ ($0.0002 \times 25,000 = 5$). Thus, if the production rate is 50,000 units per day, we would need to inspect 50% of the product to have one point on the chart per day.

This method of monitoring the quality is not effective and cannot be considered as real-time process control. Such a chart can be successfully used for reporting purposes where every point can represent a day, week, or even a month, but not for on-line process control.

Now suppose we want to sacrifice the accuracy of the approximation and use a control chart with a significantly smaller sample size, just to get hourly

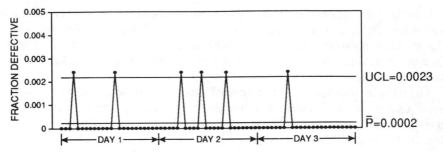

Figure 7.28   A p-chart with an hourly sample size of 400 for three days. The chart is based on the following assumptions:
1. The process average is 200 ppm (0.0002 fraction nonconformance).
2. In this case, the fraction conformance (yield) will be $1 - 0.0002 = 0.9998 = 99.98\%$.
3. With a sample size of $n = 400$, the probability of a sample containing zero nonconforming units (zero defects) is

$$Pn(0) = (0.9998)^{400} = 0.923 \approx 92\%$$

So we assume that a real chart in such conditions will have approximately 92 percent of the subgroups with zero defects. In our case, 66 points on the chart will lay on the zero level. The chart will indicate a change by frequently having terminating points outside the control limits. The chart is usually maintained directly at the workplace by the operator. It is also important to remember that whenever technical changes in the process are made, such as readjustments, changes of material or procedures, etc., notes should be made on the chart.

feedback from the process and to have a sample big enough to represent the process average after three days (three shifts per day). Accordingly, we will take a sample of 400 units per hour. And, after the three days, we will have a control chart consisting of 72 points. Of this, 66 points will have zero defects and only 6 points will have one or more nonconforming parts that will lay outside the control limits (see Figure 7.28).

It is difficult, if not impossible, to make a conclusion about the process quality from such a pattern. So what is the alternative? A cumulative count control (CCC) chart has become a very good substitute for the traditional control charts for attribute data when working in a low ppm environment. The following is a description of this relatively new technique.

7.7.2   Cumulative Count Control Chart (CCC Chart)

The progress in quality improvement generated new ideas for how to control "close to perfection" processes. Thomas W. Calvin[15] developed a technique

TABLE 7.9   Formulas for Calculating the Cumulative Count Control Chart Parameters (Based on the Geometric Distribution)

Control Chart Parameters	(1) For any value of α	(2) For $\alpha = 0.10$	(3) For $\alpha = 0.05$	(4) For $\alpha = 0.0027$
1. Center line (CL)	$CL = 0.7\bar{n}$	$CL = 0.7\bar{n}$	$CL = 0.7\bar{n}$	$CL = 0.7\bar{n}$
2. Lower control limit (LCL)	$LCL = (\alpha/2)\bar{n}$	$LCL = 0.05\bar{n}$	$LCL = 0.025\bar{n}$	$LCL = 0.00135\bar{n}$
3. Upper control limit (UCL)	$UCL = -\bar{n}\ln(\alpha/2)$	$UCL = 3.00\bar{n}$	$UCL = 3.7\bar{n}$	$UCL = 6.6\bar{n}$

$\bar{n} = 1/\bar{p}$ is the expected average number of units that have to be inspected before finding a nonconforming unit, $\bar{p}$ is the estimated average fraction defective for the process, and α is the probability of a count less than the lower control limit (LCL) or greater than the upper control limit (UCL).
Note: Column (1) is used for calculating the control chart parameter for any value of α. Columns (2) to (4) are applied when we decide to use specific values of α.

that can be used to control processes with low ppm quality levels. The technique uses the approach of plotting on a chart the number of conforming units instead of nonconforming units. The concept itself is very effective because in a low ppm environment sometimes the process runs several weeks, or even months, before finding one nonconforming unit. This is why AMD adapted the idea, developed it further, and applied it in almost all areas as a substitute for the traditional p chart.

7.7.2.1   Introduction

The cumulative count control (CCC) chart is based on the geometric distribution, which allows us to determine what the cumulative count of conforming units should be before one nonconforming unit can be expected to occur. Based on this geometric distribution, the control chart parameters can be calculated by using the formulas found in Table 7.9. Thus, given the fraction nonconformity, the center line and the control limits can easily be derived for different probability limits (α). The control chart is constructed on a semilogarithmic paper with a logarithmic vertical scale for cumulative count and a linear horizontal scale for the time period (days, weeks, etc.).

After the control chart is constructed (see the procedure in Section 7.7.2.3), we plot the cumulative count as follows. During inspection, we maintain a cumulative count of conforming items. Counting may be stopped and the latest cumulative count (n') is noted whenever inspection of a sample is completed. When inspection of the next sample starts, the count continues. However, when a defective item is encountered, n' is also noted but the count number will revert to 0, becoming 1 again for the next conforming item inspected, and so on. This will be clearer when demonstrated by an example (see Section 7.7.2.4).

7.7.2.2 Interpretation

The interpretation will be easier to understand using a real example, which will be described later. Here we will only give the major rules for reading the chart.

1. If the cumulative line is interrupted (due to finding a nonconforming item) before passing the lower control limit, it is an indication that the process is out of control.
2. If five or more interrupted lines are consecutively below the center line, it is also an indication of process instability.
3. Any line exceeding the upper control limit, or five or more interrupted lines passing the center line consecutively, indicates a significant improvement in the process.

It takes a while to get used to the new technique, but after applying it a few times, the approach becomes very simple. The cumulative count control chart is an "optimistic" technique because it highlights the number of conforming units. Out of control situations are immediately visible, and the pattern is more meaningful. In areas where the process is running on a low ppm quality level, this chart is a good substitute for p charts and other attribute control charts.

At AMD the cumulative count control chart is popular. To clarify the methodology and the interpretation, we will use an improvised example so we can demonstrate all aspects in one chart.

7.7.2.3 Procedure for Constructing a CCC Chart

Procedure	Example
Step 1	*Step 1*
Make a preliminary estimate of the process fraction defective ($\bar{p}$) to be controlled by the CCC chart.	Let's assume that the historical data show that the process average is $\bar{p} = 0.0002$ (200 ppm).
Step 2	*Step 2*
Choose α, the probability that a point will fall outside the limits even when the process is under control.	Let $\alpha = 0.10$.

Procedure	Example
Note: The traditional p chart has an $\alpha = 0.0027$. When applying a CCC chart, an α of 0.10 is also frequently used, which means that there is a 10% probability that a point will fall outside the limits even when the process is running normally.	
Step 3	*Step 3*
Determine the average run length ($\bar{n}$), which gives the expected number of items that have to be inspected before a new nonconforming item is found: $$\bar{n} = \frac{1}{\bar{p}}$$	From Step 1 $\bar{p} = 0.0002$, so $$\bar{n} = \frac{1}{0.0002} = 5000$$
Step 4	*Step 4*
Determine the center line (CL), which is the median value ($\tilde{n}$): $$CL = \tilde{n} = 0.7\bar{n}$$	From Step 3 $\bar{n} = 5000$, so $$CL = (0.7)(5000) = 3500$$
Step 5	*Step 5*
Calculate the lower control limits (LCL) and upper control limits (UCL): $$LCL = \left(\frac{\alpha}{2}\right)\bar{n}$$ $$UCL = -\bar{n}\ln\left(\frac{\alpha}{2}\right)$$	From Step 3 $\bar{n} = 5000$ and from Step 2 $\alpha = 0.10$, so $$LCL = \left(\frac{0.10}{2}\right)5000 = 250$$ $$UCL = -5000\ln\left(\frac{0.10}{2}\right) = 14{,}978.66$$ say, 15,000.

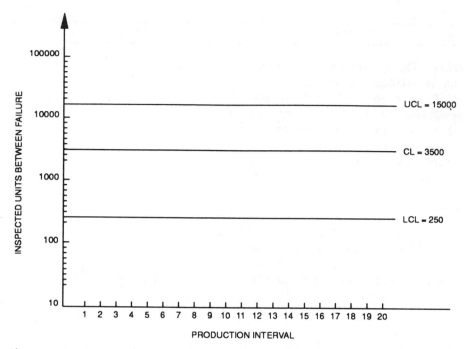

Figure 7.29 An example of cumulative count control chart ($\bar{p}=0.0002$ and $\alpha=0.10$). $\bar{p}=0.0002$; $\alpha=0.10$; $\bar{n}=1/\bar{p}=1/0.0002=5000$; CL $=0.7\bar{n}=0.7(5000)=3500$; UCL $=-\bar{n}\ln(\alpha/2)=(-5000)\ln(0.10/2)=14{,}979$, say 15,000; LCL $=(\alpha/2)\bar{n}=(0.10/2)(5000)=250$.

Procedure	Example
Step 6	*Step 6*
Construct a control chart using the parameters calculated above and semilogarithmic paper. The logarithmic vertical scale is for the number of counts (n). The linear horizontal scale is for the plotting sequence of stopping values n'.	The cumulative count control chart is constructed as shown in Figure 7.29.

The control chart is now ready to be introduced to the manufacturing floor. However, since the chart was constructed based on an estimated process average ($\bar{p}$), sometimes in the starting period the segments will frequently terminate even before passing the lower control limit. This is an indication that the preliminary estimate of the process average was too low. The reverse can also happen, when the terminating point of the segment frequently passes the center line or even the upper control limit. This is an indication that the process average estimate was too high. In any of these cases, the control chart parameters should be recalculated based on the new data collected. When the terminating points of the segments are distributed equally around the center line (CL), the tentative estimate of the process average is correct and the control chart can be continued until recognizable changes in the process average occur. The chart will indicate a change by having frequently terminating points outside the control limits.

The chart is usually monitored directly at the workplace by the operator. It is also important to remember that whenever technical changes in the process are made, such as readjustments, changes of material or procedures, etc., notes should be made on the chart.

7.7.2.4 An Improvised Example of Plotting the CCC Chart

According to the procedure, a sample of 1000 units is taken from the process every day and inspected according to the specifications. Starting with the first sample, the inspection results show that all parts conform to the specification. So a point is plotted on the chart that will represent the first result (see point 1 on Figure 7.30). The next day, another sample is inspected, and again no nonconformities are found, so the cumulative count now is 2000. A second point is plotted and connected with the previous point to form a cumulative line. On the third sample, after inspecting only 500 units, a nonconforming unit is found. In this case, only 500 units are added to the cumulative line (see dot 3), and the cumulative count is interrupted and a new count is started. Measuring the remaining 500 units, no nonconformities are found so we plot the result on the chart.

The second cumulative count is continued by taking successive samples and only at the end of the inspection of sample 8 is a nonconformity discovered. So with sample 9 again a new count is started. On sample 10 a nonconforming unit occurred before passing the lower control limit (see Figure 7.30). This is a signal that the process is no longer in a state of statistical control.

After the cause of the process disturbance has been eliminated, the process continues and the inspection starts again. At this time, no cumulative counts have been terminated below LCL, but five consecutive times the cumulative counts passed the lower control limit and did not pass the center line. This is another type of signal of process disturbance. At that time, it was found that the supplier was conducting experiments to reduce the cost of the material. A decision was made to use material from a second supplier temporarily while the prime supplier investigated the cause.

The control chart now shows an opposite situation. Four consecutive times the cumulative count has passed the center line and the fifth time the count passed the upper control limit. This is an obvious change in the process average. It was decided to continue the process a little bit longer to confirm that the supposition is correct, and then the control chart parameters are recalculated according to the new process average.

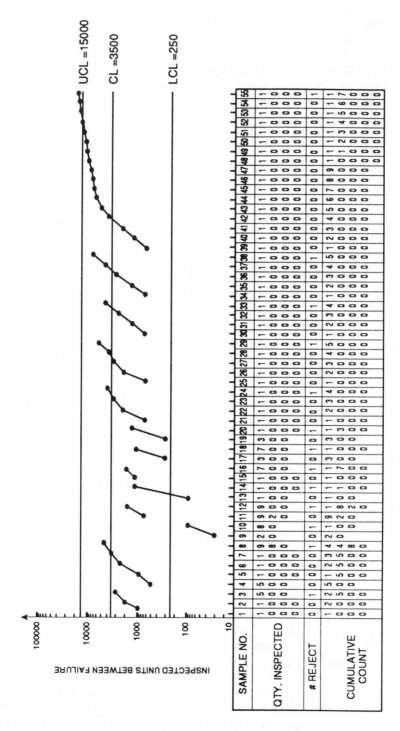

Figure 7.30 A CCC-chart based on an improvised example.

This is an illustration of different situations improvised to give the reader a feeling for how to plot and interpret the CCC chart. The following example is more simple and realistic.

Case History 7.2: Applying the CCC Chart to a Process Close to Zero Defects—Wire Bonding Process

An automatic line for integrated circuits assembly has a series of machines that are capable of self-control. This means that when something goes wrong with the process, the machine automatically stops. This automatic line is a real example of building six sigma quality into the manufacturing process.

However, to "keep a hand on the pulse" of the process, CCC charts should be introduced. Below is an example of how the chart is applied to an automatic wire bonding machine with a production rate of around 1000 units per hour. Figure 7.31 is a constructed control chart based on the previous month's process average, which was 1000 ppm ($\bar{p} = 0.001$).

The operator who is responsible for the wire bond process takes a minimum sample of 10 units on a regular basis for visual inspection. A tally of the units inspected is kept by a special counter, which is adjusted based on the machine operation. The operator increments the counter by the number of units inspected. The count is then plotted at a convenient time and after a point has been plotted, the operator presses a button that resets the counter. The cumulative count control chart is computed manually, and the operator terminates the line sequence every time a defective unit is encountered. The location of the terminated point indicates if an out-of-control situation has arisen, for which appropriate corrective action should then be taken.

In this particular case, Figure 7.31 demonstrates that an improvement in the process average has occurred, which is reflected by the cumulative line exceeding the upper control limit. Based on this signal, the engineer recalculated the control chart parameters. First he estimated the new process average by using the past and current data. The new estimate of the process average was 164 ppm ($\bar{p} = 0.000164$). Having this information, the engineer determined the mean, median, and the control limits by using the formulas as shown in Table 7.9 [see column (2) for $\alpha = 0.10$]:

$$\bar{n} = \frac{1}{\bar{p}} = \frac{1}{0.000164} = 6098$$

$$\mathrm{CL} = 0.7\bar{n} = 0.7(6098) = 4268$$

$$\mathrm{UCL} = 3.0\bar{n} = 3(6098) = 18294$$

$$\mathrm{LCL} = 0.05\bar{n} = 0.05(6098) = 305$$

Figure 7.32 is a control chart with the recalculated parameters. For illustration purposes, we replaced the data from Figure 7.31 to demon-

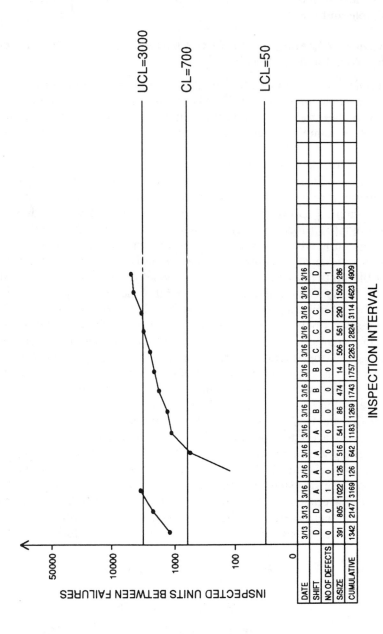

Figure 7.31 A cumulative count control chart showing process improvement.

UCL=3000
CL=700
LCL=50

INSPECTED UNITS BETWEEN FAILURES

INSPECTION INTERVAL

DATE	3/13	3/13	3/16	3/16	3/16	3/16	3/16	3/16	3/16	3/16	3/16	3/16	3/16	3/16
SHIFT	D	D	A	A	A	A	B	B	B	C	C	C	D	D
NO OF DEFECTS	0	0	1	0	0	0	0	0	0	0	0	0	0	1
S/SIZE	391	805	1022	126	516	541	86	474	14	506	561	290	1509	286
CUMULATIVE	1342	2147	3169	126	642	1183	1269	1743	1757	2263	2824	3114	4623	4909

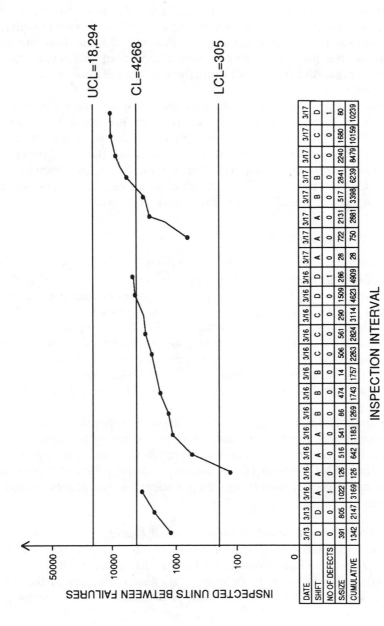

INSPECTION INTERVAL

DATE	3/13	3/13	3/16	3/16	3/16	3/16	3/16	3/16	3/16	3/16	3/16	3/16	3/16	3/16	3/17	3/17	3/17	3/17	3/17	3/17	3/17	3/17
SHIFT	D	D	A	A	A	A	B	B	B	C	C	C	D	D	A	A	A	B	B	C	C	D
NO OF DEFECTS	0	0	1	0	0	0	0	0	0	0	0	0	1	0	0	0	0	0	0	0	0	1
S.SIZE	391	805	1022	126	516	541	86	474	14	506	561	290	1509	286	28	722	2131	517	2841	2240	1680	80
CUMULATIVE	1342	2147	3169	126	642	1183	1269	1743	1757	2263	2824	3114	4623	4909	28	750	2881	3398	6239	8479	10159	10239

Figure 7.32 A cumulative count control chart with recalculated control limits.

strate how the chart would look using the second estimate of the process average. As one can see, the new cumulative count is interrupted between the center line and the upper control limit. As time passes, if five consecutive cumulative counts exceed the center line or one cumulative count exceeds the upper control limit, this will indicate that the new estimate of the process average is still higher than the actual "true" process average. And the control chart parameters will need to be recalculated again.

In this example, for demonstration purposes, we used the manual procedure of monitoring the CCC chart. This procedure was applied initially when this type of chart was first introduced. Then, after the operators and supervisors accepted the approach, all the calculations and charting were done on a computer, which is located on the assembly line close to all the operators. The CCC chart is an excellent technique for monitoring low ppm quality levels and high-speed processes.

7.7.3 The Pre-Control Technique

In 1991 AMD held its Second International Operators Symposium where workers from the process shared their experiences and the results of applying statistical principles. At this symposium a large group of operators received SPC awards for their participation in process improvement. One of the operators, Ho Chwee Gaik, accepting the award said, "I am proud to receive this. It will motivate me to do even a better job. Thank you. But what is really motivating me is the satisfaction I experience from knowing how I am doing my job, and being able to prevent a problem before the process starts producing bad parts. I've been able to do this since I started using statistical tools."

Giving an operator the opportunity to know how he/she is doing and clearly showing what can be done to correct a problem is the greatest thing that management can and should do if it wants the operator to do a great job. From the time AMD introduced pre-control and control chart techniques, employees had a greater opportunity to participate in the process improvement activity.

7.7.3.1 What Is Pre-Control and How Does It Work?
Pre-control, sometimes called "stoplight control," is a simple technique that allows the operator to have direct control of his/her process. This technique uses the normal distribution to determine significant changes in the mean and in the variance. The tolerance band is divided into three zones (see Figure 7.33):

Zone 1—the target zone (also called the green zone).
Zone 2—the cautionary zone (also called the yellow zone).
Zone 3—same as Zone 2.

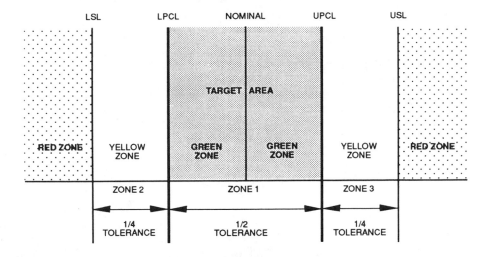

LSL = LOWER SPECIFICATION LIMIT

LPCL = LOWER PRE-CONTROL LINE

USL = UPPER SPECIFICATION LIMIT

UPCL = UPPER PRE-CONTROL LINE

Figure 7.33 Pre-control design for bilateral tolerance.

The area outside the tolerance limit is labeled the "red zone." The cautionary zones are designated just inside each tolerance extreme. These zones are used as follows. The operator measures two units periodically. If both measurements fall in the cautionary zones or either falls outside the tolerance, it is a signal for the operator to adjust the process. Otherwise the process should be left alone. This instantaneous warning allows the operator to adjust the process before nonconforming units can be produced. This is why the technique is called pre-control. Below is a step-by-step procedure of constructing and applying the pre-control technique.

Step 1 Divide the allowable specification spread into four equal parts, draw vertical lines as shown in Figure 7.33, and label the zones green, yellow, or red.

Step 2 Qualify the process by measuring five consecutive units. If all the measurements fall within the target (green) zone, the process is qualified (see Figure 7.34 as a guideline). If the process cannot be qualified, it should be diagnosed to identify the cause. Process qualification should be conducted any time a change is made in the system; e.g., a new setup, new material, etc.

Step 3 After the process is qualified, produce some units without inspection (or testing). Instead, monitor the process by taking periodic samples of

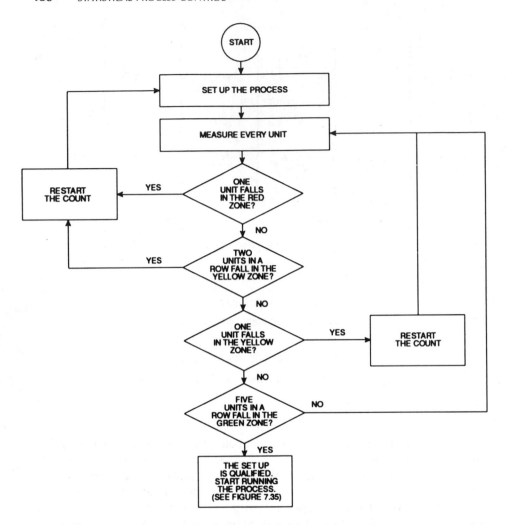

NOTE: WHEN THERE IS A CHANGE IN THE PROCESS, SUCH AS NEW
MATERIAL, NEW OPERATOR, ETC., THE PROCESS SHOULD BE REQUALIFIED.

Figure 7.34   The flowchart for the setup qualification.

two units (units A and B as a pair). Take action only if both A and B are in
the cautionary (yellow) zone or either is outside the tolerance limit (i.e., in
the red zone) (see Figure 7.35 as a guideline).

Now that we have become familiar with pre-control, we are interested in
knowing where this technique can be applied, what the risk of using pre-control is, how frequently we should sample, and other details about this
algorithm.

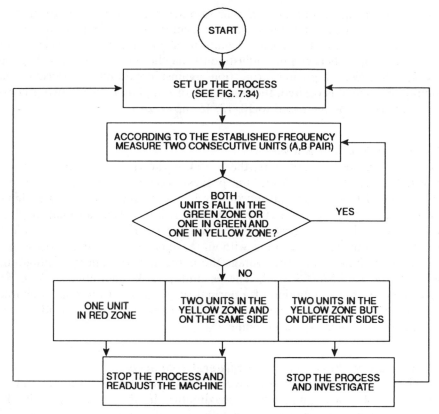

Figure 7.35 The flow chart for running the process by using pre-control.

7.7.3.2 Where Pre-Control Can Be Applied

Pre-control can be successfully applied for any discrete or continuous process where the operator can measure the quality parameter of interest (dimension, strength, etc.) and can adjust the process to change that parameter. To apply pre-control, there are no special requirements for the process such as assumption of normality or process capability. However, pre-control is most effective when we know about the process and we know how to take corrective action when the process is "going red."

7.7.3.3 Frequency of Sampling

The desired frequency for taking samples of two units (A, B) becomes apparent after using pre-control for a week or two (depending on the production rate). According to Shainin,[16] "Usually, one or two samples per eight-hour shift suffice for the typically stable quality characteristic during the period following setup approval." The original pre-control calculations indicated that a frequency of 25 A, B pair samples between two typical process adjustments (stoppages) would ensure an average of 1% or less production out of tolerance. But the latest calculations show that the frequency of six A, B pair samples between two adjustments may be sufficient to provide good results. However, if greater protection against continuing a process that should be stopped is desired, the frequency of sampling between two process adjustments can be increased to 10, 12, 15, up to 24.

7.7.3.4 The α and β Risks

With the pre-control technique, the α risk is the risk of having a signal to adjust the process when in fact the process should not be adjusted. The β risk is the risk of producing nonconforming units and not getting a signal. Using pre-control there is a 2% (worst case) risk of getting a signal to adjust the process when no adjustments are needed (α risk). As for the risk of producing nonconforming units without a signal to stop the process, the calculations show that on the average the percentage of nonconforming units will be 1% to 1.5% (β risk). The α and β risks mentioned above are calculated on the assumption of a normal distribution of the quality characteristic with its mean μ at the nominal (target) and ± 3 sigma coincident with the upper and lower tolerance limits.

7.7.3.5 Using Pre-Control for Unilateral Tolerances

In the electronics industry, we have process variables that require only a one-sided tolerance (minimum or maximum). In this case, to apply the pre-control technique we proceed as follows:

1. Determine the best unit by making a frequency distribution from at least 50 units taken from the process.
2. Place one pre-control line one-quarter of the distance between the tolerance limit and the best unit (see Figure 7.36).
3. Color the areas as follows:
   a. The area from the pre-control line to the best unit line (green).
   b. The area from the tolerance limit to the pre-control line (yellow).
   c. The area beyond the tolerance limit (red).

Everything mentioned about process setup, running the process, etc., for bilateral tolerances remains the same for unilateral tolerances when using pre-control.

7.7.3.6 Pre-Control and Six Sigma Quality

The probability of not stopping the process when producing nonconforming units is controlled by the size of the cautionary (yellow) zones and frequency of taking samples. Theoretically, if taking six pairs between typical adjust-

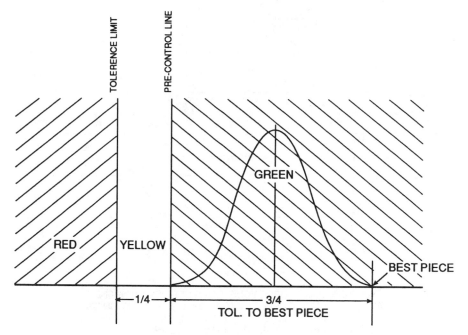

Figure 7.36   Pre-control for unilateral tolerance.

ments, the probability of producing nonconforming units will not exceed 1%. This is a relatively low risk. However, to apply pre-control in a low ppm environment, and particularly to control a six sigma quality process, we need the β risk to be close to zero. In this case, to avoid modification of the existing pre-control technique and still be able to apply it in the six sigma quality program, use the 50% concept. Simply stated, with this concept only 50% of the customer's specification limits are used during the manufacturing process. This concept can only be applied when the process approaches a Cp = 2 quality. Applying the 50% concept with the pre-control technique allows us to run the process with an average quality below one ppm. In this section we will demonstrate this with a case history that uses pre-control in a six sigma quality environment.

7.7.3.7 Charting Pre-Control

If a control chart cannot be implemented without actually plotting the data on a special graph, pre-control can be applied without monitoring the measurement results. However, there are some situations where the customer requires evidence of the process quality. The pre-control chart can be used for this (see Figure 7.37). As previously mentioned, the chart is color coded as follows: The target area (the middle half of the tolerance) is green,

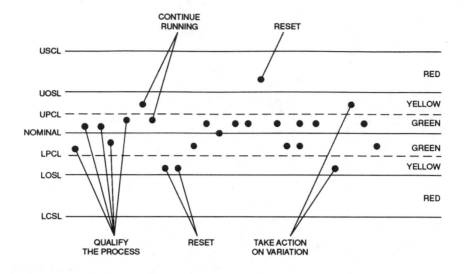

Figure 7.37 An improvised example of a pre-control chart.

the upper and lower quarters of the tolerance are yellow, and the area outside the tolerance limits is red. For the six sigma pre-control chart, the red is the upper and lower zones between the customer specification limits and the operational specification limits. The operator doesn't need to make any calculations. The rules are simple.

1. Five consecutive greens—the setup is qualified.
2. Two greens or one green and one yellow—continue.
3. Two yellows or one red—stop.

The operator makes slash-mark entries in the color-coded form. Sometimes management also requires that actual readings be entered on the chart so that capability indexes can be calculated. These additional requirements are optional and are used depending on the circumstances.

7.7.3.8 The AMD Experience of Introducing Pre-Control

Considering that pre-control is an operator-oriented technique, it cannot be introduced without the operator's cooperation. In any company or plant you always have some employees who are just waiting to be involved in new activities. And if you are a good leader, you know them.

At AMD it had already become a rule at the International Operators Symposium, that, in addition to the operators reports of what they had achieved, some new ideas and new initiatives should be on the agenda. In preparation for the last symposium, the working committee selected a small group of employees, representatives from different operations, to whom the technique was explained. The operators were then asked to come up with their own ideas where and how they could apply pre-control. In the second meeting, the selected operators were given a specially developed set of written rules and forms, and an engineer-instructor was assigned to every three operators to help them introduce the technique. After 10 working days of running the process using pre-control, a third meeting was held to discuss the results. At this meeting the operators suggested a number of interesting ideas of how to improve the working procedure to make it easier to apply. This group of operators were the presenters at the Operators Symposium who first introduced pre-control at AMD; and from this the approach was spread to all AMD workers. Operators became interested in this approach and they started to request participation. At that time, a special eight-hour course was given for those who wanted to apply pre-control in their area. This is how pre-control became an important tool at AMD. Below is a case history of how the pre-control approach has been applied to control a six sigma process capability.

Case History 7.3: The First Trial of Applying Pre-Control

A certain amount of epoxy adhesive paste is used to attach the die to the lead frame when assembling an electronic device. It has been proven that there is a strong correlation between the paste thickness control and the quality of the die attach. This is why controlling the paste thickness has become a critical parameter (one that is included in the six sigma quality program).

The previous results of a process capability study showed that the paste thickness capability was approaching a Cp of 2. With this information in mind, the supervisor decided to apply the pre-control technique to control the paste thickness quality. To be sure that the process would run on a zero defect quality level, he decided to apply the 50% concept, which means that he will use only half of the allowable specification range in the application of pre-control (see Figure 7.38). So even when the process occasionally produces one red (which will give a signal for investigation), the process will still produce zero defects. To see how the process will behave when using the 50% concept, the supervisor decided to temporarily implement a pre-control chart and record the values. Figure 7.39 shows that after 102 pairs (A and B) were inspected, there were two stoppages where only slight adjustments were necessary. In the first two weeks the

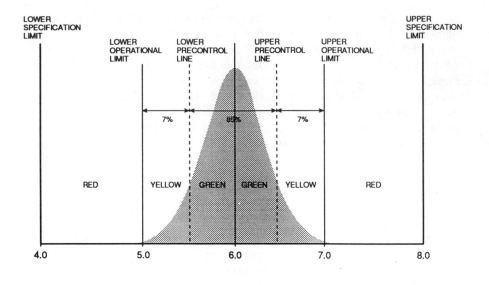

Figure 7.38   The fifty-percent concept and pre-control technique in controlling paste thickness quality.

supervisor took samples (A, B pairs) every two hours just to have an impression of how the pre-control technique was working. Once he had the results, he calculated the frequency interval of samples as follows:

1. The two yellows were received 1/10 at 10 a.m. (sample 2), and another two yellows were received 1/19 at 12:00 p.m. (sample 101). The period in between the two stoppages is

$$T_s = \text{sampling interval } (ST_2 - ST_1)$$

   where T_s is the time between two stoppages, ST_2 is the sample number of the second stoppage, and ST_1 is the sample number of the first stoppage. In our example,

$$T_s = 2 \times (101 - 2) = 198 \text{ hours}$$

2. The frequency of sampling can be calculated as follows:

$$\text{Frequency} = \frac{T_s}{24}$$

$$= \frac{198 \text{ hours}}{24}$$

$$= 8.3 \text{ hours}$$

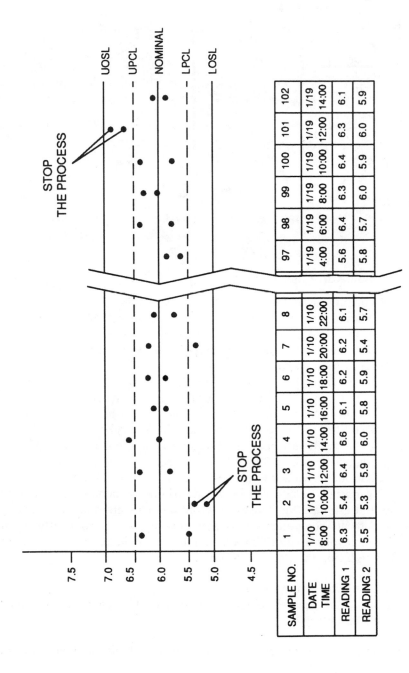

SAMPLE NO.	1	2	3	4	5	6	7	8		97	98	99	100	101	102
DATE TIME	1/10 8:00	1/10 10:00	1/10 12:00	1/10 14:00	1/10 16:00	1/10 18:00	1/10 20:00	1/10 22:00		1/19 4:00	1/19 6:00	1/19 8:00	1/19 10:00	1/19 12:00	1/19 14:00
READING 1	6.3	5.4	6.4	6.6	6.1	6.2	6.2	6.1		5.6	6.4	6.3	6.4	6.3	6.1
READING 2	5.5	5.3	5.9	6.0	5.8	5.9	5.4	5.7		5.8	5.7	6.0	5.9	6.0	5.9

Figure 7.39 Pre-control chart for paste thickness.

137

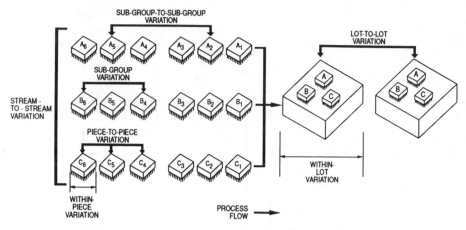

Figure 7.40 Major sources of product variation.

7.7.4 The Multi-Vari Chart Method

The most frequently used tool for process capability studies is the control chart which allows us to investigate the "consecutive-piece" and "time-to-time" sources of variation. However, in situations where we suspect "within-piece," "measurement error," or other components of variation, the multi-vari chart can be successfully applied.

The multi-vari chart is a graphical method of presenting the process spread in such a way that allows a visual expression of the degree of contribution of lot-to-lot variation, time-to-time, within-piece, and consecutive-piece variabilities to the total spread (see Figure 7.40).

By adding the multi-vari chart to the statistical "toolbox," the process capability study becomes broader and more effective. The application of the multi-vari chart let us view the process spread as a composite of several "sources of variation" which can be divided into components in order to understand and estimate the causes of variation. This approach of studying the variation by its components makes it possible to determine the most influential source of variation, so we can make engineering efforts more successful in reducing the process variation.

7.7.4.1 Description of the Chart

The multi-vari chart can be compared with a control chart of individual measurements except:

1. The upper and lower limits are specification limits and not control limits.
2. Every unit has a number of measurements that let us determine the "within-piece" variation.

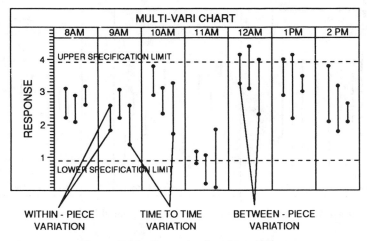

Figure 7.41   Example of multi-vari chart.

For example, the plating thickness of a unit can be measured in several locations to determine the "within-piece" thickness variations. A molded part can be measured in two or more locations to determine the impact of the mold cavity on the total process variation. The result of the reading of one unit is plotted on the chart as a single vertical line which represents the range of those readings. In other words, the line connects the minimum and maximum values of the reading of an individual unit. As for the usual control chart, a series of successive parts is taken from the process from time to time and plotted on the multi-vari chart. The only difference is that instead of a "dot," a vertical line is plotted for every unit (see Figure 7.41).

7.7.4.2 Interpretation of the Chart

By comparing the lengths of the vertical lines to the allowable tolerance range, the magnitude of the "within-piece" variation can be visualized (see Figure 7.42). When the average length amounts to one-half of the tolerance or more, this is an indication that actions should be taken to reduce "within-piece" variability.

Observing the scatter among successive parts, we can visualize the capability of the process. If it is desired, we can also have a tentative measure of the capability by using conventional formulas. The process capability (PC) would be

$$PC = 6\left(\frac{\bar{R}}{d_2}\right)$$

where $\bar{R}$ is the average range of the consecutive maximum (or minimum)

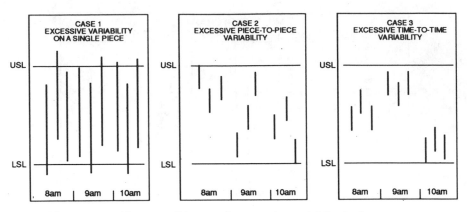

Figure 7.42   Three possible cases for excessive variability in the process.

readings, and d_2 is a constant factor which depends on the sample size. The tentative capability index (C_P) would be

$$C_P = \frac{\text{USL} - \text{LSL}}{6\left(\dfrac{\bar{R}}{d_2}\right)}$$

Note: This estimate should be treated as an approximation because we don't know if the process is in a state of statistical control from this chart alone.

When the process capability does not meet the requirements, actions should be taken to reduce the "consecutive-piece" variability by determining and eliminating the causes which affect the ability of the process to reproduce its cycle. Such causes may be: variation in temperature or pressure, looseness in fixtures, etc. Examination of the positions of the plotted lines from one sample cluster to another allow us to make conclusions about the extent of contribution of time-to-time variability to the overall spread.

Presence of drifts or trends in the process can be caused by incoming lot-to-lot differences in material, tool wear, excessive process readjustments, etc. So the multi-vari chart allows us to observe the overall process as variability within a single part, variability between parts, and also time-to-time variability.

Studying the variation by its components makes it easier to work on reduction of the overall process variation. A case history which will demonstrate the application of the multi-vari chart follows.

Case History 7.4: Studying the Components of Variation in the Tin-Plating Process

The thickness of the tin plating is one of the most important elements of plating quality. It's dependent on a number of variables, such as temperature at various locations of the plating bath, amount of tin deposited at the cathode of the plating barrel, variations in bath-to-bath concentration of the plating solution, etc. Drifts or trends in the process can also be caused by the depletion of tin at the anode as chemical reactions take place, daily environment temperature fluctuations gradually building up a tin deposit at the cathode of the plating barrel, etc. The manufacturing workers constantly monitor the process by a number of different control charts. But a process engineer decided, in addition, to use a multi-vari chart to obtain a visual picture of the relationship between some of the components of variation. There was also a need to confirm management's suspicions that there may be a big difference in the bath variation.

To obtain the data, the engineer took five measurements from five parts randomly selected from each plating bath at four hourly intervals. The data were tabulated as shown in Table 7.10. The multi-vari charts were then constructed (see Figure 7.43) so that for each bath the vertical line represents the "within-parts" variability. There are five such vertical lines representing the "between-parts" variability for each time period. The different time periods represent "time-to-time" variability. The multi-vari charts for different plating baths can also be compared with each other for "between-baths" variability.

Interpretation

1. For the overall period of analysis, no parts outside the specification have been observed.
2. The average "within-parts" variation consumes only 15% of the specification range, which is acceptable. However, in bath 3, time period 3, an increase in "within-parts" variation is observed. Also "between-parts" variation has been slightly increased as the vertical lines within the period were dispersed further apart. An investigation revealed that the fillers used for electrical conductivity were already too thick. Changing the fillers remedied the problem.
3. The "between-parts" variation seemed to be more than that of "time-to-time" or "between-lots" variation.
4. Bath 1 shows that the process is centered on the specification nominal. However, a slight shift of the process is observed in the upper specification limit when observing all three baths together.

The results also show that management's suspicions that there may be a significant variation from "bath to bath" is not confirmed by this investigation.

TABLE 7.10   Tin Plating Results from Three Baths

Plating Bath 1				Plating Bath 2				Plating Bath 3			
Time Period	Sample Number	X min	X max	Time Period	Sample Number	X min	X max	Time Period	Sample Number	X min	X max
1	1	493	571	1	1	540	611	1	1	517	563
1	2	510	572	1	2	496	586	1	2	522	565
1	3	513	591	1	3	555	592	1	3	504	547
1	4	566	653	1	4	621	684	1	4	523	564
1	5	585	687	1	5	582	663	1	5	450	490
2	1	601	668	2	1	602	637	2	1	500	560
2	2	592	682	2	2	621	687	2	2	531	610
2	3	635	688	2	3	628	698	2	3	490	574
2	4	461	542	2	4	493	549	2	4	539	610
2	5	537	592	2	5	553	599	2	5	509	588
3	1	550	584	3	1	607	639	3	1	651	736
3	2	627	636	3	2	689	738	3	2	457	566
3	3	571	672	3	3	670	760	3	3	563	669
3	4	462	511	3	4	575	676	3	4	635	744
3	5	575	654	3	5	508	576	3	5	534	675
4	1	597	622	4	1	650	686	4	1	590	742
4	2	582	638	4	2	653	709	4	2	495	689
4	3	642	679	4	3	531	594	4	3	505	693
4	4	545	598	4	4	571	655	4	4	472	644
4	5	561	664	4	5	601	641	4	5	622	772

7.7.5   The Control Chart for Individual Measurements with Limits from a Moving Range

The control chart of individual measurement with limits calculated based on the moving range can be considered a special technique for two major reasons: The first is because it can be applied not only for variable data but also for attribute data; and the second reason is that the special way of calculating the control limits makes the chart less sensitive to small process changes but perfectly able to indicate trends, large shifts, and drifts. This is why the chart is used when we want to have an overall picture of a macroprocess, or when we are dealing with large lots or batches.

Sometimes a measurement (X) may represent a full day, a shift, a week, or even several months. For example, in a low ppm environment where nonconforming units are rarely found, we need to accumulate a very large sample (for several days or even weeks) to have some nonconforming parts. Or, let's say we want to control the accident rate in a company to see the trend of safety improvement. If you want to accumulate a result of 4 or 5 accidents, you might have to wait a long time. There are also manufacturing parameters such as temperature, pressure, voltage, and moisture content, where it doesn't make sense to take a subgroup sample size. Or, for example,

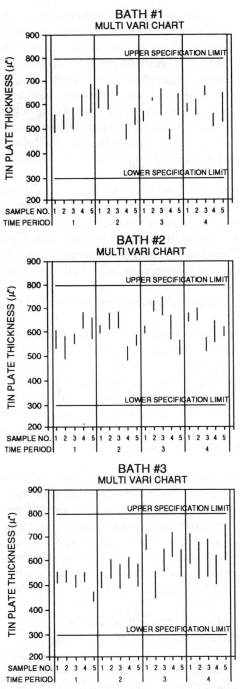

Figure 7.43 Multi-vari charts for three baths.

if you want to control shipments, orders, productivity, scrap, rework, or other accounting data, what kind of chart would be appropriate for such situations?

The answer is a control chart for individual measurements with control limits based on a moving average. This type of chart can be used for variable data and, in special cases, for attribute data. When applying to variable data, the control chart for individual measurements can be used for process control. However, considering that this chart is significantly less sensitive than an $\overline{X}$-R chart or an $\overline{X}$-S chart, it should be used only when individual observations are available. As for its application to attribute data, this type of chart is used mainly for reporting purposes. For attribute data, it is more a management chart than a control chart.

On some occasions this type of chart is used to determine the process capability for manufacturing support and nonmanufacturing processes: for example, if we want to determine the capability of a group of inspectors, or we want to determine the quality level of a maintenance group, or we want to control the quality level of the purchasing department. In these and other similar situations, the control chart for individual measurements is legitimate.

7.7.5.1   Using the Individual Measurements Control Chart for Attribute Data

When p charts are used at the end of the production process (where a number of characteristics are combined) and when the sample size is very large, it is difficult to apply a conventional p chart. For this, it is more appropriate to use a control chart for individual measurements. We consider the fraction defective (or percent defective) an individual measurement, similar to the countable or variable data. The moving range method can be applied to calculate the control limits. Control charts with limits calculated in this way will not reveal small changes in the process, but will indicate large general trends.

In relation to attribute data, the control chart with limits based on the moving range can be used mainly for reporting purposes when we want to have an overall picture of a plant or manufacturing process, or where a large number of variables are combined on one chart, making the sample size very large. This chart can also be used to report the percentage of scrap, rework or retest, percentage of product returns from the customer, etc.

The following examples will illustrate the procedure of applying the control chart for individual measurements using variable and attribute data.

7.7.5.2   Examples of Applying Control Charts for Individual Measurements

Example 1: Using Variable Data   Silver glass paste comes from the supplier in lots. Because this paste is very important to the integrated circuits assembly process, its quality is inspected at the Receiving Inspection Chemical Laboratory and analyzed on a number of different parameters. To

TABLE 7.11   Results of Silver Glass Paste Monitoring for Two Parameters

Lot Number	(1) Percentage Solid	(2) MR	(3) Average Adhesion	(4) MR
1	81.0	0.1	87.1	8.9
2	80.9	1.0	78.2	19.0
3	81.9	1.7	59.2	16.7
4	80.2	1.1	75.9	0.6
5	81.3	0.1	76.5	1.8
6	81.2	0.23	78.3	5.9
7	80.97	0.83	72.4	4.8
8	81.8	0.5	77.2	4.5
9	81.3	0.01	72.7	15.7
10	81.31	1.06	57.0	18.7
11	82.37	2.17	75.7	6.8
12	80.2	1.29	82.5	21.0
13	81.49	0.29	61.5	15.7
14	81.2	0	77.2	4.5
15	81.2	0.57	72.7	6.1
16	81.77	0.03	78.8	14.1
17	81.8	0.9	64.7	11.0
18	80.9	0.9	75.7	5.3
19	81.8	0.2	70.4	17.1
20	82.0	1.5	87.5	7.2
21	80.5	1.13	80.3	2.6
22	81.63	0.59	77.7	12.8
23	81.04	0.15	64.9	24.6
24	81.19	0.97	89.5	25.0
25	80.22		64.5	

monitor the quality of this incoming material, a control chart of individual measurements with control limits calculated by the moving range is used.

At the beginning of introduction, the control chart for individual measurements was plotted, together with the moving range chart. Table 7.11 shows the results of monitoring two parameters (percentage solid and average adhesive) from 25 lots. Figure 7.44 represents the control chart for percentage solid. To construct the control chart, the following steps were performed.

Step 1 Calculate the average of all the individual measurements. This will be the center line of the chart:

$$\bar{X} = \frac{X_1 + X_2 + \cdots + X_k}{k}$$

$$= \frac{81 + 80.9 + \cdots + 80.22}{25}$$

$$= 81.2$$

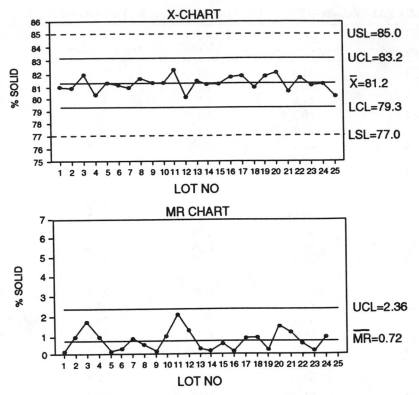

Figure 7.44 X-MR chart for percent solid analysis.

where $X_1, \ldots, X_k$ are the individual measurements, and k is the number of individual measurements.

Step 2 Compute the moving ranges (MR) by taking the absolute value of the difference between the first and second measurements, then the second and third measurements, etc.

As an illustration, the moving range for samples 1 and 2 are

$$MR_1 = |81 - 80.9| = 0.1$$

For samples 2 and 3:

$$MR_2 = |80.9 - 81.9| = 1.0$$

The results of the moving range calculations are recorded on the datasheet [see column (2) in Table 7.11]. *Note:* The total number of differences, or "moving ranges," should be one less than the total number of individual measurements.

Step 3 Find the average of the moving ranges by the following formula:

$$\overline{MR} = \frac{MR_1 + MR_2 + \cdots + MR_{k-1}}{k-1}$$

$$= \frac{0.1 + 1.0 + \cdots + 0.97}{25 - 1}$$

$$= 0.72$$

Step 4 Calculate the upper and lower control limits as follows:

$$\text{UCL} = \overline{X} + 2.66(\overline{MR})$$

$$= 81.2 + 2.66(0.72)$$

$$= 83.2$$

$$\text{LCL} = \overline{X} - 2.66(\overline{MR})$$

$$= 81.2 - 2.66(0.72)$$

$$= 79.3$$

Interpretation

As one can see from Figure 7.44, the range is in control and the individual measurements are also within control limits. Considering that every point represents one individual measurement, we made two extra lines on the chart that represent the specification limits. So now we can say that the percentage solid parameter of the paste came from a controlled process, and the product conforms to the specification requirements. Figure 7.45 represents the same type of chart to control the average adhesive strength (in pounds), which is

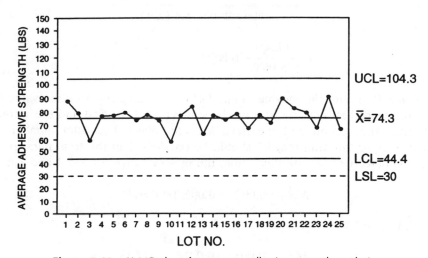

Figure 7.45   *X*-MR chart for average adhesive strength analysis.

another parameter of the same paste. With the parameter, the moving range chart is not monitored, but it is calculated to determine the control limits. The control chart for individuals with the moving range is frequently used in the Receiving Inspection Chemical Laboratory.

Example 2: Using Attribute Data   Frame attach is a process of connecting the frame onto the base of a ceramic DIP (dual-in-line packages). This process has a number of quality characteristics that can be visually inspected 100%. Table 7.12 shows the inspection results for 22 lots. Taking into consideration that the lots come to the inspection area in very large sizes, it was decided to use a p chart with control limits calculated from the moving range. To construct the chart, the following steps have been performed.

Step 1 Compute the fraction defective (p) for every lot as follows:

$$p = \frac{\text{number of defectives }(np)}{\text{sample size }(n)}$$

As an example for lot 1:

$$p = \frac{79}{30,348} = 0.0026$$

The results of fraction defective calculation are recorded in column (4) of Table 7.12.

Step 2 Calculate the average fraction defective ($\bar{p}$) by the following formula:

$$\bar{p} = \frac{\text{total number of defectives }(\Sigma np)}{\text{total sample size }(\Sigma n)}$$

$$= \frac{1924}{805,003} = 0.0024$$

Step 3 Compute the moving ranges (MR) by taking the absolute value of the difference between the fraction defectives (p) of the first and the second lot, then the second and the third lot, etc. *Note:* The total number of differences or "moving ranges" should be one less than the total number of fraction defectives. As an illustration, the moving range for lots 1 and 2 is

$$MR_1 = |0.0026 - 0.0057| = 0.0031$$

For lots 2 and 3:

$$MR_2 = |0.0057 - 0.0014| = 0.0043$$

TABLE 7.12   Datasheet for the Individual Measurement Chart for Frame Attach Visual Inspection Results

(1) Lot Number	(2) Sample Size (n)	(3) Number of Defectives (np)	(4) p	(5) MR
1	30,348	79	0.0026	0.0031
2	9,090	52	0.0057	0.0043
3	41,077	58	0.0014	0.0014
4	34,358	98	0.0029	0.0018
5	13,775	15	0.0011	0.0005
6	51,073	79	0.0015	0.0021
7	21,645	79	0.0036	0.0013
8	96,820	227	0.0023	0.0012
9	44,936	52	0.0012	0.0011
10	10,924	25	0.0023	0.0009
11	9,063	29	0.0032	0.0047
12	5,320	42	0.0079	0.0028
13	19,513	99	0.0051	0.0042
14	35,265	31	0.0009	0.0021
15	52,342	156	0.0030	0.0008
16	45,583	99	0.0022	0.0016
17	10,729	41	0.0038	0.0033
18	38,089	21	0.0006	0.0010
19	24,684	39	0.0016	0.0039
20	94,234	521	0.0055	0.0052
21	96,870	31	0.0003	0.0023
22	19,265	51	0.0026	
	$\Sigma n = 805,003$	$\Sigma np = 1,924$		

The results of the moving range calculation are recorded on the datasheet [see column (5) in Table 7.12]. *Note:* The total number of differences or "moving ranges" should be one less than the total number of individual measurements.

Step 4 Find the average of the moving ranges by the following formula:

$$\overline{MR} = \frac{MR_1 + MR_2 + \cdots + MR_{k-1}}{k-1}$$

$$= \frac{0.0031 + 0.0043 + \cdots + 0.0023}{22-1}$$

$$= 0.0024$$

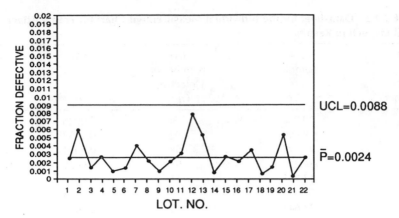

Figure 7.46   A *P*-MR chart for frame attach inspection results.

Step 5 Calculate the upper and lower control limits as follows:

$$UCL = \bar{p} + 2.66(\overline{MR})$$

$$= 0.0024 + 2.66(0.0024)$$

$$= 0.0088$$

$$LCL = \bar{p} - 2.66(\overline{MR})$$

$$= 0.0024 - 2.66(0.0024)$$

$$= -0.0040$$

Since the lower control limit is negative, it is not considered.

Step 6 The control chart for individual measurement is constructed as shown in Figure 7.46.

Interpretation
It is observed that all the points are inside the control limits and the pattern appears to be normal. Based on the control chart, we can make a conclusion that the frame attach process is stable and in a state of statistical control.

7.7.6   The $\bar{X}$-R Chart with Modified Control Limits
In the electronics industry, there are situations where the spread of natural process variation is considerably less than the allowable specification spread. This kind of situation grows from day to day because of the activities of reducing the variation around the target and the introduction of the six sigma

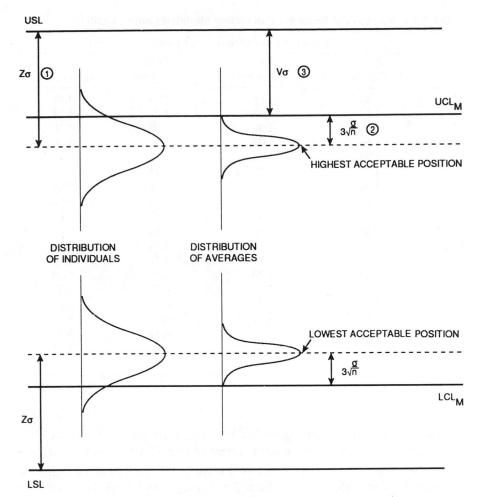

USL

$Z\sigma$ ①

$V\sigma$ ③

UCL_M

$3\frac{\sigma}{\sqrt{n}}$ ②

HIGHEST ACCEPTABLE POSITION

DISTRIBUTION
OF INDIVIDUALS

DISTRIBUTION
OF AVERAGES

LOWEST ACCEPTABLE POSITION

$3\frac{\sigma}{\sqrt{n}}$

LCL_M

$Z\sigma$

LSL

Figure 7.47   Configuration for modified control limits.

quality concept. In such cases sometimes it is more economical to control these parameters by using *reject limits* instead of *control limits*.

When reject limits are applied to control a process, they are called *modified control limits*.[17] The modified control limits are designed to allow the process average to shift just high enough or just low enough to produce all parts within the specification range. Figure 7.47 illustrates the development of such limits.

Now let's assume that the process is in its highest acceptable position with the process average exactly $Z\sigma$ below the upper specification limit (USL) where Z is the number of sigma between USL and the highest acceptable

TABLE 7.13   Values of *V* Factor for Calculating Modified Control Limits

| | Values of *V* for Different ppm (or Sigma) Levels | | | |
n	$Z=3$ (1350 ppm)	$Z=4$ (32 ppm)	$Z=5$ (0.3 ppm)	$Z=6$ (0.001 ppm)
2	0.88	1.88	2.88	3.88
3	1.27	2.27	3.27	4.28
4	1.50	2.50	3.50	4.50
5	1.66	2.66	3.66	4.66
6	1.78	2.78	3.78	4.78
7	1.87	2.87	3.87	4.87
8	1.94	2.94	3.94	4.94
9	2.00	3.00	4.00	5.00
10	2.05	3.05	4.05	5.05
11	2.10	3.10	4.10	5.10
12	2.13	2.14	2.15	2.16
13	2.17	3.17	4.17	5.17
14	2.20	3.20	4.20	5.20
15	2.23	3.23	4.23	5.23
16	2.25	3.25	4.25	5.25
17	2.27	3.27	4.27	5.27
18	2.29	3.29	4.29	5.29
19	2.31	3.31	4.31	5.31
20	2.33	3.33	4.33	5.33

process average (see ① in Figure 7.47). In this case, on an $\bar{X}$ chart with a subgroup size of *n*, the upper modified control limit (UCL_M) will be $3\sigma/\sqrt{n}$ above the process average (see ② in Figure 7.47). This limit can also be designated as the upper reject limit for averages ($URL_{\bar{x}}$). Now we can express the distance of UCL_M below the USL as ($Z\sigma - 3\sigma/\sqrt{n}$), which can be transformed to ($Z - 3/\sqrt{n}$)σ. This factor ($Z - 3/\sqrt{n}$) is designated as *V* (see ③ in Figure 7.47).

Table 7.13 gives values of *V* corresponding to different sample sizes (*n*) and multiples of sigma (*Z*). Using the same approach of calculation, it can be shown that the lower modified control limit (LCL_M) is $V\sigma$ above the lower specification limit. Mathematically, the modified control limits can be determined by the following formulas:

$$UCL_M = USL - V\sigma$$

$$LCL_M = LSL + V\sigma$$

By applying modified control limits we can avoid the extra cost of stopping the process when there is a process shift, especially if the shift is not significant enough to cause the production of nonconforming units. This is important when the plants are working on reducing costs and cycle time.

In Figure 7.47 one can see that the process average in this case varies from the position marked "highest acceptable position" of the process to that marked "lowest acceptable position." However, it is important to note here that (1) when we use modified limits only, we lose the opportunity of disclosing the presence of or absence of statistical control in the manufacturing process, and (2) the protection given by the modified (reject) limits depends on a good estimate of the standard deviation (σ). Because of this, it is important to make sure that the range chart of the process remains in a state of statistical control. Below is an example of applying the chart with modified control limits.

Case History 7.5: Using Modified Control Limits for the Wire Bonding Process

Wire bond is a critical variable in the integrated circuit assembly process. To control the bond strength, a destructive method is used where the operator tests $n = 5$ wires from a unit, taken randomly from the process every four hours. The results are plotted on an $\overline{X}$-R chart (see Figure 7.48).

Figure 7.48 shows that the $\overline{X}$ chart demonstrates a process that is not in a state of statistical control. This is because the chart represents a number of different types of devices that differ from each other by the wire length, which impacts the measurement results. If we used only one type of device on one machine, the same process would demonstrate a state of statistical control (see Figure 7.49). However, in real life, it is sometimes difficult to specialize the bonding machine for a particular device because of the large variety of devices and limited number of machines. An alternative would be to have separate control charts for every device type. Unfortunately, this is not a practical solution because of the quantity and frequency of changes in the device type.

The investigation of a large number of different types of devices shows that the difference in the spread between the types of devices is insignificant. The only difference here is the average value (see Figure 7.50). Taking into consideration that every individual product has a normal distribution and also that the spread of every type of device is significantly smaller than the allowable specification range (USL $= 11.0$, LSL $= 3.0$) which is the same for all the family of devices, a control chart with modified control limits is decided to be used which will permit a larger range of variation for the average in comparison with the conventional ± 3 sigma limits.

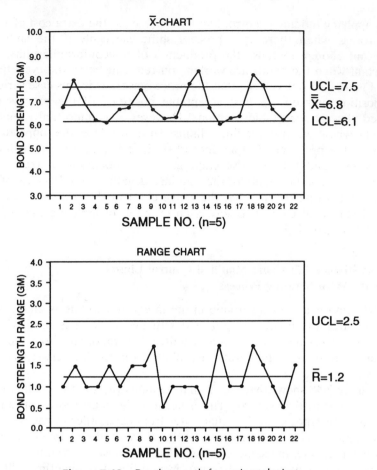

Figure 7.48   Bond strength for various devices.

In this application, it was decided to control the process at the six sigma quality level. The calculation of the modified control limits are as follows:

$$UCL_M = USL - V\sigma$$

$$LCL_M = LSL + V\sigma$$

In our example, USL = 11 and LSL = 3. From Table 7.13, for $Z = 6$, $n = 5$:

$$V = 4.66$$

The value of σ is estimated by $\bar{R}/d_2$ where (from Figure 7.48) $\bar{R} = 1.2$,

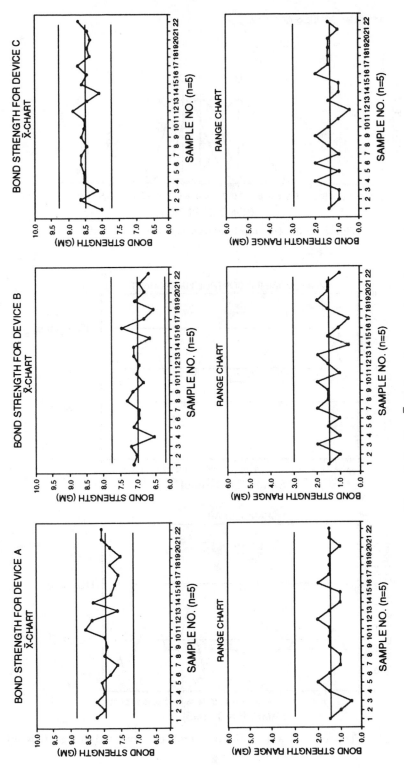

Figure 7.49 The $\bar{X}$-R charts for separate devices.

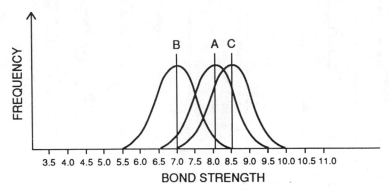

Figure 7.50 The distributions for devices *A*, *B*, and *C*.

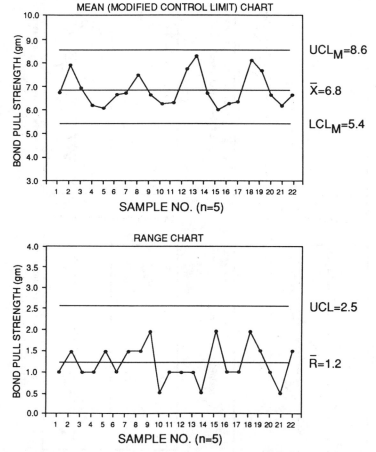

Figure 7.51 A control chart with modified control limits for bond strength (for various devices).

for $n = 5$, $d_2 = 2.326$; therefore,

$$\sigma = \frac{\bar{R}}{d_2} = \frac{1.2}{2.326} = 0.52$$

The modified control limits are determined as follows:

$$\text{UCL}_M = \text{USL} - V\sigma$$

$$= 11 - (4.66)(0.52)$$

$$= 8.6$$

$$\text{LCL}_M = \text{LSL} + V\sigma$$

$$= 3 + (4.66)(0.52)$$

$$= 5.4$$

The control chart with modified control limits is constructed as shown in Figure 7.51.

This example illustrates that modified control limits may be used when the inherent variability of the manufacturing process is much smaller compared with the width of the specification tolerance zone. Therefore, the process mean may shift from the central target value without defective product being made. In our application, relaxing the control of the means without jeopardizing the quality is an economical approach.

REFERENCES

1. Western Electric Co., "Statistical Quality Control Handbook" (1956), p. 3.
2. Walter A. Shewhart, "Economic Control of Quality Manufactured Product" (1931), Van Nostrand, p. 6.
3. Walter A. Shewhart, "Economic Control of Quality Manufactured Product" (1931), Van Nostrand, p. 5.
4. Walter A. Shewhart, "Industrial Quality Control" (1947), p. 23.
5. "Glossary and Tables for Statistical Quality Control," ASQC, Statistical Division.
6. Walter A. Shewhart, "Statistical Methods from the Viewpoint of Quality Control" (1939), Washington, D.C. Graduate School, Department of Agriculture, Chapter 1, Dover Publications, Inc. (1986).
7. A. J. Duncan, "Quality Control and Industrial Statistics" (1974), Richard D. Irwin, Inc., p. 94.
8. Lloyd S. Nelson, "The Shewhart Control Chart–Tests for Special Causes," Journal of Quality Technology, Vol. 16, No. 4 (October 1984), pp. 237–239.

9. Western Electric Co., "Statistical Quality Control Handbook" (1956), Section I, Part B and Section II, Part F, p. 27, Figure 38.

10. Western Electric Co., "Statistical Quality Control Handbook" (1956), p. 172.

11. Walter A. Shewhart, "Economic Control of Quality Manufactured Product" (1931), Van Nostrand, p. 314.

12. A. J. Duncan, "Quality Control and Industrial Statistics" (1974), Richard D. Irwin, Inc., p. 449.

13. William B. Rica, "Control Charts in Factory Management" (1947), John Wiley & Sons, Inc., p. 82.

14. A. J. Duncan, "Quality Control and Industrial Statistics" (1974), Richard D. Irwin, Inc., p. 449.

15. Thomas W. Calvin, "Quality Control Techniques for 'Zero Defects'" Statistical Quality Control, edited by Warren H. Klippely, (1984), Society of Manufacturing Engineers, pp. 110–118.

16. Dorian Shainin, "Techniques for Maintaining a Zero Defects Program," American Management Association Bulletin, Vol. 71, (1965), p. 18.

17. D. Hill, "Modified Control Limits," Applied Statistics, Vol. 5, No. 1, (1956), pp. 12–19.

Chapter **8**

Measurement and Inspection Capability Studies

In the survey Quality Pulse[1] conducted at the Quality Expo '87, the attendees were asked to fill out a questionnaire. One of the true/false questions was: "Our management makes sure we never release or ship products known to be substandard in quality."

One hundred percent of the administrators agreed with this statement, but only 56% of those in quality control and 67% of those in engineering agreed. Why did the administrators feel more confident with the shipped quality than the quality control or engineering personnel? This is because a technician or inspector keeps a hand "on the pulse" of the quality more than an administrator. These people know better how the testers are calibrated and what the inspectors' accuracy is.

Management makes decisions about the shipped quality from the available information on the yields, test, or inspection results, but how accurate are these results? How much can we rely upon them? To answer these questions, we need to know more about the test system capability which includes not only calibration, but also repeatability, reproducibility, and stability. We also need to know what our inspectors' capability is, and what should be done to improve their performance.

In this chapter the reader will be introduced to a series of topics related to measurement and inspection activities.

8.1 CONCEPTS AND TERMINOLOGY

8.1.1 Treating Measurement as a Production Process

How important is SPC to a testing or inspection process? Some people think that SPC is a technique that belongs more to a production process. Frequently statements are made like this, "SPC is important when you want to control a manufacturing process. We are a testing organization and we don't produce anything, so SPC is not applicable to us."

Here again we have a terminology problem. What is a process? By definition, any set of conditions or set of causes working together to produce an outcome is called a *process*. This can be a manufacturing or assembly process, and this can also be a measurement process where the outcome is the reading of an instrument.

It is very important to regard measurement as a production process where the "raw material" is the units to be measured and the "product" is the measurements that the measurement process yields (see Figure 8.1). Treating measurement as a production process is valuable because it not only identifies several potential sources of variability, but it also shows the possibility and necessity of using Shewhart's control charts to control the measurement process.

8.1.2   What Do You See when You Measure?

When you look at the numbers that the instrument is "throwing out," you believe in them. You make decisions based on these numbers, and you think that your decisions are right. The more expensive the measurement equipment is, the stronger your belief is in the figures it generates. According to procedures, you periodically calibrate the instrument, and your confidence in the figures grows again.

In fact, no matter how expensive the instrument is, or how good the calibration results are, the value you get by measuring a product consists of two major components: the *true value* of the product and the value of the *measurement error*. The relationship between these components can easily be seen if we consider a right triangle (see Figure 8.2) in which the shorter leg is proportional to the standard deviation of the *measurement error* (σ_e), the longer leg is proportional to the standard deviation of the *product* (σ_p), and the hypotenuse is proportional to the standard deviation of the *observed measurements* (σ_o). Mathematically, this can be expressed as

$$\sigma_o = \sqrt{\sigma_p^2 + \sigma_e^2}$$

and

$$\sigma_p = \sqrt{\sigma_o^2 - \sigma_e^2}$$

If we can separate the measurement error (σ_e) from the production variation (σ_p), then we can perform the following two analyses to see whether the measuring system is adequate.

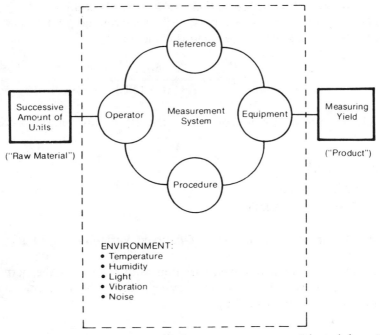

Figure 8.1 Measurement as a production process concept. Adapted from Karl, F. Speitel, "Measurement Assurance," Chapter 8.2 of "Handbook of Industrial Engineering," edited by Gavriel Salvendy (1982), John Wiley & Sons, p. 8.2.2[2].

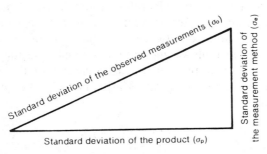

Figure 8.2 The relationship between measurement components. Adapted with permission from W. Edwards Deming "Sample Design in Business Research" (1960), John Wiley & Sons, Inc., p. 62[3].

8.1.2.1   *Precision / Tolerance Ration (P / T Ratio) Analysis*

The precision/tolerance ratio represents the percentage of tolerance consumed by the measurement error, and is computed by

$$P/T \text{ ratio} = \frac{6\sigma_e}{(\text{USL} - \text{LSL})}$$

Generally the criteria for acceptance of the P/T ratio of the measuring system are

Under 10%—acceptable.

10% to 25%—marginal.

Over 25%—not acceptable.

8.1.2.2   *Measurement Error (σ_e)/Observed Variation (σ_o) Ratio (E/O Ratio) Analysis*

The measurement error/observed variation ratio represents the percentage of total observed variation consumed by the measurement error, and can mathematically be represented as

$$E/O \text{ ratio} = \left(\frac{\sigma_e}{\sigma_o}\right) \times 100$$

According to Juran's *Quality Control Handbook*, if this ratio is less than 10%, then the effect of measurement error upon the product variation (σ_p) will be less than 1% and is negligible.[4] The allowable error depends on the manufacturing and economic conditions.

8.1.3   Accuracy and Precision

Accuracy or systematic error (SE) refers to the deviation of the measurement from the true value. The accuracy of a measurement can be quantified by determining the difference between the true value of the characteristic and the average value of multiple measurements made on the characteristic. So the difference between the true measurement and the observed average represents the instrument accuracy. Mathematically, the systematic error can be described as

$$\text{SE} = \overline{X}_P - \mu_R$$

where SE is the systematic error, $\overline{X}_P$ is the average value of the measurement process, and μ_R is the expected reference level.

Precision of a measurement process is the degree of agreement among individual measurements of the same sample, and is associated with repeata-

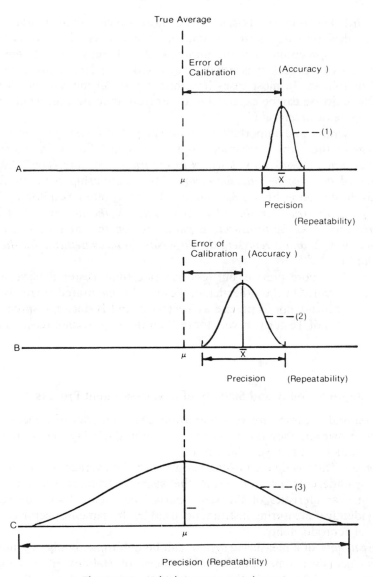

Figure 8.3   Which instrument is better?

bility; i.e., the measurement is precise if upon repetition the individual measurements taken on a single characteristic cluster close together. If the variability is small, then the measurement is precise, but not necessarily accurate (see Figure 8.3).

The *index of precision* is calculated as a multiple of the standard deviation of repeated measurements made on one unit. The measurement process with a standard deviation σ_{e1} is said to be more precise than another with a standard deviation σ_{e2}, if σ_{e1} is smaller than σ_{e2}. (In fact, σ_e is really a measure of imprecision since the imprecision is directly proportional to σ_e.)

The index of precision is usually expressed as $\pm 3\sigma_e$, and can be interpreted as follows. In most cases less than 1% of all measurements from a statistical universe can be expected to differ from their average value by more than 3 sigma of precision ($3\sigma_e$).

It is essential and important to understand the distinction between the meanings of the terms "accuracy" and "precision." Stated in the simplest form, *"accuracy" deals with closeness to the truth* and *"precision" deals with closeness of the repeated measurements*. The relationship between the two terms is illustrated in Figure 8.3, and can be interpreted as follows:

Let μ be the true (or the reference) value, $\overline{X}$ the average of individual measurements made by measuring a parameter on the same device, and the distributions 1, 2, and 3 represent the dispersion in measurements for the testers A, B, and C.

Tester A is more precise, but less accurate than Tester B because of its calibration error. On the other hand, Tester C is calibrated correctly, but is neither as accurate nor as precise as Testers A and B since the spread in the measurements of Tester C covers more than the corresponding spreads of A or B.

8.1.4 Reproducibility and Stability of a Measurement Process

Precision and accuracy are very important characteristics of a measurement process. However, they don't address the reproducibility and stability of a process, which are also very important.

Reproducibility is defined as the variation in the average of measurements made by different operators using the same instrument when measuring identical characteristics of the same parts (see Figure 8.4). The variation among identical measuring instruments used by the same inspector is another source of reproducibility.

The *stability* of a measuring system can be determined only by studying its performance over time. For a quick check of stability, it is enough to determine the difference in the average of at least two sets of measurements obtained with the same instrument on the same part taken at different time (see Figure 8.5).

The Shewhart control chart is the standard statistical technique which can be used successfully to control measurement process stability. Keeping a control chart on every instrument will give a signal when some assignable causes have entered the measuring process and shifted the process level or the variability. When the control chart has a natural pattern, this means that the measurement process is in a state of statistical control. In other words, this means that the measurement process is operating in a predictable

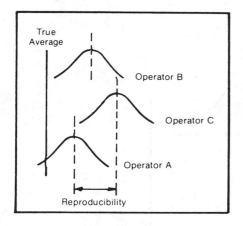

Figure 8.4 Instrument reproducibility. Reprinted with permission from "Statistical Process Control Manual," ASQC (1986), p. 3-2[5].

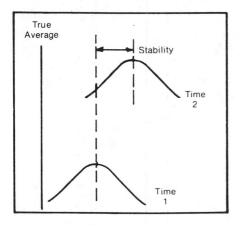

Figure 8.5 Instrument stability. Reprinted with permission from "Statistical Process Control Manual," ASQC (1986), p. 3-2[6].

manner within set limits based on the inherent random variability (often referred to as noise).

8.1.5 Overall Remarks

There are three things that need to be emphasized.

1. A measurement operation can and should be treated as a production process.
2. If this is true, we need to remember that to evaluate the capability of the measurement process, we need it brought into a state of statistical control. Unless this is achieved, we do not know the true capability of the measurement system.

The Instrument Capability Schematic

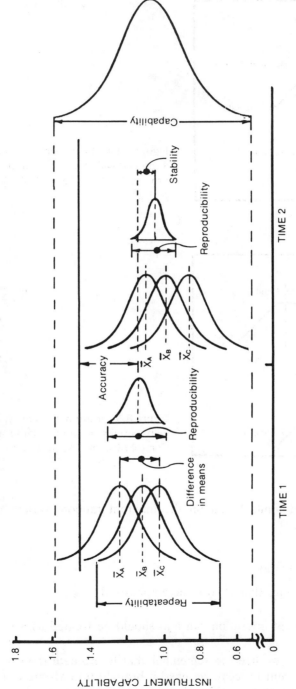

Figure 8.6 The instrument capability schematic. Adapted with permission from Harvey C. Charbonneau and Gordon L. Webster, "Industry Quality Control," Prentice-Hall, Inc. (1978), p. 124[7].

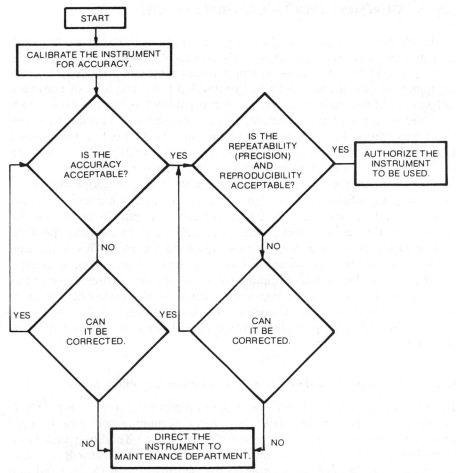

Figure 8.7 Instrument evaluation flow chart. Adapted with permission from "Statistical Quality Control Manual," ASQC (1986), p. 3-0[8].

3. The overall capability of an instrument is represented by its accuracy, repeatability, reproducibility, and stability (see Figure 8.6). To maintain a good measurement process, we need to evaluate the instrument capability as shown in Figure 8.7. Also an inspection capability study should be done on a continuous basis. The detailed methodology of inspection capability studies is described in Section 8.3.

Now that the reader has become more familiar with the terminology related to the test and inspection process, we can demonstrate some procedures and case histories related to test and inspection capability studies.

8.2 MEASUREMENT PROCESS CAPABILITY STUDIES

In the electronics industry, we have a large variety of test systems that work on different principles and measure thousands of parameters. Most of the test equipment is very expensive, which forces the organizations to use the equipment in two or three shifts and to develop tight schedules of maximum utilization. At the same time, the measured parameters require great precision and accuracy of measurement, so the organizations introduce complicated procedures for periodic calibration and check. Finding the optimal frequency of calibration is not only a matter of quality, but it is also an economical issue of equipment utilization and cycle time reduction.

Now in the ppm environment, and also when trying to adopt Motorola's philosophy of introducing six sigma quality, a careful evaluation of the precision and accuracy of all test equipment is required. Improving the precision of the measurement process is equivalent to opening the specification requirements without jeopardizing the product quality. This is again an economical issue because sometimes we find ourselves proposing to invest a lot of money to buy technical equipment to satisfy the customer's specification instead of improving the measuring system. All this makes the measurement capability study a very important area for improvement. In this section we will describe some procedures which helped AMD to improve its measurement system.

8.2.1 The Simplest Method for a Measurement Capability Study

In this subsection we will describe a measurement capability study (MCS) demonstrating the simplest procedure for determining the measurement process stability, accuracy, and repeatability (precision). This study took place at a plating process where the tin thickness of a semiconductor lead frame was measured. To perform the study, only one instrument (X-ray) and one operator were randomly selected from all the instruments and operators available by using the random table. The purpose of this study was to determine if the measurement process capability would satisfy the new process requirements, and if the frequency of recalibration could be reduced.

8.2.1.1 Determining the Measurement Process Stability
According to the existing operating procedure, in parallel with measuring the manufacturing product, every two hours the operator performs an additional measurement using a reference standard to test the measuring instrument. And, if the measurement result deviates from the reference standard, the operator readjusts the instrument. To determine the stability of the measurement process, we decided to continue the procedure and plot the results of measuring the reference standard on a control chart of individual measurements with a moving range (*X-MR* chart). This control chart is shown in Figure 8.8.

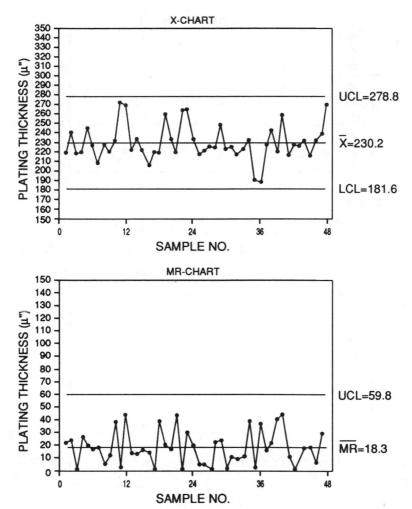

Figure 8.8   An *X-MR* chart showing the results of monitoring the reference value every two hours.

By analyzing the pattern of the chart, one can see that the variations of the X chart are very large and there are some points that give an out-of-control signal. However, after slight instrument readjustments, the process runs a relatively long time in a state of statistical control. The range chart shows reasonable control all the time, even when the X chart gives signals that the process is out of control.

Based on the control chart information, as an experiment, it was decided to change the frequency of checking the instrument from every two hours to one time per shift. The results are reflected in Figure 8.9. As one can see, the

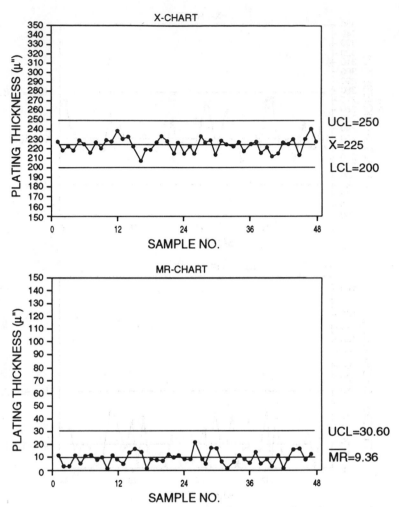

Figure 8.9   An *X-MR* chart showing the results of measuring the reference value every shift.

variability of the measuring process was significantly reduced by leaving the process alone. Then we can conclude that knowing when to "touch" the process and when to "leave it alone" is very important.

Note here that we applied a reference standard to monitor the control chart, so we can use the same data to determine the constant error (the accuracy). However, when a reference standard is not available, you can replicate the measurement of production samples to check the consistency of the measurement process. The range of these replicate measurements should be plotted on an ordinary range chart. The operator, while performing

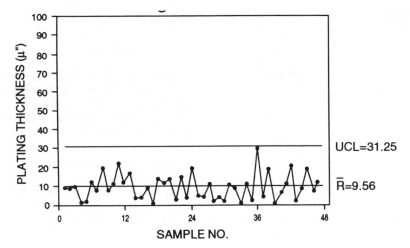

Figure 8.10 A range chart of the results from measuring each unit twice (one time per shift).

routine measurements of production units, in every shift takes one unit from the production line and measures it two times. The difference of this duplication is plotted on a range chart (see Figure 8.10). In this chart, the average range ($\bar{R}$) is almost the same as when we used the reference value to check one time per shift. When the measurement procedure is destructive, every effort must be made to obtain and maintain consistency by measuring two samples that are very close relative to the conditions of how they were made. For example, if we used the procedure described above to monitor the consistency for wire bond quality, we would take two measurements from two consecutive devices on the same lead number.

8.2.1.2 Estimating the Index of Precision (Repeatability)
Once we know that the measurement process is in a state of statistical control, we can use the range chart to estimate the standard deviation of the measurement process (S_e). Since each point of the range chart is a result of the difference between the two readings taken from the same reference standard at the same location, the overall range ($\bar{R}$) represents the consistency (the precision) of the measurement process. In an ideal situation the instrument would be capable of recognizing that it is measuring the same value, and the range of every point would be zero, meaning that the instrument is perfect—zero imprecision. The standard deviation of the measurement process can be determined as follows:

$$S_e = \frac{\bar{R}}{d_2} = \frac{9.36}{1.128} = 8.30$$

TABLE 8.1   Interpretation of the Index of Precision

Standard Deviation	Index of Precision	Interpretation
2	$\pm 16.60\ \mu''$ $(2S_e)$	In most cases about 5% of the measurements from the measurement process can be expected to differ from their average value by more than $2S_e$, which means $212 \pm 16.60\ \mu''$.
3	$\pm 24.90\ \mu''$ $(3S_e)$	In most cases less than 1% of all measurements from the measurement process can be expected to differ from their average value by more than $3S_e$, which means $212 \pm 24.90\ \mu''$.

where $\overline{R}$ is the average range taken from the control chart, d_2 is a constant factor for estimating the standard deviation of individual measurements when the number of observations is 2 (see Appendix B), and S_e is the estimated standard deviation of the instrument error. The index of precision can be represented as ± 2 or ± 3 standard deviations of the measurement error. In our example, the index of precision can be presented and interpreted as in Table 8.1.

8.2.1.3   Estimating the Accuracy (Systematic Error)

In our study we used a reference standard that was calibrated in a higher level laboratory. So we can say that we know the "true" value of the measuring unit which is 212 microinches (μ''). Using the same data from Figure 8.10, which represents measurements taken from the same value 48 times, we made a histogram (see Figure 8.11) and determined the $\overline{X}$ and S_e. The difference between the average value ($\overline{X}_P$) and the true value (μ_R) of the reference would be an estimate of the systematic error. In our example, the systematic error (SE) can be calculated and interpreted as follows:

$$SE = \overline{X}_P - \mu_R$$
$$= 225 - 212$$
$$= 13$$

where SE is the systematic error, $\overline{X}_P$ is the estimated average of n observations made on the same unit, and μ_R is the accepted reference value.

It would be quite unusual for $\overline{X}_P$ to be further than $2S_e$ from the reference value, or about $16.60\ \mu''$ in this example. Thus, the value 225, can be attributed to sampling variability. Maybe before trying to improve the tester calibration, it would be wise to take a second sample of 50 measurements.

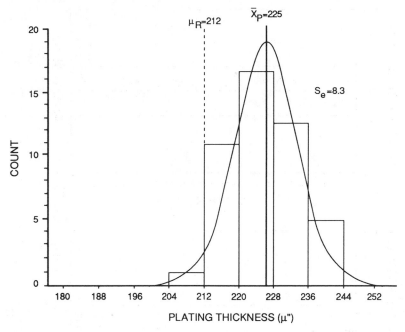

Figure 8.11 A histogram of the results of measuring the reference value every shift.

8.2.1.4 Determining the Precision / Tolerance Ratio

Having estimated the measurement system standard deviation, we can determine the precision/tolerance ratio (P/T ratio), which reflects the relationship between the measurement precision (repeatability) and product tolerances. To do this, we proceed as follows:

$$P/T \text{ ratio} = \frac{\pm 3S_e}{\text{USL} - \text{LSL}} = \frac{6S_e}{\text{total tolerance}}$$

where USL and LSL are upper and lower specification limits, respectively. Substituting the actual values from our study, we have

$$P/T \text{ ratio} = \frac{6(8.30)}{650 - 150} = 0.0996 = 9.96\%$$

Applying the "tenth rule" that the instrument should be able to divide the tolerance into about 10 parts, we can say that the instrument is acceptable. But how will the measurement error inflate the product variation? Figure 8.12 is an $\bar{X}$-R chart of the tin plating thickness, which represents 125 measurements with a subgroup sample size of $n = 5$, taken from the manu-

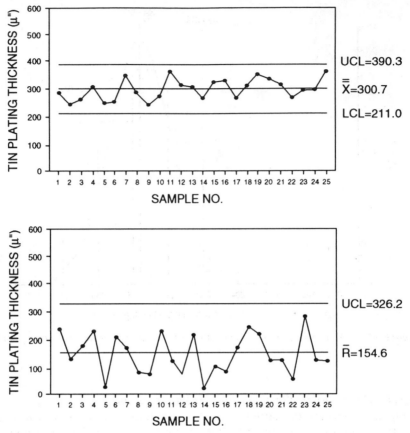

Figure 8.12 $\bar{X}$-R chart for tin-plating thickness.

facturing process. The chart shows that the process is in a state of statistical control, so the average range ($\bar{R}$) from the range chart can be used to estimate the observed (or overall) standard deviation (S_o) of the manufacturing process. To do this, we proceed as follows:

$$S_o = \frac{\bar{R}}{d_2} = \frac{154.6}{2.326} = 66.466$$

This standard deviation ($S_o = 66.466$) includes two components of variation: (1) the variation of the product and (2) the variation of the measuring method. This can be mathematically represented as

$$S_o = \sqrt{S_p^2 + S_e^2}$$

where S_o is the standard deviation of the observed data, S_p is the standard deviation of the product, and S_e is the standard deviation of the measuring method. Now to solve for the standard deviation of the product (S_p), we have

$$S_p = \sqrt{S_o^2 - S_e^2}$$

Substituting the result from our study, we have

$$S_p = \sqrt{(66.46)^2 - (8.30)^2} = 65.95$$

Therefore, the ratio of the standard deviation of the product (S_p) to the overall standard deviation (S_o) is

$$\frac{S_p}{S_o} = \frac{65.95}{66.466} = 0.992$$

This indicates that the instrument variation inflated the product variation less than 1%, which is negligible. The results of this example almost exactly reflect the statement in Juran's *Quality Control Handbook* that if the standard deviation of the measurement error (σ_e) is less than one-tenth the standard deviation of the observed measurements (σ_o), then the effect upon the standard deviation of the product (σ_p) will be less than 1%.

8.2.1.5 Summary
Now let's summarize the information we obtained from the measurement capability study.

1. By introducing an *X-MR* chart, we concluded that the measurement system was stable and that we could significantly reduce the frequency of checking and recalibrating the system, which will not only increase the productivity and reduce the cycle time, but will also reduce the measurement error.
2. The systematic error is not significant and may be due to the sampling error.
3. The P/T ratio is also acceptable because it consumes less than 10% of the allowable specification spread.
4. The product value is not significantly inflated by the measurement error because the S_e is slightly more than 10% of the S_o $(S_e/S_o = 8.30/66.466 = 0.125 = 12.5\%)$. From this information we can say that the measurement system totally conforms to the requirements.

8.2.2 The Influence of the Instrument Imprecision on the Reporting Results of Process Capability

When we measure the capability of a process, most of the time we assume that the readings obtained from the measuring instruments represent the true value of the product quality. However, in real life we have situations where we are not satisfied with the process capability index (Cp or Cpk) without knowing that this is because of the instrument imprecision. If we knew, we would have obtained a more precise instrument and improved customer satisfaction without having to make any effort to improve the manufacturing process. The case history below is an example that reflects this issue.

Case History 8.1: Relationship Between Process Capability and Precision of a Measurement Method

In integrated circuit manufacturing, employing TAB (tape automated bonding) bumping is the process of adding raised metal contacts to the wafer chip bond pads. This provides a platform for bonding on and off of etched copper leads to semiconductor device. Bump height is one of the critical process parameters which influences the bond quality.

This is why it was decided to perform a capability study on the bumping process before releasing it to the manufacturing floor. To do this, the process was properly set up, test equipment was calibrated, and the operator and inspector were given the necessary instructions. In other words, a condition for a classical process capability study was prepared. The results of the study follow.

The histogram and control chart (see Figures 8.13 and 8.14) show that the process is stable and has a normal distribution. But the capability index (Cp = 0.83, Cpk = 0.67) indicates that the process is not capable of meeting the specification requirements.

We could go in two different directions at this stage. The easiest way would be to relax the specification range, but this is not preferable because it would jeopardize the bond quality. The second, more desirable option is to work on reducing the process variations. But how? And where do we start?

Let's take a second look at the histogram (see Figure 8.13). Does it really represent the process spread? To some extent, yes, but not exactly. As we know, the variability observed in measured values of a dimension (in our case bump height) is due in part to the variability of the product and in part to the variability inherent in the method of measuring. Expressing this in a formula, we have

$$\sigma_o = \sqrt{\sigma_p^2 + \sigma_e^2}$$

where σ_o is the observed variability, σ_p is the product variability, and σ_e is

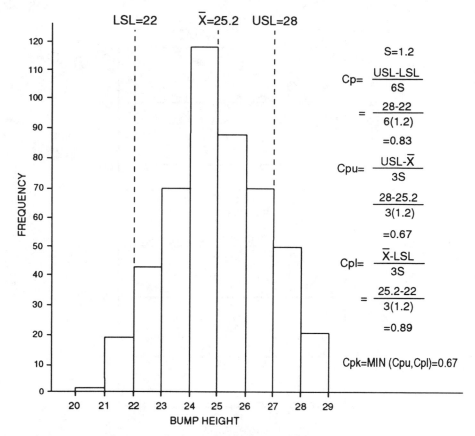

Figure 8.13 A histogram for bump height by using the X-Y-Z measuring system.

the measurement variability. So the spread reflected in the histogram (Figure 8.13) is a result of the product variation plus the variation inherent in the measurement method.

Once we have the results of the capability study, we become suspicious of the precision of the "X-Y-Z" measurement microscope (which impacts the performance of the measurements in the process capability study). To justify these suspicions a simple measure of precision was performed by measuring the same bump at the same phase 50 times (see Table 8.2). The results of the repeated measurement (repeatability) allow us to calculate the standard deviation of the error of measurement, which in this case is $\sigma_e = 0.56$. Having the standard deviation of the observed measurements ($\sigma_o = 1.2$, see Figure 8.13) and the sigma of the measurement error (σ_e), we can easily determine the product variability (σ_p) (or phrased differently, the "true manufacturing process variation").

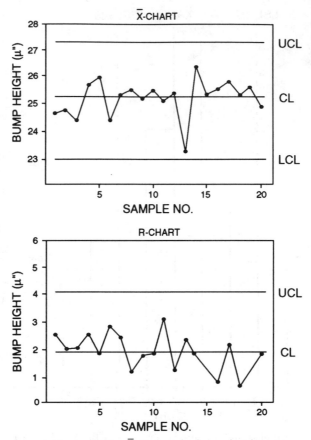

Figure 8.14 $\bar{X}$-R chart for bump height.

Now we can express the measurement error (1) in relation to the total variability and (2) in relation to the specification range.

1. Measurement error in relation to observed variability.

$$\frac{\sigma_e}{\sigma_o} = \frac{0.56}{1.2} = 0.47$$

which means that the measurement error consumed 47% of the total variability. In other words, if we had a perfect measurement system, the histogram that represents the process spread would be 47% narrower.

TABLE 8.2 Results of Measuring the Same Bump 50 Times (Using the X-Y-Z Measuring System)

Sample Number	Bump Height	Sample Number	Bump Height
1	24.00	26	25.00
2	25.00	27	24.00
3	24.00	28	23.00
4	24.00	29	24.00
5	25.00	30	24.00
6	24.00	31	24.00
7	24.00	32	24.00
8	24.00	33	25.00
9	25.00	34	23.00
10	25.00	35	24.00
11	24.00	36	24.00
12	24.00	37	24.00
13	23.00	38	23.00
14	24.00	39	24.00
15	24.00	40	24.00
16	24.00	41	24.00
17	24.00	42	23.00
18	23.00	43	24.00
19	24.00	44	24.00
20	23.00	45	24.00
21	24.00	46	25.00
22	24.00	47	24.00
23	24.00	48	24.00
24	23.00	49	24.00
25	24.00	50	25.00

2. Measurement error in relation to the specification or precision/tolerance ratio (P/T ratio):

$$P/T \text{ ratio} = \frac{6\sigma_e}{\text{tolerance range}} \times 100$$

$$= \frac{6(0.56)}{28-22} \times 100$$

$$= 56\%$$

This ratio shows that the measurement error itself consumes 56% of the specification range. So, reducing the spread of the measurement variation is equivalent to opening the specification range without jeopardizing the quality. According to the "tenth rule" for the P/T ratio, $\pm 3\sigma_e$ should consume no more than 10% of the total tolerance range.

**TABLE 8.3   A Comparison of Precision Study Results
for the Three Measurements Systems**

	"X-Y-Z"	"A-B-C"	"Q-R-S"
σ_e	0.56	0.042	0.068
σ_e / σ_t	47%	3.5%	5.6%
P/T ratio	56%	4.2%	6.8%

Two other measuring instruments became available to replace the existing instruments: an "A-B-C" surface profile measuring system and a "Q-R-S" laser beam surface profiler. Are they better than the "X-Y-Z?" If yes, which of them should we use?

To select the best measuring instrument, the same procedures that were used earlier have been applied to the two new instruments. Comparing the results of the test of precision (see Table 8.3), we can conclude that both instruments, are significantly more precise than the "X-Y-Z." Although the "A-B-C" has a slightly smaller measurement error, the "X-Y-Z" was replaced with the "Q-R-S" due to other engineering and economic reasons.

The capability study was repeated and this time the "Q-R-S" was used as the measuring instrument. It was easy to predict that the capability index would improve (see Figure 8.15). And it is obvious that this was not an indication of process improvement, but a result of improving the measurement precision. This example demonstrates the importance of knowing the measurement capability before we make conclusions about process capability.

8.2.3   Separation of Instrument Errors from Product Variation

As we know, the variation of the observed measurements is a combination of two elements: (1) the variation in the product itself and (2) the variation due to the measurement error. The relationship of those two elements can be expressed as follows:

$$\sigma_o^2 = \sigma_p^2 + \sigma_e^2$$

where σ_o^2 is the variance of the observed measurements, σ_p^2 is the variance of the product, and σ_e^2 is the variance of the measurement error.

If we want to know the variation of the instrument or the variation of the product itself, we need to know how to separate these elements. When you have only one instrument and you cannot make repeated measurements on the same unit, there is no way you can determine how much variation is due to the product and how much is due to the error of the measurement. To do this, you need at least two instruments.

The following are some of AMD's procedures for separating product variation from instrument variation. These procedures are based on Frank E.

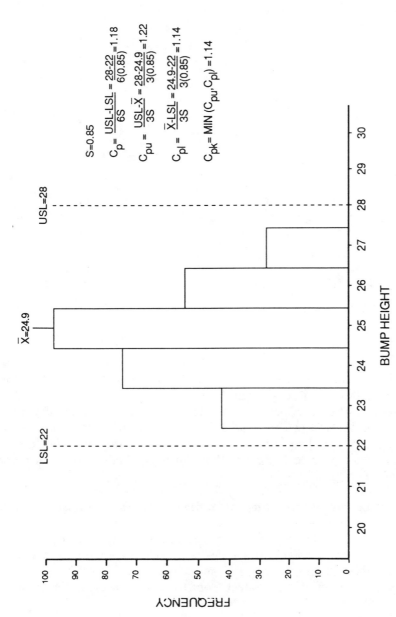

Figure 8.15 A histogram for bump height by using the "Q-R-S" measuring system.

$S = 0.85$

$$C_p = \frac{USL - LSL}{6S} = \frac{28 - 22}{6(0.85)} = 1.18$$

$$C_{pu} = \frac{USL - \bar{X}}{3S} = \frac{28 - 24.9}{3(0.85)} = 1.22$$

$$C_{pl} = \frac{\bar{X} - LSL}{3S} = \frac{24.9 - 22}{3(0.85)} = 1.14$$

$$C_{pk} = MIN (C_{pu}, C_{pl}) = 1.14$$

TABLE 8.4   Measurement Results for 50 Devices on Two Testers

Device Number	(1) X_A	(2) X_B
1	1.3	1.8
2	1.3	1.8
3	1.5	1.8
4	1.3	1.8
5	1.5	1.8
6	1.8	2.3
7	1.3	1.5
8	1.3	1.5
9	1.5	1.8
10	1.8	1.9
⋮	⋮	⋮
46	1.8	1.9
47	1.3	1.3
48	1.3	1.3
49	1.5	1.8
50	1.3	1.5

$$\bar{X}_A = 1.406 \qquad \bar{X}_B = 1.684$$
$$\sigma_{oA} = 0.2185 \qquad \sigma_{oB} = 0.2595$$
$$\sigma_{oA}^2 = 0.048739 \qquad \sigma_{oB}^2 = 0.068718$$

Grubb's methods[9] and are used mainly when it is impossible or undesirable to perform repeated measurements on the same device; e.g., when using destructive methods of testing, when a device is sensitive to repeated measurements, etc. The procedures will be demonstrated by using actual data extracted from AMD's project results. We will start with the most simple procedure, when only two instruments are used, and then we will demonstrate a more complicated procedure that can be applied for three or more instruments.

8.2.3.1   Procedure of Separating Instrument Error from Product Variation Using Two Instruments

Two instruments are selected and calibrated in accordance with the existing procedure. A sample of 50 electronic devices is selected from a product line and tested on both instruments for one parameter (see Table 8.4). To separate the variation of the product from the variation of the instrument, proceed as follows.

1. Determine the averages, standard deviations, and variances for the measurements of each instrument (see the bottom portion of Table 8.4).

TABLE 8.5   Deviations of Every Measurement from Its Average Value

Device Number	(1) $X_A - \bar{X}_A$	(2) $X_B - \bar{X}_B$	(3) $(X_A - \bar{X}_A)(X_B - \bar{X}_B)$
1	−0.106	0.116	−0.0123
2	−0.106	0.116	−0.0123
3	0.094	0.116	0.0109
4	−0.106	0.116	−0.0123
5	0.094	0.116	0.0109
6	0.394	0.616	0.2427
7	−0.106	−0.184	0.0195
8	−0.106	−0.184	0.0195
9	0.094	0.116	0.0109
10	0.394	0.216	0.0851
⋮	⋮	⋮	⋮
46	0.394	0.216	0.0851
47	−0.106	−0.384	0.0407
48	−0.106	−0.384	0.0407
49	0.094	0.116	0.0109
50	−0.106	−0.184	0.0195

$$\sum (X_A - \bar{X}_A)(X_B - \bar{X}_B) = 2.26478$$

2. Determine the deviation of every measurement from its average value:

$$\text{For Instrument A} := X_A - \bar{X}_A$$

$$\text{For Instrument B} := X_B - \bar{X}_B$$

[see columns (1) and (2) in Table 8.5].

3. Multiply these deviations:

$$(X_A - \bar{X}_A)(X_B - \bar{X}_B)$$

[see column (3) in Table 8.5].

4. Sum the results of the multiplication and divide by $(n-1)$. This is the variance of the product (σ_p^2):

$$\sigma_p^2 = \frac{\Sigma(X_A - \bar{X}_A)(X_B - \bar{X}_B)}{n-1} = \frac{2.26478}{50-1} = 0.04622$$

(see Table 8.5). Therefore, the standard deviation of the product (σ_p) is

$$\sigma_p = \sqrt{\sigma_p^2} = \sqrt{0.04622} = 0.215$$

5. Determine the variance in measurement error by subtracting the variance of the product (σ_p^2) obtained in step 4 from the variance of the observed values (σ_o^2) for each instrument (see Table 8.4):

   For Instrument A:

$$\sigma_{eA}^2 = \sigma_{oA}^2 - \sigma_p^2$$

$$= 0.048739 - 0.04622$$

$$= 0.002519$$

For Instrument B:

$$\sigma_{eB}^2 = \sigma_{oB}^2 - \sigma_p^2$$

$$= 0.068718 - 0.04622$$

$$= 0.022498$$

6. Calculate the standard deviation of measurement error:

   For Instrument A:

$$\sigma_{eA} = \sqrt{\sigma_{eA}^2} = \sqrt{0.002519} = 0.05019$$

For Instrument B:

$$\sigma_{eB} = \sqrt{\sigma_{eB}^2} = \sqrt{0.022498} = 0.15$$

Index of Precision

Now that we know the variation in the measuring equipment, we can estimate the index of precision $(\pm 3\sigma_e)$:

For Instrument A:

$$\pm 3\sigma_{eA} = \pm(3 \times 0.05019) = \pm 0.15057$$

For Instrument B:

$$\pm 3\sigma_{eB} = \pm(3 \times 0.15) = \pm 0.45$$

Interpretation of an Instrument's Capability Based on the Index of Precision

Comparing the two instruments, we can say that Instrument A has greater precision because of its smaller value of index of precision $(\pm 3\sigma_e)$. We also can say that in most experiences less than 1% of all measurements taken from the process can be expected to differ from their average value by more than $3\sigma_e$.

Determining the Precision / Tolerance Ratio (P / T Ratio)

1. The ratio of the estimated measurement precision (repeatability) to the total tolerance (the difference between the upper and lower specification limits) can be expressed as follows:

$$P/T \text{ ratio} = \frac{\pm 3\sigma_e}{\text{USL} - \text{LSL}} = \frac{6\sigma_e}{\text{total tolerance}}$$

2. Taking into consideration that our example we have a situation of unilateral tolerance where only USL is given (USL = 6), the P/T ratio formula can be transformed as

$$P/T \text{ ratio} = \frac{3\sigma_e}{\text{USL} - \overline{X}}$$

Substituting the above results in the formula, we have
For Instrument A:

$$P/T \text{ ratio} = \frac{3\sigma_{eA}}{\text{USL} - \overline{X}_A} = \frac{3 \times 0.05019}{6 - 1.4060} = 3.3\%$$

For Instrument B:

$$P/R \text{ ratio} = \frac{3\sigma_{eB}}{\text{USL} - \overline{X}_B} = \frac{3 \times 0.15}{6 - 1.684} = 10.4\%$$

Interpretation of the P / T Ratio

The P/T ratio shows the relation of the $\pm 3\sigma_e$ spread to the allowable specification range. In our example it shows that instrument A consumes only 3.3% of the specification range and Instrument B consumes slightly more than 10%. However, according to the existing "tenth rule" which says that the instrument may consume 10% or less of the range to be negligible, this means that both of the instruments are acceptable.

Determining the Ratio of the Measurement Error to Total Variation

According to Juran,[4] if the σ_e is less than one-tenth of σ_o, then the effect upon σ_p will be less than 1%. To determine this relationship, we proceed as follows:
For Instrument A:

$$\frac{\sigma_{eA}}{\sigma_{oA}} = \left(\frac{0.05019}{0.2185} \right) \times 100 = 23\%$$

TABLE 8.6   Repeated Measurements of 50 Units on Three Testers

Row	X_1	X_2	X_3
1	12.00	12.20	12.20
2	12.50	12.50	12.50
3	12.00	12.50	12.20
4	12.00	12.20	12.00
5	12.20	12.20	12.20
6	12.20	12.50	12.50
⋮	⋮	⋮	⋮
46	12.00	12.00	12.20
47	12.00	12.00	12.20
48	12.20	12.20	12.00
49	12.50	12.80	12.50
50	11.70	12.80	11.70

$$\sum X_1 = 602.6 \qquad \sum X_2 = 608.4 \qquad \sum X_3 = 607.5$$
$$\sum \overline{X}_1 = 12.05 \qquad \overline{X}_2 = 12.17 \qquad \overline{X}_3 = 12.15$$
$$\sigma_{x1}^2 = 0.06 \qquad \sigma_{x2}^2 = 0.07 \qquad \sigma_{x3}^2 = 0.06$$
$$\sigma_{x1} = 0.24 \qquad \sigma_{x2} = 0.27 \qquad \sigma_{x3} = 0.24$$

For Instrument B:

$$\frac{\sigma_{eB}}{\sigma_{oB}} = \left(\frac{0.15}{0.2595} \right) \times 100 = 58\%$$

This means that the measurement variation in the instrument, especially in Instrument B, has inflated the variation in the product.

8.2.3.2 Procedure to Determine the Instrument Error and Product Variation (by Using N Instruments, N ≥ 3)

Grubbs suggested a method that can be used to separate the instrument variation from the product variation by using three or more instruments. The following is a step-by-step procedure which utilizes three testers to illustrate the calculations.

Step 1 Select N testers (in our example, $N = 3$) that are capable of measuring the parameter to be used for the experiment. Label the testers as $X_1, X_2, \ldots, X_N$.

Step 2 Select a random sample of 50 devices from the line and label them from 1 to 50.

Step 3 Measure the sample devices on each tester and record the readings (see Table 8.6 for an example).

Step 4 Compute the sum, average, variance, and standard deviation for each tester (see Table 8.6).

TABLE 8.7   The Results of Cross Product Calculation

$X_1 X_2$	$X_1 X_3$	$X_2 X_3$
146.40	146.40	148.84
156.25	156.25	156.25
150.00	146.40	152.50
146.40	144.00	146.40
148.84	148.84	148.84
152.50	152.50	156.25
160.00	156.25	160.00
146.40	144.00	146.40
146.40	146.40	148.84
152.50	156.25	152.50
⋮	⋮	⋮
144.00	146.40	146.40
144.00	146.40	146.40
148.84	146.40	146.40
160.00	156.25	160.00
149.76	136.89	149.76
$\sum X_1 X_2 = 7334.4$	$\sum X_1 X_3 = 7324.1$	$\sum X_2 X_3 = 7393.9$

Step 5 Compute the sum of the cross products of each tester with other testers (see Table 8.7 as an example).

Step 6 Calculate the covariance of each tester with other testers and then find the sum of all the covariances. The covariance can be computed according to the formula:

$$\text{Covariance } X_i \cdot X_j = \frac{1}{n-1}\left[\sum X_i X_j - \frac{1}{n} \sum X_i \sum X_j \right] \qquad (8.1)$$

where n is the number of devices and $X_i X_j$ is the test set for testers i and j, respectively. For example:

$$\text{Covariance } X_1 \cdot X_2 = \frac{1}{n-1}\left[\sum X_1 X_2 - \frac{1}{n} \sum X_1 \sum X_2 \right]$$

From Tables 8.6 and 8.7 we can easily get

$$\text{Covariance } X_1 \cdot X_2 = \frac{1}{50-1}\left[(7334.4) - \frac{1}{50}(602.6)(608.4) \right]$$

$$= \frac{1}{50-1}[7334.4 - 7332.4]$$

$$= 0.04$$

Proceed in this way, and the results are

$$\text{Covariance } X_1 \cdot X_2 = 0.04$$
$$\text{Covariance } X_1 \cdot X_3 = 0.05$$
$$\text{Covariance } X_2 \cdot X_3 = 0.04 \tag{8.2}$$
$$\text{Sum of the covariances} = 0.13$$

Step 7 Compute the variance and standard deviation of the product. The average of the covariance is the estimated variance of the product:

$$\sigma_p^2 = \frac{2}{N(N-1)} (\text{sum of the covariances}) \tag{8.3}$$

where N is the number of testers. Substituting the result (8.2) in (8.3) we have

$$\sigma_p^2 = \frac{2}{3(3-1)} (0.13) = 0.043$$

$$\sigma_p = 0.208$$

Step 8 Compute the variance and the standard deviation of the test error of each tester by using the following formula:

$$\sigma_{ei}^2 = \sigma_{Xi}^2 - \frac{2}{N-1} \left[\begin{array}{c} \text{sum of the covariances} \\ \text{involving tester } i \end{array} \right]$$

$$+ \frac{2}{(N-1)(N-2)} \left[\begin{array}{c} \text{sum of the covariances} \\ \text{not involving tester } i \end{array} \right] \tag{8.4}$$

Applying formula (8.4) to the data, we have

$$\sigma_{e1}^2 = \sigma_{X1}^2 - \frac{2}{N-1} [\text{Covariance } X_1 \cdot X_2 + \text{Covariance } X_1 \cdot X_3]$$

$$+ \frac{2}{(N-1)(N-2)} [\text{Covariance } X_2 \cdot X_3]$$

$$= 0.06 - \frac{2}{3-1} [0.04 + 0.05] + \frac{2}{(3-1)(3-2)} [0.04]$$

$$= 0.010$$

and $\sigma_{e1} = 0.10$.

In the same way, we can get σ_{e2} and σ_{e3}, and the results are as follows:

$$\sigma_{e1}^2 = 0.01 \qquad \sigma_{e1} = 0.10$$
$$\sigma_{e2}^2 = 0.04 \qquad \sigma_{e2} = 0.20$$
$$\sigma_{e3}^2 = 0.01 \qquad \sigma_{e3} = 0.10$$

TABLE 8.8   Summary of the Test Capability Study Results

Testers	(1) σ_e	(2) $3\sigma_e$	(3) $\bar{X}$	(4) P / T Ratio
1	0.10	0.30	12.05	8%
2	0.20	0.60	12.17	16%
3	0.10	0.30	12.15	8%

Index of Precision
The index of precision will tell us how precise a tester is. The higher the instrument precision, the smaller the index is. The index is usually expressed as $\pm 3\sigma_e$. Column 2 of Table 8.8 shows the index of precision for the three testers. It is obvious that tester 2 is worse than the other two testers.

Precision / Tolerance (P / T Ratio)
The P/T ratio shows the percentage of the tolerance range consumed by the tester error and is determined by

$$P/T \text{ ratio} = \frac{6\sigma_e}{\text{USL} - \text{LSL}} \times 100$$

In our case, we have a unilateral specification (USL = 16); therefore, we apply

$$P/T \text{ ratio} = \frac{3\sigma_e}{\text{USL} - \bar{X}} \times 100$$

Column 4 of Table 8.8 shows the P/T ratios for the three testers. The P/T ratios of testers 1 and 3 are less than 10%, which is in the acceptable range. The P/T ratio for tester 2 is less than 25%, which indicates that it is in the marginally acceptable range.

Conclusion
The procedure described above illustrates how we can split the components of variations (tester precision and product variation). A better understanding of the testers can be derived by knowing the instrument precision and product variation, and corrective action can be taken to prevent a bad tester from being on the test floor.

8.3   INSPECTION CAPABILITY STUDIES

In recent years there has been a tendency toward inspection automation, which reduces or totally eliminates the human error in inspection. However, many people are still involved in inspection activities.

Tightening customer requirements sometimes forces a supplier to perform 100% visual inspection on a number of cosmetic parameters to ensure that the customer receives "close to perfection" products. But, at the same time, 100% inspection in high quantities causes inspector fatigue, which increases error in the inspection process. To improve the quality of the inspection procedure, operations expend a lot of resources for inspector training and improving the organization of the inspector's workplace.

One of the important elements related to inspector capability improvement is the inspector capability study. The results from these studies allow us to determine more specifically in what areas the inspector's knowledge should be improved; how to better organize the inspector's workplace; and they also allow the operator and supervisor to observe the progress of improvement and take action when necessary. Listed below are procedures based on John L. Hradesky's methods, [10] which have worked effectively at AMD.

8.3.1   Procedure for Inspection Capability Study (ICS) Using Variable Data

Most variable data are obtained by operators according to a measuring procedure using a particular instrument. The procedure described below will allow us to not only evaluate the inspector capability, but also evaluate the instrument precision and accuracy. This will help management determine whether the capability of the inspector or instrument needs to be improved. We will use real examples to describe step by step the application of the procedure.

8.3.1.1   Example 1: One Instrument and Multiple Inspectors Involved

The data used to illustrate this example are from an inspection capability study that was performed to investigate epoxy thickness quality. The manufacturing management suspected that the instrument being studied was not precise (as is almost always the case), but the maintenance manager suspected that the operators weren't being trained well enough. This is a typical situation where the results of a capability study can provide better insight.

We will divide the study into two parts: The first part is the preparation and data collection, and the second part is the analysis and conclusion. Both parts are equally important, but we want to emphasize the importance of preparation since people sometimes start a study without spending enough time on planning. The following is a step by step procedure illustrated with data from a real example.

Preparation for the ICS and Data Collection

Step 1 Determine the number of instruments and inspectors to be involved in the study. In this particular study, it was decided to involve one instrument and two inspectors.

TABLE 8.9   Sample Sizes for Inspection Capability Studies Involving Variable Data

Number of Inspectors	Number of Instruments	Minimum Number of Parts	Minimum Number of Measurements per Part
1	1	10	5
1	2	15	3
2	1		
2	2		
1 or 2	3 or more	10	2
3 or more	1 or 2		
3 or more	3 or more		

Step 2 Refer to Table 8.9 to determine the minimum number of units and minimum number of measurements per unit that are required for the study. In our example, when one instrument and two inspectors are involved in the study, the table shows that at least 15 units are required and each unit should be measured at least three times.

Step 3 Select from the process one set of units ($n = 15$) to be measured by each operator. Label each unit and randomize the order in which the parts are given to the operators.

Step 4 Calibrate the instrument(s), revise the inspection procedure, and make sure the inspectors know how to inspect the units, and that they understand the purpose of the study.

Step 5 Determine which quality characteristic is to be used in the study. Make sure that every operator uses the same location on the units to take the measurement. In our example we use die attach epoxy thickness as the measurement parameter. The specification range requirement for this parameter is UCL = 3.00 mils and LCL = 0.5 mil.

Step 6 Conduct the study by having the operators measure each part once in a random order and then record the readings on the datasheet (see Table 8.10). Repeat this cycle until the determined number of readings per part are obtained. In our example we need to repeat the cycle three times. *Note:* It is important to record the measurement results on a separate sheet to avoid the human inclination to reduce the variation between the readings. See the measurement results in Table 8.10.

Calculation and Analysis of ICS

To perform the calculation and analysis of the ICS, a worksheet is used (as shown in Figure 8.16). The procedure of analysis is as follows:

Step 1 Perform the range evaluation (see Section 1 in Figure 8.16).

1. Compute the range (R) of the repeated readings for each part and record them in the row identified as the "Range" (see Table 8.10).
2. Compute the average range $(\bar{R})$ for each inspector, find the overall average range $(\bar{\bar{R}})$, and record it on Section 1, Range Evaluation, of the worksheet in Figure 8.16. In our example:

$$\bar{R}_1 = \frac{(0.10 + 0.10 + 0.20 + \cdots + 0.00 + 0.20)}{15} = 0.14$$

$$\bar{R}_2 = \frac{(0.30 + 0.20 + 0.30 + \cdots + 0.20 + 0.10)}{15} = 0.19$$

$$\bar{\bar{R}} = \frac{(\bar{R}_1 + \bar{R}_2)}{2} = \frac{(0.14 + 0.19)}{2} = 0.165$$

3. Compute the upper control limit for the range (UCL_R) as shown in Section 1, Upper Control Limit Range, of Figure 8.16:

$$\text{UCL}_R = D_4 \bar{\bar{R}}$$

where D_4 is a constant factor (see Table 8.11) that depends on the number of repeat readings (n) on each part. In our example, $n = 3$, so

TABLE 8.10 Epoxy Thickness Measurement Results from Inspector 1

Inspector 1

Trial	1	2	3	4	5	6	7	8	9	10	11	12	13	14	15
1	1.07	0.97	0.97	0.97	0.97	0.87	0.87	0.97	1.07	1.07	1.07	0.97	0.97	0.97	0.97
2	0.97	0.97	0.97	1.07	0.97	0.87	0.97	0.87	0.87	1.17	1.07	0.97	0.97	0.97	0.87
3	0.97	1.07	1.17	0.97	1.27	1.17	0.97	0.97	0.87	0.97	1.07	0.87	1.07	0.97	1.07
Range	0.10	0.10	0.20	0.10	0.30	0.30	0.10	0.10	0.20	0.20	0.00	0.10	0.10	0.00	0.20

$$\bar{X}_1 = 0.994 \qquad \bar{R}_1 = 0.14$$

Inspector 2

Trial	1	2	3	4	5	6	7	8	9	10	11	12	13	14	15
1	1.07	1.07	0.87	0.87	0.77	0.97	0.97	0.97	0.97	0.77	0.87	0.97	1.17	0.87	0.97
2	0.77	0.87	0.77	1.17	0.87	1.07	0.77	0.77	0.87	0.97	0.97	0.97	1.27	0.87	0.87
3	0.97	0.97	1.07	1.07	0.77	0.87	0.87	0.87	0.87	0.97	0.87	1.07	0.97	1.07	0.97
Range	0.30	0.20	0.30	0.30	0.10	0.20	0.20	0.20	0.10	0.20	0.10	0.10	0.30	0.20	0.10

$$\bar{X}_2 = 0.939 \qquad \bar{R}_2 = 0.19$$

$D_4 = 2.574$, and

$$UCL_R = (2.574)(0.165) = 0.425$$

If any range exceeds the UCL_R, proceed as follows.

a. Remeasure the out-of-control results to see if there is a recording error. If a recording error is found, remove the affected data from the computations and recalculate the average range and UCL_R.

Part number _____ Date _____

Dimension or characteristic Epoxy Thickness Measurement

Specification or tolerance   1.75 ± 1.25 (Total tolerance = 2.5 mils)

Section 1: Range Evaluation:

Original	Revised	Upper Control Limit Range (UCL$_R$)

$\bar{R}_1 = $ _0.14_   _____ $n = $ _3_ $D_4 = $ _2.574_

$\bar{R}_2 = $ _0.19_   _____ $UCL_R = D_4\bar{R} = $ _0.425_

$\bar{R}_3 = $ _____   _____   Number of points above $UCL_R = $ _0_

$\Sigma\bar{R} = $ _0.33_   _____   Number of points discarded = _0_

$\bar{\bar{R}} = $ _0.165_   _____

Section 2: Repeatability Evaluation:

$$S_R = \left(\frac{1}{d_2}\right)\bar{R}$$

$n = $ _3_ $1/d_2 = $ _0.592_ $S_R = $ _0.098_

Repeatability = $6 \times S_R = $ _0.586_

$P/T_R = [(6 \times S_R)/(\text{total tolerance})] \times 100\% = $ _23%_

Section 3: Reproducibility Evaluation (Inspectors or Instruments)

$$R_M = \bar{X}_l - \bar{X}_s$$

$\bar{X}_l = $ _0.994_ $\bar{X}_s = $ _0.939_ $R_M = $ _0.055_

$$S_r = DR_M$$

$K = $ _2_ $D = $ _0.709_ $S_r = $ _0.039_

Reproducibility = $6 \times S_r = $ _0.234_

$P/T_r = [(6 \times S_r)/(\text{total tolerance})] \times 100\% = $ _9%_

Section 4: Inspection Capability Evaluation:

$$S_o = \sqrt{(S_R)^2 + (S_r)^2}$$

$S_o = $ _0.105_

Inspection capability = $6 \times S_o = 6 \times 0.105 - 0.63$

$P/T_o = [(6 \times S_o)/(\text{total tolerance})] \times 100\% = $ _25%_

Figure 8.16 Variable Data Worksheet (for Example 1)

b. If there is more than one range exceeding the UCL_R, the method should be revised and the study repeated. If only one range exceeds the UCL_R due to a measurement method error, remove the affected data on that part from the computations and recompute the average range and UCL_R. In our example, all ranges are in control.

Step 2 Perform the repeatability evaluation (see Section 2 in Figure 8.16).

1. Compute the standard deviation of the repeatability (S_R) as shown in Section 2 of the worksheet (see Figure 8.16) by the following formula:

$$S_R = \left(\frac{1}{d_2}\right)\overline{\overline{R}}$$

where $1/d_2$ is a constant factor (see Table 8.11) that depends on the number of repeated measurements (n). In our example, $n = 3$, so $1/d_2 = 0.592$ and

$$S_R = (0.592)(0.165) = 0.098$$

2. Compute the repeatability by the following formula:

$$\text{Repeatability} = 6 \times S_R = 6 \times 0.098 = 0.586$$

3. Compute the percentage tolerance consumed by repeatability (P/T_R) by the following formula:

$$P/T_R = \frac{6 \times S_R}{\text{total tolerance}} \times 100$$

In our example, the total tolerance = 2.5 mils; therefore,

$$P/T_R = \frac{6 \times 0.098}{2.5} \times 100 = 23\%$$

Step 3 Perform the reproducibility evaluation (see Section 3 in Figure 8.16). *Note:* If only one inspector and instrument is involved in the study, the reproducibility cannot be calculated.

TABLE 8.11 Repeatability Factors for Inspection Capability Studies Involving Variable Data

n	$1/d_2$	D_4
2	0.885	3.268
3	0.592	2.574
4	0.485	2.282
5	0.429	2.114

TABLE 8.12   Reproducibility Factors for Inspection Capability Studies Involving Variable Data

K	D
2	0.709
3	0.524
4	0.446
5	0.403

1. Find the largest $(\overline{X}_L)$ and smallest $(\overline{X}_S)$ averages. Compute the difference of these two averages (R_M). In our example, we only have two inspectors, so

$$\overline{X}_L = \overline{X}_1 = 0.994 \qquad \text{and} \qquad \overline{X}_S = \overline{X}_2 = 0.939$$

therefore,

$$R_M = \overline{X}_L - \overline{X}_S = 0.994 - 0.939 = 0.055$$

2. Compute the standard deviation of reproducibility (S_r) by the following formula:

$$S_r = DR_M = (0.709)(0.055) = 0.039$$

where D is a factor dependent on the number of inspectors (or instruments) used (K) (see Table 8.12).

3. Compute the reproducibility by the following formula:

$$\text{Reproducibility} = 6 \times S_r = 6 \times 0.039 = 0.234$$

4. Compute the percentage tolerance consumed by reproducibility (P/T_r):

$$P/T_r = \frac{6 \times S_r}{\text{total tolerance}} \times 100$$

$$= \frac{0.234}{2.5} \times 100$$

$$= 9\%$$

Step 4 Perform the inspection capability evaluation (see Section 4 in Figure 8.16).

1. Compute the standard deviation for the overall inspection capability (S_o). The standard deviation of inspection capability is composed of the standard deviations of repeatability (S_R) and reproducibility (S_r). Therefore, the standard deviation of inspection capability (S_o) is determined by

$$S_o = \sqrt{S_R^2 + S_r^2}$$

$$= \sqrt{(0.098)^2 + (0.039)^2}$$

$$= 0.105$$

2. Compute the inspection capability (IC):

$$IC = 6 \times S_o = 6 \times 0.105 = 0.63$$

3. Compute the percentage tolerance consumed by the ICS (P/T_o):

$$P/T_o = \frac{6 \times S_o}{\text{total tolerance}} \times 100$$

$$= \frac{6 \times 0.105}{2.5} \times 100$$

$$= 25\%$$

Note: If more than one instrument is used in the study, the standard deviation of the reproducibility must also be computed for the instruments and then added to the final overall inspection capability.

Interpretation

Generally, the criteria for acceptance of the overall P/T ratio consumed by inspection capability (P/T_o) is as shown in Table 8.13. It is just a working criteria which is based on a rule of thumb. In our example, since the P/T_o is 25%, the inspection capability for this measuring system is marginally acceptable. And, as we can see the percentage tolerance consumed by repeatability

TABLE 8.13 Criteria for the Acceptance of the Overall Percentage Tolerance Consumed by Inspection Capability (P/T_o)

PTCC Value	Study Result
10% or less	Acceptable
Between 10% and 25%	Marginal
Greater than 25%	Unacceptable

(P / T_R) is 23% and the percentage tolerance consumed by inspector reproducibility (P / T_r) is 9%. So, to improve the inspection capability system, corrective actions should be taken to improve the instrument repeatability (precision).

8.3.1.2 Example 2: Multiple Instruments and Multiple Inspectors Involved
The purpose of this example is to show the slight difference in the procedure of an inspection capability study when using more than one instrument and inspector. Taking into consideration that the "preparation part" of the procedure is the same as for one instrument, we will focus only on the second part "calculation and analysis" in this example.

Situation
In manufacturing plastic leadless chip carrier (PLCC) devices, there is a critical parameter that is measured by a Micro-Vu profile projector. The instrument together with the operator form a measuring system. The PLCC manufacturing line utilizes a number of such measuring systems. For the purpose of IC study only, two instruments and two inspectors have been randomly selected as representatives. The specification requirements for this parameter are

$$\text{USL} = 495 \text{ mils}$$

$$\text{LSL} = 485 \text{ mils}$$

Calculations and Analysis
Step 1 Determine the sample size and the number of repetitions. According to Table 8.9 when two inspectors and two instruments are involved in the study, the sample size should be at least 10 ($n = 10$) and every unit should be measured at least twice.

Step 2 Perform the measurements by using all possible instrument-inspector combinations. Each inspector uses the first instrument (I_1) to measure the 10 units randomly twice, and uses the second instrument (I_2) to measure the same 10 units twice more. Table 8.14 represents the average range ($\bar{R}$) and the average reading ($\bar{X}$) for each instrument-inspector combination, derived from the measurement results, which are not shown here.

TABLE 8.14 The average Range and Average Reading from the Four Instrument-Inspector Combinations

Inspector	Instrument	$\bar{R}$	$\bar{X}$
1	A	0.38	490.410
1	B	0.69	490.695
2	A	0.36	491.490
2	B	0.42	491.430

Part number _____ Date _____
Dimension or characteristic E-Dimension
Specification or tolerance   490 ± 5 mils (Total tolerance = 10)

Section 1: Range Evaluation:

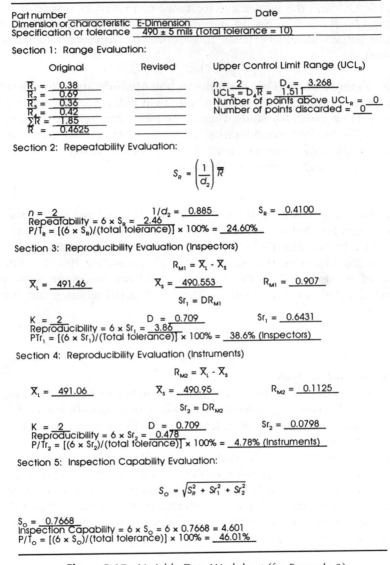

Original	Revised	Upper Control Limit Range (UCL$_R$)
$\bar{R}_1$ = 0.38		n = 2 D_4 = 3.268
$\bar{R}_2$ = 0.69		UCL_R = $D_4\bar{R}$ = 1.511
$\bar{R}_3$ = 0.36		Number of points above UCL$_R$ = 0
$\bar{R}_4$ = 0.42		Number of points discarded = 0
$\sum\bar{R}$ = 1.85		
$\bar{\bar{R}}$ = 0.4625		

Section 2: Repeatability Evaluation:

$$S_R = \left(\frac{1}{d_2}\right)\bar{\bar{R}}$$

n = 2 $1/d_2$ = 0.885 S_R = 0.4100
Repeatability = 6 × S_R = 2.46
P/T$_R$ = [(6 × S_R)/(total tolerance)] × 100% = 24.60%

Section 3: Reproducibility Evaluation (Inspectors)

$$R_{M1} = \bar{X}_L - \bar{X}_S$$

$\bar{X}_L$ = 491.46 $\bar{X}_S$ = 490.553 R_{M1} = 0.907

$$Sr_1 = DR_{M1}$$

K = 2   D = 0.709 Sr_1 = 0.6431
Reproducibility = 6 × Sr_1 = 3.86
PTr$_1$ = [(6 × Sr_1)/(Total tolerance)] × 100% = 38.6% (Inspectors)

Section 4: Reproducibility Evaluation (Instruments)

$$R_{M2} = \bar{X}_L - \bar{X}_S$$

$\bar{X}_L$ = 491.06 $\bar{X}_S$ = 490.95 R_{M2} = 0.1125

$$Sr_2 = DR_{M2}$$

K = 2   D = 0.709 Sr_2 = 0.0798
Reproducibility = 6 × Sr_2 = 0.478
P/Tr$_2$ = [(6 × Sr_2)/(total tolerance)] × 100% = 4.78% (Instruments)

Section 5: Inspection Capability Evaluation:

$$S_O = \sqrt{S_R^2 + Sr_1^2 + Sr_2^2}$$

S_O = 0.7668
Inspection Capability = 6 × S_O = 6 × 0.7668 = 4.601
P/T$_O$ = [(6 × S_O)/(total tolerance)] × 100% = 46.01%

Figure 8.17   Variable Data Worksheet (for Example 2)

Step 3 Perform the range evaluation (see Section 1 in Figure 8.17).

1. Determine the overall average range ($\bar{\bar{R}}$) and record it in Section 1, Range Evaluation, of the worksheet in Figure 8.17:

$$\bar{\bar{R}} = \frac{\left(\bar{R}_1 + \bar{R}_2 + \bar{R}_3 + \bar{R}_4 \right)}{4}$$

$$= \frac{(0.38 + 0.69 + 0.36 + 0.42)}{4}$$

$$= 0.4625$$

2. Compute the upper control limit for the range (UCL_R) as shown in Section 1, Upper Control Limit Range, of Figure 8.17:

$$UCL_R = D_4 \bar{\bar{R}}$$

$$= (3.268)(0.4625)$$

$$= 1.511$$

where D_4 is a constant factor from Table 8.11 when $n = 2$ repeat measurements. In our example, all the ranges are below the UCL_R.

Step 4 Perform the repeatability evaluation (see Section 2 in Figure 8.17).

1. Compute the standard deviation of the repeatability (S_R):

$$S_R = \left(\frac{1}{d_2} \right) \bar{\bar{R}}$$

$$= (0.885)(0.4625)$$

$$= 0.4100$$

where $1/d_2 = 0.885$ is from Table 8.11 with $n = 2$ repeat measurements.

2. Determine the repeatability:

$$\text{Repeatability} = 6 \times S_R$$

$$= 6 \times 0.41$$

$$= 2.46$$

3. Determine the percentage tolerance consumed by repeatability (P / T_R). The specification is 490 ± 5 mils; therefore, the total tolerance $= 10$ mils, and

$$P / T_R = \frac{6 \times S_R}{\text{total tolerance}}$$

$$= \frac{6 \times 0.4100}{10} \times 100$$

$$= 24.6\%$$

Step 5 Perform the reproducibility evaluation for the inspectors (see Section 3 in Figure 8.17).

1. Find the average readings of each inspector based on the measurement results (see Table 8.14):

$$\bar{X}_1 = \frac{(490.410 + 490.695)}{2} = 490.553$$

$$\bar{X}_2 = \frac{(491.490 + 491.430)}{2} = 491.46$$

2. Determine the range from the average readings of the inspectors:

$$R_{M1} = \bar{X}_L - \bar{X}_S$$

$$= \bar{X}_2 - \bar{X}_1$$

$$= 491.46 - 490.553$$

$$= 0.907$$

3. Compute the standard deviation of reproducibility of the instrument:

$$S_{r1} = D R_{M1}$$

$$= (0.709)(0.907)$$

$$= 0.6431$$

where $D = 0.709$ from Table 8.12 with $K = 2$ inspectors.

4. Determine the reproducibility of the inspectors:

$$\text{Reproducibility} = 6 \times S_{r1}$$

$$= (6)(0.6431)$$

$$= 3.86$$

5. Determine the percentage tolerance consumed by the instrument's reproducibility (P/T_{r1}):

$$P/T_{r1} = \frac{\text{reproducibility}}{\text{total tolerance}} \times 100$$

$$= \frac{3.86}{10} \times 100$$

$$= 38.6\%$$

Step 6 Perform the reproducibility evaluation for the instruments (see Section 4 in Figure 8.17).

1. Find the average readings of each instrument based on the measurement results (see Table 8.14):

$$\overline{X}_A = \frac{(490.410 + 491.490)}{2} = 490.9500$$

$$\overline{X}_B = \frac{(490.695 + 491.430)}{2} = 491.0625$$

2. Determine the range from the average readings of the instruments:

$$R_{M2} = \overline{X}_L - \overline{X}_S$$

$$= \overline{X}_B - \overline{X}_A$$

$$= 491.0625 - 490.95$$

$$= 0.1125$$

3. Compute the standard deviation of reproducibility of the instruments:

$$S_{r2} = DR_{M2}$$

$$= (0.709)(0.1125)$$

$$= 0.0798$$

4. Determine the reproducibility of the instruments:

$$\text{Reproducibility} = 6 \times S_{r2}$$

$$= (6)(0.0798)$$

$$= 0.478$$

5. Determine the percentage tolerance consumed by the instrument's reproducibility:

$$P/T_{r2} = \frac{\text{reproducibility}}{\text{total tolerance}} \times 100$$

$$= \frac{0.478}{10} \times 100$$

$$= 4.78\%$$

Step 7 Perform the inspection capability evaluation (see Section 5 in Figure 8.17).

1. To evaluate the overall inspection capability, we must find the standard deviation for the total inspection capability (S_o), which includes the standard deviation of repeatability plus the standard deviation of reproducibility for both the inspectors and instruments. Therefore,

$$S_o = \sqrt{(S_R)^2 + (S_{r1})^2 + (S_{r2})^2}$$

$$= \sqrt{(0.4100)^2 + (0.6431)^2 + (0.0798)^2}$$

$$= 0.7668$$

2. Now we can compute the inspection capability (IC):

$$IC = 6 \times S_o$$

$$= (6)(0.7668)$$

$$= 4.601$$

3. The percentage tolerance consumed by overall inspection (P/T_o) is

$$P/T_o = \frac{IC}{\text{total tolerance}} \times 100$$

$$= \frac{4.601}{10} \times 100$$

$$= 46.01\%$$

Conclusion
Based on the inspection capability results (see Table 8.15), we can conclude that the precision of the evaluated systems consumes 46% of the allowable tolerance range. In other words the imprecision of the measuring system

TABLE 8.15 The Results of the Inspection Capability Study for Example 2

Overall* precision of the system (P/T_o)	46%
Precision related to the repeatability (P/T_R)	24.6%
Inspector precision related to the reproducibility (P/T_{r1})	38.6%
Instrument precision related to the reproducibility (P/T_{r2})	4.78%

*The overall precision in this case is expressed as a result combined from repeatability and reproducibility from inspector and instrument.

consumes almost half of the specification range, which can cause the rejection of a lot of conforming units. Based on these results, the measuring system is unacceptable (see Table 8.13). The reliability of the measuring system ($P/T_R = 24.6\%$) is marginally acceptable, which means that there is room for improvement. As for the inspector reproducibility ($P/T_{r1} = 38.6\%$), we can say that the results between the operators are not correlated very well. A separate inspector study would be preferable, which can demonstrate the necessity of inspector retraining. The reproducibility of the instruments ($P/T_{r2} = 4.78\%$) is very small and can be neglected. Overall, we can say that if the selected two systems represent the overall measurement system of PLCC quality, an engineering investigation should be made to resolve the problem.

Six Months Later

The above results were received at the beginning of the learning course, and highlighted the interest in instrument and inspector capability studies. Six months later a second study was performed that involved all available instruments and inspectors. Table 8.16 shows the results from the second study, which doesn't need commentary.

TABLE 8.16 The results of the Second Inspection Capability Study (Six Months Later)

Overall precision of the system (P/T_o)	15.8%
Precision related to the repeatability (P/T_R)	11.2%
Inspector precision related to the reproducibility (P/T_{r1})	12.5%
Instrument precision related to the reproducibility (P/T_{r2})	4.01%

8.3.2 Procedure for Inspection Capability Study (ICS) Using Attribute Data

In the electronics industry much of the cosmetic product quality is controlled by visual inspection. Because of this, the parts are attributed to either good or bad purely by the inspector's judgment. The quality of inspection will directly impact the quality of the product. Therefore, it is very important to control the inspection activities and know when and where to take action to improve the inspector's capability.

In this section we will introduce a step by step procedure for the inspection capability study by using attribute data. But before describing the procedure, we will discuss the preparation activities and become familiar with some special terms related to this subject.

Preparation Activities

Before conducting the actual study, a special "expert committee" is selected. The committee comprised of a group of people whose expertise in a particular inspection activity is used as a reference to compare the inspector's results. They are usually people who have a strong practical experience in the job under investigation. The committee receives the inspection procedure, selects devices from the process to form a "reference lot" which will include all types of blemishes and nonconformities which the inspector can meet in the real inspection activity. The "standard lot" is completed in such a way as to have one-third conforming units, one-third nonconfirming units, and one-third marginal units. The marginal units are selected so they have 50% marginal conformity over 50% marginal nonconformity. The sample size and number of replicants needed to perform the study are determined by using Table 8.17 which depends on the number of inspectors. The last step before starting the study is to describe to the inspectors the established quality criteria, methods of inspection, and the purpose of the study.

8.3.2.1 Example: Inspection Conditions

One of the visual quality characteristics in semiconductor manufacturing utilizing hermetic packages is "seal quality." To perform a study of this visual inspection, seven inspectors were randomly selected. A "reference lot" of 30

TABLE 8.17 Sample Sizes for the Inspection Capability Study Using Attribute Data

Number of Appraisers	Minimum Number of Parts	Minimum Number of Inspections per Unit
1	24	5
2	18	4
3 or more	12	3

TABLE 8.18   Data for the Attribute Data Analysis Example

Operator's Code	Conforming Correct (1)	Nonconforming Correct (2)	Total Correct (3)	False Alarm (4)	Misses (5)	Grand Total (6)
A	75	68	143	0	7	150
B	73	70	143	2	5	150
C	67	64	131	8	11	150
D	74	65	139	1	10	150
E	71	64	135	4	11	150
F	73	71	144	2	4	150
G	69	69	138	6	6	150
Overall	503	469	972	23	55	1050

Notes:

Conforming Correct—Column (1) is the number of conforming units identified correctly by the inspector. Since there are 15 conforming units, each inspected 5 times, 75 opportunities exist for correct identification of the conforming units for each person. For example, Inspector A correctly identified the conforming units 75 times.

Nonconforming Correct—Column (2) is the number of nonconforming units identified correctly by the inspector. Since there are 15 nonconforming units, each inspected 5 times, 75 opportunities exist for correct identification of the nonconforming units for each person. For example, Inspector A correctly identified the nonconforming units 68 times.

Total Correct—Column (3) is the sum of columns (1) and (2), and is the numerator of the formula for computing O_E. For example, the "Total Correct" for inspector A is $75 + 68 = 143$.

False Alarm—Column (4) is the number of false alarms for each inspector. Inspector A had zero false alarms, which indicates this inspector did not reject any conforming units.

Misses—Column (5) is the number of misses for each inspector. Inspector A had 7 misses, which means this inspector did not correctly identify the nonconforming units 7 times.

Grand Total—Column (6) is the sum of columns (3), (4), and (5), and should equal the number of units inspected times the number of inspections per unit. In the example, 30 units were inspected 5 times for a total of 150.

units (15 conforming and 15 noncomforming units) was prepared for inspection. Each inspector inspected the same 30 units in a random order, once per day, for five consecutive days.

Evaluation of the Inspection Results

Table 8.18 represents the inspection results examined by the members of the expert committee. Based on these results, the following indexes which will reflect the inspector capability can be computed.

1. The operator effectiveness index (O_E): This index reflects the ability of an inspector to correctly identify conforming and nonconforming units, and it is expressed as a number between 0 and 1, where 1 is perfection. The index can be computed as follows:

$$O_E = \frac{\text{number of units correctly identified}}{\text{total opportunities to be correct}}$$

Note: The "total opportunities to be correct" is a function of the number of units used and the number of replicants made. For example:
a. 30 units are inspected five times, so there are 150 opportunities to be correct ($5 \times 30 = 150$).
b. Inspector A correctly identified the samples 143 times; Therefore,

$$O_E = \frac{143}{150} = 0.95$$

2. The false alarm index (I_{FA}): This index compares the "number of false alarms" to the "number of opportunities of having false alarms" and it can be computed as follows:

$$I_{FA} = \frac{\text{number of false alarms}}{\text{number of opportunities for a false alarm}}$$

The "number of opportunities for a false alarm" is a function of the "number of conforming units" used in the study and the "number of times each unit is inspected." For example:
a. If there are 15 conforming units and each unit is inspected five times, then there are $5 \times 15 = 75$ opportunities for a false alarm.
b. Inspector A made zero false alarms; therefore,

$$I_{FA} = \frac{0}{75} = 0$$

3. The index of a miss (I_{MISS}): This index compares the "number of misses" to the "number of opportunities for a miss," and it can be calculated as follows:

$$I_{MISS} = \frac{\text{number of misses}}{\text{number of opportunities for a miss}}$$

The "number of opportunities for a miss" is a function of the "number of nonconforming units used in the study" and the "number of times each unit is inspected." For example:
a. If 15 nonconforming units are used and each unit is inspected five times, then the number of missed opportunities would be 75 ($5 \times 15 = 75$).

b. Inspector A made seven misses; therefore,

$$I_{\text{MISS}} = \frac{7}{75} = 0.09$$

4. The bias index (I_B): Bias is a measure of an inspector's tendency to classify a unit as conforming or nonconforming, and is a function of I_{MISS} and I_{FA}. Bias values are equal to or greater than 0 and have the following meaning:

$I_B = 1$ implies no bias.

$I_B > 1$ implies bias toward rejecting units.

$I_B < 1$ implies bias toward accepting units.

Bias is computed by the equation:

$$I_B = \frac{B_{\text{FA}}}{B_{\text{MISS}}}$$

where B_{FA} is a factor found in Table 8.19 and is a function of I_{FA}, and B_{MISS} is a factor found in Table 8.19 and is a function of I_{MISS}. For example, for Inspector B, $I_{\text{FA}} = 0.03$ and $I_{\text{MISS}} = 0.07$. The factor from Table 8.19 corresponding to I_{FA} is $B_{\text{FA}} = 0.0681$; and the factor corresponding to I_{MISS} is $B_{\text{MISS}} = 0.1334$. The bias value is then

$$I_B = \frac{B_{\text{FA}}}{B_{\text{MISS}}} = \frac{0.0681}{0.1334} = 0.51$$

which indicates that Inspector B has a bias toward accepting units. Table 8.20 lists special cases encountered in computing the bias value when I_{FA} or I_{MISS} are 0 or greater than 0.50.

The overall ICS evaluation is shown in Table 8.21.

Interpretation

Table 8.22 shows the recommended evaluation criteria for the four computed measures O_E, I_{FA}, I_{MISS}, and I_B. The results of Table 8.21 show that the effectiveness (O_E) of all the inspectors is in the acceptable range except Inspector C. This indicates that this inspector needs special attention. The I_{FA}'s of Inspector C and Inspector G are in the unacceptable and marginal range, which indicates that both inspectors should be retrained to improve their ability to identify conforming units. The I_{MISS} of all the inspectors is equal to or more than 0.05, which strongly indicates that all the inspectors

MEASUREMENT AND INSPECTION CAPABILITY STUDIES

TABLE 8.19 Bias Factor Table for the Inspection Capability Study Involving Attribute Data

I_{FA} or I_{MISS}	B_{FA} or B_{MISS}	I_{FA} or I_{MISS}	B_{FA} or B_{MISS}
0.01	0.0264	0.26	0.3251
0.02	0.0488	0.27	0.3312
0.03	0.0681	0.28	0.3372
0.04	0.0863	0.29	0.3429
0.05	0.1040	0.30	0.3485
0.06	0.1200	0.31	0.3538
0.07	0.1334	0.32	0.3572
0.08	0.1497	0.33	0.3621
0.09	0.1626	0.34	0.3668
0.10	0.1758	0.35	0.3712
0.11	0.1872	0.36	0.3739
0.12	0.1989	0.37	0.3778
0.13	0.2107	0.38	0.3814
0.14	0.2227	0.39	0.3836
0.15	0.2323	0.40	0.3867
0.16	0.2444	0.41	0.3885
0.17	0.2541	0.42	0.3910
0.18	0.2613	0.43	0.3925
0.19	0.2709	0.44	0.3945
0.20	0.2803	0.45	0.3961
0.21	0.2874	0.46	0.3970
0.22	0.2966	0.47	0.3977
0.23	0.3034	0.48	0.3984
0.24	0.3101	0.49	0.3989
0.25	0.3187	0.50	0.3989

TABLE 8.20 Special Cases in Computing Bias

I_{FA}	I_{MISS}	I_B	Decision or Action
0	More than 0	0	Unacceptable
More than 0	0	No value	Use O_E, I_{FA}, and I_{MISS} directly
0	0	No value	This is the same as $I_B = 1$, since $I_{FA} = I_{MISS}$; acceptable
More than 0.5	0.5 or less	More than 1.5	Unacceptable
0.5 or less	More than 0.5	Less than 0.5	Unacceptable
More than 0.5	More than 0.5	No value	Bias unimportant; study is unacceptable based on I_{MISS} and I_{FA} being more than 0.5

TABLE 8.21 ICS Results for the Data Analysis Example

Operator's Code	Effectiveness (O_E)	Probability of a False Alarm I_{FA}	Probability of a Miss I_{MISS}	Bias I_B
A	0.95	0	0.09	0
B	0.95	0.03	0.07	0.51
C	0.87	0.11	0.15	0.81
D	0.93	0.01	0.13	0.31
E	0.90	0.05	0.15	0.45
F	0.96	0.03	0.05	0.65
G	0.92	0.08	0.08	1.00

TABLE 8.22 Evaluation Criteria for the Inspection Capability Study Involving Attribute Data

Parameter	Acceptable	Marginal	Unacceptable
O_E	0.9 or more	0.8–0.9	Less than 0.8
I_{FA}	0.5 or less	0.05–0.10	More than 0.10
I_{MISS}	0.02 or less	0.02–0.05	More than 0.05
I_B	0.80–1.20	0.50–0.80 or 1.2–1.5	Less than 0.50 or more than 1.5

have the tendency to accept nonconforming units and should be retrained to identify them since misidentification of the nonconforming units will cause returns and customer complaints. After corrective action is completed, the ICS should be redone to verify improvement of the inspection capability.

REFERENCES

1. Scott Montgomery, Quality (February 1988), p. 70.
2. Karl F. Speitel, "Measurement Assurance," Chapter 8.2 of "Handbook of Industrial Engineering," edited by Gavriel Salvendy, (1982), John Wiley & Sons, Inc., p. 8.2.2.
3. W. Edwards Deming, "Sampling Design in Business Research" (1960), John Wiley & Sons, Inc. p. 62.
4. J. M. Juran and Frank M. Gryna, "Juran's Quality Handbook," 4th ed. (1988), McGraw-Hill Book Co., p. 18.66.
5. ASQC Automotive Division, "Statistical Quality Control Manual" (1986), ASQC, pp. 3-2, Figure 3-3.

6. ASQC Automotive Division, "Statistical Quality Control Manual" (1986), ASQC, p. 3-2, Figure 3-4.

7. Harvey C. Charbonneau and Gordon L. Webster, "Industrial Quality Control" (1978), Prentice-Hall, Inc., p. 124, Figure 8A1-1.

8. ASQC Automotive Division, "Statistical Quality Control Manual" (1986), ASQC, p. 3-0.

9. Frank E. Grubbs, "On Estimating Precision of Measuring Instruments and Product Variability," Journal of the American Statistical Association, Vol. 95 (1948), pp. 243–264.

10. John L. Hradesky, "Productivity and Quality Improvements, (1988), McGraw-Hill Book Co., Chapter 5, pp. 73–99.

Chapter 9

Process Capability Study

A competitor advertises his product at a higher quality level. Can your company match this level or do even better? A customer is planning to place a large order, but the specification requirements are tighter than usual. Can we make it? We need to reduce the cost, and one way of doing this is by reducing the inspection activities. Can we do it without jeopardizing the quality? A customer requires that the product be delivered within a specific cycle time. Can our process meet this requirement? To decide where to buy a product, a potential customer wants to know your capability index. Are you ready to answer?

These questions can be concentrated into one general question: Is your process adequate? To find the answer, we need to perform a process capability analysis.

There are a broad variety of methods and techniques used to evaluate and improve the process capability. In this chapter we describe those that are most frequently used in the electronics industry, and illustrate them by using case histories and examples from the AMD plants.

9.1 PROCESS CAPABILITY

Before describing the methods of determining the process capability, let's define what a process is and what we mean by "process capability." In general, any repeated event or any combination of conditions which work together to produce a given result is a process. A gardener performs the process of maintaining the yard, a teacher performs the process of educating students, a truck driver performs the process of delivering goods, a planner performs the process of planning and forecasting. All these diverse activities have a similarity, they have methods, machines, and/or tools/materials; they have an input, output, and feedback.

A gardener, for example, has a method of feeding the flowers and grass and a machine to cut the grass. He uses fertilizers and gasoline as materials. His input is labor and materials, his output is the result of his work, and his feedback is the customer satisfaction, suggestions, or complaints.

We brought up this example to make a point, that process capability analysis can and should be made not only in a manufacturing plant, but also in any other areas where we have a process and we decide to improve it. This can be maintenance, planning, forecasting, physical distribution, training, etc.

Going further into generalizing the meaning of a process, it can also be said that any meaningful activity that utilizes time can be viewed as a process. For example, playing golf, baseball, or music should be treated as a very important process since it helps us recover the energy and power lost during the working hours.

From a manufacturing point of view, a process is defined as a combination of labor, machines, tools, and methods used in a specific environment to manufacture a given product. The term "process" refers to a system of causes which introduce variation in the quality of the product. The causes of this variation should be studied and eliminated (or reduced) if we want to improve the quality of the process output (the product). We will discuss how to do this later in the chapter.

Before taking actions for process improvement, we need to have a *process benchmark*, which means we need to know the existing process capability. *Process capability* refers to the normal behavior of a process when operating in a state of statistical control (see the definition on Section 7.2), and can be defined as the *minimum spread* of a specific measured variation that will include 99.73% of the measurements from a given process.

It is customary to consider that a process is capable if 99.73% of the production will fall in six sigma limits. In other words, if a process is centered on the nominal specification and follows a normal distribution, 99.73% of the production will fall within the ± 3 sigma specification. If the data show no evidence of instability, the process capability (PC) can be estimated as

$$PC = 6\sigma$$

This is the most widely adopted standardized formula for process capability, where σ is the standard deviation of the process under a state of statistical control.[1]

Figure 9.1 illustrates a process that has a normal distribution with a mean μ and standard deviation σ. The upper and lower natural tolerance limits (UNTL and LNTL) of the process fall at $\mu + 3\sigma$ and $\mu - 3\sigma$, respectively. For a normal distribution, the natural tolerance limits (NTL) include 99.73% of the variables, which means that only 0.27% (or 2700 parts per million) of the process output will fall outside the natural tolerance limits.

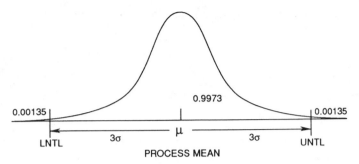

Figure 9.1   For a normal distribution, the natural tolerance limits include 99.73% of the product.

Having a standardized formula like $PC = 6\sigma$ is of great value for quality planning. This formula gives in a quantified form the natural limits of a process when running in a state of statistical control. It is important to note that process capability, in terms of 6σ of variation, is absolutely unrelated to the product tolerance. Whether a process will or will not meet the specification requirements depends on a number of factors, but mainly on the allowable specification range.

9.2   THE MEASURE OF PROCESS CAPABILITY

In recent years Cp and Cpk indexes have become very popular as a measure of process capability in relation to the specification requirements. The terms Cp and Cpk are frequently heard at different meetings, a lot of publications are dedicated to this subject, and customers have started requesting specific values of Cp and Cpk from their suppliers. In other words, Cp and Cpk create more interest today than all other types of indexes. Why? Is this a new approach of measuring the process? Are there other reasons that make these indexes so popular? Do they deserve this kind of attention? If yes, why? To answer these questions, we need to go back in time and analyze the background of these indexes.

9.2.1   The Process Capability Ratio (PCR)

In Juran's *Quality Control Handbook*,[2] there is a rule of thumb that describes the relationship of the process capability to the tolerance. According to this rule, the process capability must be no greater than 75% of the bilateral tolerance and no greater than 88% of the unilateral product tolerance. It allows a higher ratio for unilateral specification limits because in the bilateral specification case the process capability is allowed to consume 75% of the

TABLE 9.1   The Rule of Thumb for Adequate Process Capability Ratios

	Bilateral Tolerance	Unilateral Tolerance
Existing Processes	75%	88%
New Processes	67%	83%

total tolerance range and leave 25% as the guardband, which means $25\% \div 2 = 12.5\%$ on each side. Therefore, for the unilateral specification case, the process capability is allowed to consume $75 + 12.5\% = 87.5\% \approx 88\%$ of the total tolerance. This was the basis for developing an index known as the process capability ratio (PCR)[3]. For bilateral tolerances,

$$\text{PCR} = \frac{6\sigma \text{ variation}}{\text{total tolerance}}$$

For unilateral tolerances,

$$\text{PCR} = \frac{3\sigma \text{ variation}}{|\text{specification limit} - \overline{X}|}$$

where $\overline{X}$ is the process average. This rule of thumb has been further developed for different types of processes (see Table 9.1 and Figure 9.2).

The purpose of this rule of thumb is to provide some kind of a safety factor that will compensate for not centering the process between the upper and lower tolerance limits, or for other reasons that may cause a shift of the process from the nominal. Juran's *Quality Control Handbook*[4] also describes the experience of a Japanese company and that of the automotive industry. They used two forms of calculating capability indexes, defined as follows.

1. Six standard deviations as computed from the frequency distribution. Expressing this in a formula, we have

$$\text{Process capability} = 6\sigma = 6s = 6\sqrt{\frac{\Sigma(X - \overline{X})^2}{n-1}}$$

2. Eight standard deviations as computed from average ranges. Expressing this in a formula, we have

$$\text{Process capability} = 8\sigma = 8s = 8\frac{\overline{R}}{d_2}$$

The relationship of these two values derived in different ways is used to judge whether time-to-time variation is excessive. Here the process capability index

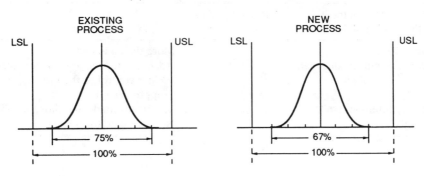

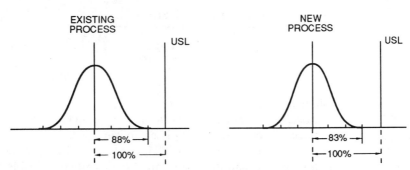

Figure 9.2   Graphical representation of the rule of thumb for adequate process capability ratios.

(PCI) is defined as

$$\text{PCI} = \frac{\text{tolerance width}}{\text{process capability}}$$

This means that when we use a histogram to determine the standard deviation, the process capability index will be derived by

$$\text{PCI} = \frac{\text{USL} - \text{LSL}}{6s} = \frac{\text{USL} - \text{LSL}}{6\sqrt{\dfrac{\Sigma(X - \bar{X})^2}{n - 1}}}$$

But when we use a control chart ($\bar{X}$-R chart) to determine the standard

TABLE 9.2 The Grades and Decision Rules Based on the Capability Index

Value of the Process Capability Index	Resulting Class of Process	Decision
Above 1.33	1	More than adequate
Under 1.33, but greater than 1	2	Adequate for the job, but requires close control as the index approaches 1
Under 1, but above 0.67	3	Marginally inadequate for the job
Under 0.67	4	Not adequate for the job

deviation, the process capability index will be derived by

$$PCI = \frac{USL - LSL}{8s} = \frac{USL - LSL}{8\dfrac{\overline{R}}{d_2}}$$

where d_2 is a constant factor.

Based on this process capability index, four grades of processes are developed and a pattern of decision making is established (see Table 9.2).

It is important here to note two things: (1) The criteria and requirements of the capability index for the manufacturing process described in Table 9.2 are based on an article which was published in 1964.[5] This reflects the Japanese requirements for process quality at that time. (2) The methodology mentioned so far in this section has been used by some companies, but this was not during the recent period of excitement about capability indexes.

9.2.2 Cp and Cpk

In July 1984, *Quality Progress*[6] published an article, "Reducing Variability: A New Approach to Quality," where the author brought up an important topic related to process variation. He mentioned an index of measuring the variation around the target. This index is the Cpk which has become so popular recently.

Later, in January 1986, the *Journal of Quality Technology*[7] published an article, "Process Capability Indexes," which presented a family of capability indexes (Cp, Cpu, Cpl, and Cpk) used to measure the process parameters. In the article, the author mentioned that this system of measurements was being used by a number of Japanese industries and that the U.S. automotive industry had *started* using these measurements in a number of areas. The formulas for the four capability indexes are shown in Table 9.3.

TABLE 9.3   Formulas for Process Capability Indexes

Process capability
  potential
$$Cp = \frac{USL - LSL}{6\sigma}$$

Process capability
  performance
$$Cpu = \frac{USL - \overline{X}}{3\sigma}$$

$$Cpl = \frac{\overline{X} - LSL}{3\sigma}$$

$$Cpk = \min\{Cpu, Cpl\}$$

or

$$Cpk = \frac{|\overline{X} - \text{nearest spec}|}{3\sigma}$$

Later in this section we will describe the Cp and Cpk in more detail, but now let's make a comparison between the indexes in Table 9.3 and the capability ratios mentioned earlier. As you can see, the Cp index is just the reciprocal of the process capability ratio (PCR) which reflects the percentage of the specification used:

$$PCR = \text{Percentage of specification used} = 100\frac{1}{Cp}$$

For example, if a process tolerance is ± 2, the total tolerance range becomes 4, and assuming the standard deviation is 0.5, the PCR and Cp values would be

$$PCR = \frac{6s}{USL - LSL} = \frac{6(0.5)}{4} \times 100 = 75\%$$

$$Cp = \frac{USL - LSL}{6s} = \frac{4}{6(0.5)} = 1.33$$

Comparing the two indexes, the traditional capability ratio appears more attractive because it reflects what percentage of the allowable specification range is consumed by the process spread and what percentage is left as a "safety zone" for the shifts and drifts of the process average from the nominal (or target). But the process capability ratio and the Cp index do not consider the location of the process mean. A process can have a very good PCR or Cp, but produce many parts outside the specification limits (see Figure 9.3). These two indexes quantify the *potential* process performance

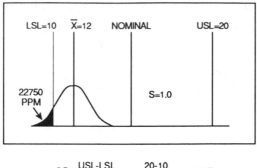

$$CP= \frac{USL\text{-}LSL}{6S} = \frac{20\text{-}10}{6(1.0)} = 1.67$$

$$PCR= \frac{6S}{USL\text{-}LSL} = \frac{6(1.0)}{20\text{-}10} = 0.6 = 60\%$$

Figure 9.3 A process with an adequate capability ratio and Cp, but producing many parts outside of the specification limit.

that can be attained only if the process average is on the nominal of the specification limits and stays there all the time (which is very rare in practice).

This is why the Cpk index has been added to the existing family of capability indexes. The Cpk utilizes the process average and can be considered as a measure of the *process performance*. This is why the traditional process capability ratio has been reciprocally transformed into a Cp, to make this ratio relative to the Cpk.

However, the Cpk is not a substitute for the Cp. Just the opposite, they work together and complement each other. It is also important to note that the absolute Cp and Cpk are not very important because their values depend on the specification limits which are arbitrary in most cases. What is important here is why they change over time. This is the reason that in measuring the progress of process improvement, a reporting system should be developed which will reflect the change of the capability indexes from month to month, from year to year, etc.

9.2.3   The Relationship Between Capability Indexes and PPM Levels

To determine the ppm levels based on capability indexes, we need to use the table of area under the normal curve or Z table (see Table 9.4). The following example is used to demonstrate the procedure of using the Z table to derive the ppm levels based on capability indexes.

Example   Given a process with Cpu = 1.15, Cpl = 1.08, determine the total ppm level.

TABLE 9.4 Table of Area Under the Normal Curve Beyond Selected Z Values*

Z	0.00	0.01	0.02	0.03	→0.04	→0.05	0.06	0.07	0.08	0.09
3.00	$1.350E-03$	$1.306E-03$	$1.264E-03$	$1.223E-03$	$1.183E-03$	$1.144E-03$	$1.107E-03$	$1.070E-03$	$1.035E-03$	$1.001E-03$
3.10	$9.676E-04$	$9.354E-04$	$9.042E-04$	$8.740E-04$	$8.477E-04$	$8.163E-04$	$7.888E-04$	$7.622E-04$	$7.364E-04$	$7.114E-04$
→ 3.20	$6.871E-04$	$6.637E-04$	$6.410E-04$	$6.190E-04$	$\boxed{5.977E-04}$	$5.770E-04$	$5.571E-04$	$5.378E-04$	$5.191E-04$	$5.010E-04$
3.30	$4.835E-04$	$4.665E-04$	$4.501E-04$	$4.343E-04$	$4.189E-04$	$4.041E-04$	$3.898E-04$	$3.759E-04$	$3.625E-04$	$3.495E-04$
→ 3.40	$3.370E-04$	$3.249E-04$	$3.132E-04$	$3.019E-04$	$2.909E-04$	$\boxed{2.804E-04}$	$2.702E-04$	$2.603E-04$	$2.508E-04$	$2.416E-04$
3.50	$2.327E-04$	$2.242E-04$	$2.159E-04$	$2.079E-04$	$2.002E-04$	$1.927E-04$	$1.855E-04$	$1.786E-04$	$1.719E-04$	$1.655E-04$
3.60	$1.592E-04$	$1.532E-04$	$1.474E-04$	$1.418E-04$	$1.364E-04$	$1.312E-04$	$1.262E-04$	$1.214E-04$	$1.167E-04$	$1.123E-04$
3.70	$1.079E-04$	$1.038E-04$	$9.974E-05$	$9.587E-05$	$9.214E-05$	$8.855E-05$	$8.509E-05$	$8.175E-05$	$7.854E-05$	$7.543E-05$
3.80	$7.248E-05$	$6.961E-05$	$6.685E-05$	$6.420E-05$	$6.165E-05$	$5.919E-05$	$5.682E-05$	$5.455E-05$	$5.236E-05$	$5.025E-05$
3.90	$4.822E-05$	$4.627E-05$	$4.440E-05$	$4.260E-05$	$4.086E-05$	$3.920E-05$	$3.760E-05$	$3.606E-05$	$3.458E-05$	$3.316E-05$
4.00	$3.179E-05$	$3.048E-05$	$2.921E-05$	$2.800E-05$	$2.684E-05$	$2.572E-05$	$2.465E-05$	$2.362E-05$	$2.263E-05$	$2.168E-05$
4.10	$2.076E-05$	$1.989E-05$	$1.905E-05$	$1.824E-05$	$1.747E-05$	$1.672E-05$	$1.601E-05$	$1.533E-05$	$1.467E-05$	$1.404E-05$
4.20	$1.344E-05$	$1.286E-05$	$1.231E-05$	$1.77E-05$	$1.126E-05$	$1.077E-05$	$1.031E-05$	$9.857E-06$	$9.426E-06$	$9.014E-06$
4.30	$8.619E-06$	$8.240E-06$	$7.878E-06$	$7.530E-06$	$7.198E-06$	$6.879E-06$	$6.574E-06$	$6.282E-06$	$6.002E-06$	$5.734E-06$
4.40	$5.478E-06$	$5.233E-06$	$4.998E-06$	$4.773E-06$	$4.558E-06$	$4.353E-06$	$4.156E-06$	$3.968E-06$	$3.787E-06$	$3.615E-06$
4.50	$3.451E-06$	$3.293E-06$	$3.143E-06$	$2.999E-06$	$2.861E-06$	$2.730E-06$	$2.604E-06$	$2.484E-06$	$2.369E-06$	$2.225E-06$
4.60	$2.154E-06$	$2.054E-06$	$1.959E-06$	$1.867E-06$	$1.780E-06$	$1.697E-06$	$1.617E-06$	$1.541E-06$	$1.469E-06$	$1.399E-06$
4.70	$1.333E-06$	$1.270E-06$	$1.210E-06$	$1.153E-06$	$1.098E-06$	$1.046E-06$	$9.956E-07$	$9.480E-07$	$9.026E-07$	$8.593E-07$
4.80	$8.151E-07$	$7.787E-07$	$7.411E-07$	$7.054E-07$	$6.712E-07$	$6.387E-07$	$6.077E-07$	$5.782E-07$	$5.500E-07$	$5.232E-07$
4.90	$4.976E-07$	$4.733E-07$	$4.501E-07$	$4.280E-07$	$4.070E-07$	$3.869E-07$	$3.678E-07$	$3.496E-07$	$3.323E-07$	$3.159E-07$

*This is only a portion of the Z table. The complete table is shown in Appendix C.

Step 1 Compute the Z values for Cpu and Cpl:

$$Z_U = 3 \times \text{Cpu} = 3 \times 1.15 = 3.45$$

$$Z_L = 3 \times \text{Cpl} = 3 \times 1.08 = 3.24$$

Step 2 Locate the corresponding fraction defective values for Z_U and Z_L in the Z table (see Table 9.4). For $Z_U = 3.45$, find 3.4 in the leftmost column of the Z table and 0.05 in the uppermost row. At the intersection of line 3.4 and column 0.05 find the value $2.804E - 04$. In the same way we can find that for $Z_L = 3.24$ the fraction defective value is $5.977E - 04$.

Step 3 Convert the values found in the table to a ppm level. The values in the table are represented in scientific notation. For example, "$E - 04$" means we should move the decimal point four positions to the left. Therefore

$$2.804E - 04 = 0.0002804$$

$$5.977E - 04 = 0.0005977$$

To convert to a ppm level, the fraction defective value should be multiplied by 1,000,000. Therefore

$$\text{Cpu} = 0.0002804 \times 1,000,000 = 280.4 \text{ ppm}$$

$$\text{Cpl} = 0.0005977 \times 1,000,000 = 597.7 \text{ ppm}$$

Step 4 Add the ppm obtained in Step 3 for Cpu and Cpl. Therefore, the total ppm quality level is

$$280.4 \text{ ppm} + 597.7 \text{ ppm} = 878.1 \text{ ppm}$$

9.2.4   Conclusion

Now we can go back to the questions posed in the beginning of this section: Why have Cp and Cpk become so popular today? There are a number of reasons, but we will mention just two of them.

1. A method, technique, or an index is like a commodity which depends on supply and demand. A statistical tool which is developed at the right time and is matched to today's demand will become popular like a good commodity. The change in people's way of thinking, i.e., that progress in quality improvement is not only for producing parts to the specification requirements but also for reducing the process variation around the target, created a demand for statistical indexes that will reflect this new philosophy of quality.

2. The intensive educational programs in statistical methodology that recently took place in manufacturing also generated a new demand in measurement techniques based on statistical principles. People finally started to recognize the law of variation.

The Cp and Cpk methodology, like any other techniques, are not without weaknesses and pitfalls. However, if using and interpreting them properly, they can be applied successfully as tools for measuring process improvement.

In this chapter we will demonstrate a number of examples and case histories where Cp and Cpk are used. This will help the reader get a better understanding of this technique.

9.3   METHODS AND TECHNIQUES FOR CONTINUOUS PROCESS IMPROVEMENT

In the last few years the "statistical toolbox" has become more complete with different types of tools required to measure, analyze, and characterize a process.

In this section we will give a brief description of some of the tools which have been used successfully at AMD. This will give an overall picture of what tools are available. In the following sections, we will demonstrate examples in real applications.

9.3.1   Types of Data: Variable and Attribute

Process capability and performance studies can be done by using variable and/or attribute data. However, whenever possible, preference should be given to variable data because they require a relatively smaller sample size to provide the same amount of information.

Variable Data
Variable data are especially valuable in studies where it is difficult or expensive to take measurements or when the test is destructive. Variable data also give a greater opportunity for determining the causes of excess variation (more about the advantages of variable data can be found in Section 7.3.6).

Attribute Data
In the electronics industry there are areas where variable data are difficult or impossible to generate for technical or economical reasons; e.g., visual/mechanical inspection capability studies. In this case, naturally, attribute data are used to perform capability or performance studies. There is also another area where attribute data are very valuable. This is where we want to study an entire assembly process by using an overall chart. We call this a "macro capability study." This kind of study can be valuable in itself, and can also provide a good measure of the results and effectiveness of performing studies based on variable data. More about the weaknesses and strengths of attribute data can be found in Section 7.3.7.

9.3.2   Mini Capability Study (MCS)

Situations exist where we just want a quick indication about the process capability. This happens, for instance, when we visit a supplier, or perform an in-house executive audit of the manufacturing processes, or in any other analogous situations. In such cases, the mini capability study (MCS) described in Juran's *Quality Control Handbook*[8] is the right technique to be used.

MCS for Variable Data

Ten measurements are taken from the process, measured, and the range determined. The process capability (PC) defined as 6σ is then calculated as twice the range of the sample (PC $= 2R$). The calculation is based on the assumption that the process is in a state of statistical control. The "study" is a quick, but highly approximate, estimate of the true process capability.

MCS for Attribute Data

There are different forms of performing a quick capability check for attribute data. At AMD we have developed a special table (see Table 9.5), which is based on the two formulas[9]:

1. When zero nonconforming units in a sample is allowed the "capability" percentage is determined by using the following formula:

$$\text{Capability \%} = 100(0.5)^{\frac{1}{n+1}}$$

where n is the sample size.
2. When one or more nonconforming units in a sample are allowed the following formula is used:

$$\text{Capability \%} = 100\left(1 - \frac{f + 0.7}{n}\right)$$

where f is the number of nonconforming units.

TABLE 9.5   Number of Defectives Allowed in a Sample of n for Different Process Capabilities

Number of Defectives	Process Capability and Sample Size			
	99.73% (2700 ppm)	99.90% (1000 ppm)	99.95% (500 ppm)	99.99% (100 ppm)
0	250	700	1400	7000
1	630	1700	3400	
2	1000	2700		
3	1370			
4	1740			

As an example to illustrate how to apply the table, let's assume that we visited a supplier and we want to make a preliminary check on its process capability. Our requirements for a particular parameter is 1000 ppm. So, we take a sample of 700 units (see Table 9.5) and inspect them. If no nonconforming units are found, the process is preliminarily considered running at a quality level no worse than 1000 ppm.

As for variable data, the MCS for attribute data is quick and easy, and gives a highly approximate estimate of the true process capability and can be used only as a preliminary check.

9.3.3 Tentative Capability Study (TCS)

Another form of deriving an approximation of the process capability is the "tentative capability study." This technique is based on the application of a histogram, tally sheet, or a normal probability plot (NPP). In a short time frame with no process adjustment, a sample of at least 50 units is taken directly from the process, measured, and plotted in the form of a histogram, tally sheet, or NPP, and the standard deviation (σ) is calculated. The tentative process capability is then regarded as 6σ. The results are "tentative" because we don't have enough evidence that the process is in a state of statistical control from this kind of study.

Without knowing if the process is stable, there may be a risk that the standard deviation is inflated by unnatural causes of variation. This means the 6σ spread may not represent the true process potential. But, the risk is relatively small because the data are taken in a very short time interval. So the risk of process drift is small. This simple method is widely used by manufacturing practitioners as a preliminary tentative measure of the process quality. Certainly this method cannot be used, for example, to design specification limits or make any other fundamental decisions about the process, but it can be used successfully to determine the priorities of what process variable needs immediate attention. This method is more accurate than a mini capability check but is still an approximation.

9.3.4 Short-Term Capability (STC) Study

The purpose of a short-term capability study is to determine the process capability potential in a given instant of time when running in a state of statistical control. In other words, this is a study to determine the best the process will ever produce under current conditions (see Figure 9.4).

STC Study Using Variable Data
The time frame needed to make an STC study depends on the production rate of the process under investigation and also on the time required to perform the measurements. The minimum time recommended for an STC study is usually two hours but can vary from one to eight hours depending on

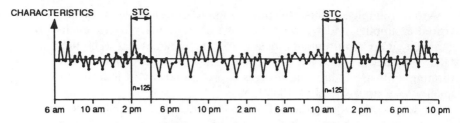

Figure 9.4   Short-term capability (STC) study.

different conditions and circumstances. It is preferred that the time frame be no more than one working shift. And the sample size is usually recommended to be a total of 125 parts ($n = 125$) so as to be able to form an $\overline{X}$-R control chart consisting of 25 subgroups ($k = 25$) of samples of 5 ($n = 5$). (When data are limited, the number of subgroups may be reduced to $k = 10$.) The same data are used to make a histogram or to plot a normal probability plot (NPP) which, together with the control chart, will show whether the process is in a state of statistical control and has a normal distribution. Evidence of normality is essential to determine the demonstrated process capability potential.

In some circumstances instead of using an $\overline{X}$-R chart, a control chart for individual measurements with control limits based on a moving range is used. When using this type of chart, data are usually limited. So a total sample of 10 consecutive measurements is permissible, provided each measurement (as a subgroup of a sample of 1) covers a representative period of time (a week, month, or quarter) to calculate the process capability ratio (PCR %), Cp or Cpk.

The standard deviation (σ) is derived from the range chart (assuming normality) by using the equation:

$$\sigma = \frac{\overline{R}}{d_2}$$

STC Study Using Attribute Data

The time frame for a short-term capability (STC) study using attribute data should be long enough to accumulate 20 to 25 consecutive subgroups (k), each consisting of a sample size (n) capable of catching one or more nonconforming parts. So the sample size (n) depends on the expected process average, and can be calculated by using

$$\frac{2.3}{\overline{p}}$$

where 2.3 is a constant factor (see Section 7.6.3) and $\overline{p}$ is the expected process average.

To perform an STC study using attribute data usually a p chart or other charts from the same family (such as np, c, or u) is used. In general, conventional 3σ control limits are used for all the charts mentioned above. However, sometimes it is recommended to use more sensitive 95% limits (roughly 2σ), particularly when the process being studied has a high-quality level (low defective rate).

In addition to the conventional attribute control charts, sometimes when the STC study is designed to be used at the end of the production process or where the sample sizes are very large, an attribute control chart with control limits based on a moving range is used (see Section 7.7.5). And, considering that by using attribute data a shift in the process is not easily revealed, a long-term capability study is preferable to demonstrate process stability.

When to Apply an STC Study

A short-term capability study is the first step in formal capability studies and can be applied whenever we want to see whether the capability of the process, as revealed by the study, is what we want it to be. If the results show that the process capability satisfies the company's requirements, we may want to perform an additional long-term capability study to confirm the results. However, experience shows that most of the time the results of the first study show that not everything is the way we would like it to be. For example, a process may demonstrate stability but at an entirely wrong level, or it may be in control and still produce parts outside the specification requirements, etc. In these types of cases, engineering efforts are required to fix the problem and then a second period or phase of capability study can be performed.

9.3.5   Long-Term Capability (LTC) Study

A long-term capability study is an investigation of the process behavior in a relatively longer time frame (see Figure 9.5) to see how the time factor influences the process variability, and also to permit a higher confidence in prediction of the process capability.

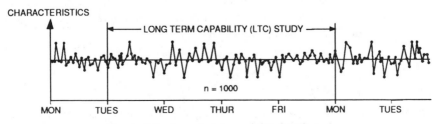

Figure 9.5   Long-term capability (LTC) study.

The time frame for this purpose is usually calculated to cover all the possible influences on the process such as normal effect of tool wear, minor variations from material lots, and all kinds of other expected variations. Usually for variable data this is a time frame of two to three shifts where at least 200 subgroups ($k = 200$) of sample size of 5 ($n = 5$) can be selected. For attribute data the time frame should be large enough to have at least 25 points on the control chart and an overall sample size which will represent the process average with a high degree of confidence and accuracy (see Section 7.6.2 to determine the sample size).

The statistical tools applied for a long-term capability study are the same as those for the short-term capability study as described earlier.

However, one important point should be made. In a long-term capability study, even when we are just interested in how different types of factors (such as tool wear, materials, and other components) impact on the process variability, the standard deviation which is used for measuring the process spread is still calculated by using the average range divided by the constant factor d_2 ($\sigma = \bar{R}/d_2$) and reflects only the "within"-subgroup variation exhibited by a process. This is the major difference between the long-term capability study and the long-term performance study (described below).

When to Apply the Long-Term Capability Study

An LTC study is used whenever we want to have greater confidence and a higher accuracy in assessing the true process capability. The results from such a study can be successfully utilized in manufacturing planning, control chart setting, and other engineering and management decision-making processes.

9.3.6 Process Performance (PP) Study

So far this section has been devoted to the formal process capability studies which are related to the determination of the inherent instantaneous variability of a process when running in a state of statistical control. Now the discussion will be expanded by describing another form of investigation called the "process performance (PP) study."

The major difference between these two forms of study is in the determination of the standard deviation. As we described earlier, in a process capability study we are concerned only with the determination of inherent instantaneous process variability which is reflected in the *within*-subgroup variation and determined by the standard deviation derived from the average range ($\bar{R}$) divided by a constant factor d_2:

$$\sigma = \frac{\bar{R}}{d_2}$$

By contrast, in a process performance study, we are concerned with two types of variation exhibited by a process: *within*-subgroups variation and *between*-

subgroups variation. The performance of a process is determined by conducting a study over an extended period of time under *normal operating conditions*. This means that the effect of different operations, tool changes, wear and adjustments, different lots of material, and other variations that may occur in the period of investigation are also included when determining the process variation. The study should be conducted over a period of five or six days. However, if the production run is shorter than five days, the data collection interval should be increased to get 200 subgroups of samples of 5 (1000 measurements) at the end of the study.

The statistical tools and methods of calculation required for this study are the same as for a formal capability study. However, the standard deviation which is used to calculate the process performance capability is derived by one of the following relative formulas:

$$\sigma = \sqrt{\frac{\sum_{i=1}^{n}(X_i - \overline{X})^2}{n-1}} \qquad \text{or} \qquad \sigma = \sqrt{\frac{n\sum X_i^2 - (\sum X_i)}{n(n-1)}}$$

where X_i is the ith individual value, $\overline{X}$ is the grand average of the subgroup averages, and n is the number of observations in the total sample.

When to Apply the Process Performance (PP) Study

A process performance study is conducted when we want to evaluate the variation in the process caused by the effects of different operations, tools, materials, and other variations that may occur over time. The value of the standard deviation derived in this study can be compared with the value of the standard deviation from a formal capability study to receive important information about the process stability. Unless the process average is stable and tightly controlled, these two values will be different; and we can judge the process stability from the extent of these differences.

9.3.7 Substitutes for Process Capability and Performance Studies

Formal capability and performance studies are based on current data analysis. The process capability study also requires several conditions which usually interfere with the production schedule. In real life, however, we don't always have the opportunity to perform such formal studies and we would like some kind of substitute to resolve the same problems.

One such substitute is described by C. C. Mentch[10] who called it "process performance evaluation (PPE)." This technique is a comprehensive investigation of a product made by the process under study, usually based on whatever historical data are available. Using historical data is the major difference between this technique and the method of process performance (PP) study described earlier in this section. Process performance evaluation involves a long time period (usually at least one full month) and demonstrates how the process has been performing in the past. Practitioners like this method even knowing that this substitute is a rough guess of the true process capability. The reason for this is because in real life we have a lot of historical data that can be used for this purpose and at the same time we have too little time to conduct formal studies.

This technique can be used when variable or attribute historical data are available. The statistical tools for this kind of evaluation are the same as when using current data. A series of PPE over a long period of time can be a practical equivalent to a formal capability or performance study.

Using Design of Experiments (DOE) in Capability Studies

In describing different forms and methods of capability study up to now, we have only been concerned with eliminating the unnatural variation, bringing the process into a state of statistical control, and determining its potential capability. But what if the determined capability value does not match the customers requirements or company requirements. Should we revise the specifications, buy new equipment, change the supplier, or maybe suspect that the test equipment has introduced too much variation? To answer this question and others like it, graphical analysis and designed experiments should be applied.

Graphical analyses, multi-variable charts, evolutionary operation (EVOP) techniques, nested design, and others tailored to the needs of design of experiments will allow us to investigate more complicated processes, by isolating and evaluating the sources of variability. This will give additional information where the engineering efforts should be directed for further process improvement.

In the following sections the reader will find the broader description and examples of application of the methods and techniques mentioned in this section.

9.4 MINI CAPABILITY CHECK

Assume you are in a situation where you need a quick preliminary estimate of the process capability. This can be, for example, when you are visiting a potential supplier, auditing your own plant, or when you make some changes in the process and you want to have a fast response on how this change influences the process capability. In situations like this, a mini capability check (MCC) is a very valuable technique to use.

Take a sample of 10 units from the process, take measurements of a critical parameter or parameters, and find the range by subtracting the minimum value from the maximum value. The capability, defined as 6σ, is

TABLE 9.6   The Values for d_2 for a Subgroup Sample Size of $n = 2$ to $n = 10$

n	d_2
2	1.128
3	1.693
4	2.059
5	2.326
6	2.534
7	2.704
8	2.847
9	2.970
10	3.078

then calculated as twice the value of the measured range:

$$\text{Process capability (a quick approximation)} = 6\sigma = 2R$$

The rationale for this simple method can be explained. We know that the estimate of

$$\hat{\sigma} = S = \frac{\overline{R}}{d_2}$$

In a special case where $n = 10$, $d_2 = 3.078$ (see Table 9.6), then

$$3S = \frac{3\overline{R}}{3.078} = \overline{R} \quad \text{or} \quad 6S = 2\overline{R}$$

So, if a quick approximation of the process capability is required, a sample of 10 units could be measured, the range calculated, and the process capability estimated as $2R$. Comparing this natural spread ($2R = 6\sigma$) with the allowable tolerance spread, we derive an approximation of Cp. Of course, this method loses the advantage of the control chart, but for a sample of 10 measurements this is much better than a guess based on pure intuition.

Case History 9.1: Using the Mini Capability Check for a Process Audit

In integrated circuit manufacturing, the die attach process is an important process step. One of the methods of attaching the die to the lead frame is by using silver glass paste whose thickness dimension is controlled by the following specification parameters: The lower specification limit (LSL) is 4 mils and the upper specification limit (USL) is 8 mils. The paste height

TABLE 9.7   Ten Measurement Results from an Audit

6.1	6.5	5.9	6.1	6.2
6.0	6.1	5.9	5.8	5.7

has a strong correlation with the die attach quality. This is why a lot of effort has been made to bring this process into a state of statistical control and improve its capability.

Analyzing the control chart for silver glass paste that was maintained for a considerably long time period, it was observed that the process was constantly stable and the capability index derived from the controlled data was very high. So, the process improvement committee (PIC) decided that it was not economically reasonable to continue to run a control chart on this quality process parameter. A process audit (PA) has been introduced where the process is checked periodically using the mini capability check (MCC). In addition, an audit was performed when some changes were made to the process; e.g., replacing an empty syringe of paste, setting up the machine for a new package or device, replacing the die pick-up tool, etc.

To illustrate the procedure, let's use the results of the audit that was performed after the paste dispensing tool was replaced. A sample of 10 units was taken from the process and the measurements of the paste height were taken. The average and range was computed from the data obtained (see Table 9.7):

$$\text{Average } (\bar{X}) = 6.03$$

$$\text{Range } (R) = 6.5 - 5.7 = 0.8$$

Substituting the range (R) value into the formula of the mini capability check (MCC), we have

$$\text{MCC} = 6S = 2R = 2(0.8) = 1.6$$

Comparing the natural spread $(6S = 2R)$ with the allowable tolerance range, we perform a rough estimate of the capability indexes, assuming that the process is in a state of statistical control:

$$\text{Cp} = \frac{\text{USL} - \text{LSL}}{2R} = \frac{8 - 4}{1.6} = 2.50$$

$$\text{Cpk} = \text{Cpu} = \frac{\text{USL} - \bar{X}}{R} = \frac{8 - 6.03}{0.8} = 2.46$$

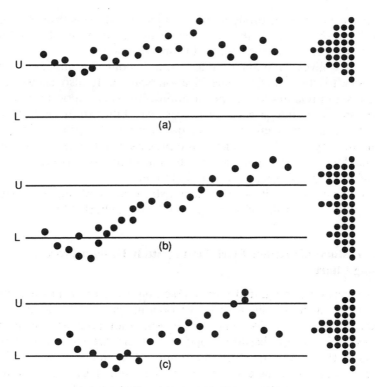

Note: U is the Upper Limit and L the Lower Limit.

Figure 9.6   Process setting chart.

Knowing that the Cp index is a high approximation, we plot the results on an audit control chart, which allows us to monitor the changes over time. The results of this particular audit show that the process is centered very close to the specification nominal, and this is why the Cp value (2.5) is very close to the Cpk value (2.46). The high capability indexing allows us to continue the process without a regular control chart just by "keeping an eye" on the process. The savings from replacing the control charts with a process audit (when the capability of the process permits) is a return on the investment in process improvement.

9.5   USING A PROCESS SETTING CHART

Some Japanese companies[11] perform preliminary process evaluations by using a process capability chart, which is a simple plot of individual measurements taken in order of production (see Figure 9.6). This process capability chart will not show you whether a process is in control or not, but will tell you how closely a process is adhering to the specification.

The chart gives preliminary information about the process spread in comparison to the allowable specification spread. It also shows how the process is set in relation to the specifications. It is very important to apply the chart before going into a classical process capability study. Experience has shown that sometimes efforts are made to bring the process into control without knowing that the average of the process is not centered between the upper and lower tolerance limits. And, having a process in control is, in itself, meaningless if specifications are not being met.

The following case history will demonstrate an application of a process capability chart. We will call it the process setting chart (PSC).

Case History 9.2: Heater Block Setting Study Using a Process Setting Chart

In the hermetic integrated circuit assembly line, there are a number of processes which contain a series of operations, each dependent on the others, with the check for quality of the operation possible only at the end. For example, in the die attach operation there are specification requirements on this operation which determine the minimum strength of the die attach. These requirements are often determined by the customer, but there are also a number of secondary operating characteristics from which the final result (die attach strength) depends.

Die attach heater block temperature is an example of a secondary quality that should be controlled to the optimum setting as closely as possible. But, before setting a control chart on this parameter, it is very important to know how closely the temperature is monitored to the specifications. Introducing a control chart will tell you if the heater block temperature is in control or not. However, it is more important to know whether the specification requirements can be met or not. So, before introducing the process control, a process capability study should be performed. Where should we start? In the assembly line in question, we have 10 heater blocks. Therefore, we decided to use the process capability chart as the check for preliminary capability study. We applied this simple technique to our blocks, analyzing the results about the interior process of temperature control. Some examples of the results of applying the process capability chart are as follows.

1. Block 1 (Figure 9.7) shows that the variation is not excessive, but the average is consistently above the nominal. An adjustment of the process average is needed.

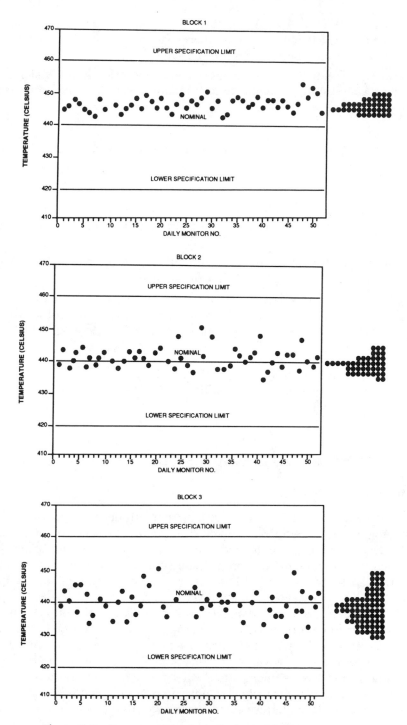

Figure 9.7 Process setting charts for heaterblock temperature.

2. Block 2 (Figure 9.7) shows that the process is set exactly on the nominal, which indicates that there will be no problem readjusting Block 1 because both blocks are the same type.

3. Block 3 (Figure 9.7) shows that extra variation exists in this system, and a capability study to eliminate any unnatural variation should be done.

Conclusion

The result of this preliminary capability check allows us to divide the blocks into three major groups:

1. Blocks which are adjusted on the nominal and have a narrow spread. For these types of blocks, control charts can be applied to control the temperature.

2. Blocks which have a narrow spread but need to be reset (then control charts can be applied to monitor the process).

3. Blocks that have excessive variation and need an additional capability study to bring the process into control. In this case, control charts will be introduced only when the special causes are eliminated.

After resolving the problems mentioned above, there is a possibility of tightening the existing specification without additional expenditures.

9.6   TENTATIVE PROCESS CAPABILITY STUDY

9.6.1   Where Do We Start?

Let's assume that it was decided to develop and introduce a process improvement program (PIP). Where should this program be started to get the best results faster and use the limited engineering and operating staff more effectively? At this point, the proper statistical approach is to perform a tentative capability study (TCS), which will help us determine the most influential areas that need to be improved.

When variable data are available, we need a sample of at least 50 measurements to perform a TCS. Considering that the data used in the study usually represent a short period (which may be only minutes), we assume that the impact on the process variation from sources such as materials, operators, adjustments, environment, etc., will be insignificant. Because of this assumption, the tentative capability study can provide a good estimate of the process potential. However, we should remember that this is still a tentative

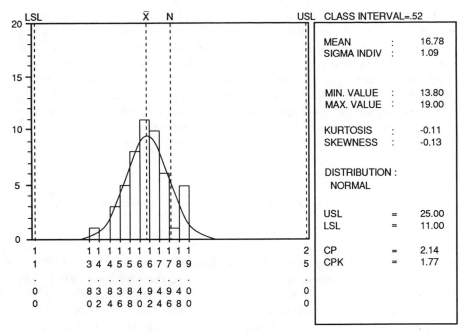

Figure 9.8 A histogram computer printout.

estimate and should only be used as preliminary information about the process. This kind of information is valuable when we want to determine, for example, the priorities of process improvement studies. The same approach can also be used when we want to make a preliminary estimate of supplier capability using the sample data taken from an incoming lot of product. Then the study can also be called *product characterization*.

To perform the tentative capability study, a simple histogram, tally sheet, or normal probability plot can be used. And since there are now personal computers in most areas and statistical software available, a tentative process capability study is very easy to perform. Just enter the data, ask the computer to make a histogram, and you will get a printout as shown in Figure 9.8. From the histogram, we can get information about the average, spread, and the distribution shape of the process.

We will also demonstrate in this section other statistical tools such as the tally sheet (TS) and normal probability plot (NPP) which can give the same information and can be performed manually. The purpose of this is to show how to perform a tentative capability study when a computer is not available. There is also another, and more important reason to demonstrate some simple "manual" techniques for process study: to provide a better understanding of the information available from a computer by using statistical software. From an investigation made by the authors of this book, 78% of our engineers couldn't clearly interpret the information about kurtosis and skewness of a distribution; this even included people who had received an introductory course on SPC. By manually performing some of the initial statistical techniques, a low computer utilization rate will be increased.

To demonstrate the procedures for performing a tentative process capability study, a case history will be used. And, we will apply three techniques: a histogram, tally sheet, and normal probability plot, using the same data so that a comparison of these methods can be made.

Case History 9.3: A Tentative Study of the Lead Separation Quality

Facts of the Case

Trim and form is an operation in the semiconductor assembly process where the leads of the lead frame are trimmed to make them of equal length and then the lead frame is formed into a certain shape. The quality of this operation depends on several dimensional characteristics such as form width, coplanarity, etc., which can be easily affected by improper handling during packaging and transportation. This is why the quality of this operation is checked several times on a number of parameters. And, to improve this operation further, an additional measurement was introduced called "lead separation." Operations required a quick measure of this parameter to get some idea of its quality level. A tentative process capability study is exactly the right technique for this situation.

After the process was set up, instruments calibrated, and the inspector given the proper instructions, a random sample of 50 units was collected and measured for lead separation (see Table 9.8). The specification requirements for this parameter are 18 ± 7 mils (i.e., USL = 25 mils, LSL = 11 mils). And by using these data, we made a histogram, tally sheet, and normal probability plot that provided a lot of information about the process setting, spread, and shape of the distribution. Based on this information, the tentative capability indexes were calculated. (Usually for a tentative capability study only one of these techniques is adequate. However we used different techniques here for demonstration purposes.)

9.6.2 Performing a Tentative Capability Study (TCS) Using a Histogram

Figure 9.8 shows a histogram computer printout. From this figure, we can see that the data are normally distributed, with a mean of $\bar{X} = 16.78$ and standard deviation $\sigma = 1.09$. The tentative capability indexes were computed

TABLE 9.8   Results of a Lead Separation Study

Observation Number	Lead Separation	Observation Number	Lead Separation
1	18.8	26	15.9
2	17.3	27	15.4
3	16.4	28	15.7
4	16.7	29	15.9
5	18.5	30	15.6
6	19.0	31	16.9
7	17.1	32	17.6
8	18.9	33	13.8
9	16.7	34	16.2
10	17.0	35	15.9
11	16.7	36	14.9
12	16.9	37	15.0
13	16.7	38	15.9
14	16.9	39	17.2
15	17.0	40	18.7
16	15.9	41	17.0
17	15.2	42	17.9
18	15.8	43	17.2
19	16.7	44	17.6
20	18.0	45	17.3
21	17.4	46	16.8
22	17.9	47	16.2
23	17.8	48	16.0
24	17.6	49	17.3
25	15.5	50	16.9

as follows:

$$Cp = \frac{USL - LSL}{6\sigma} = \frac{25 - 11}{6 \times 1.09} = 2.14$$

$$Cpu = \frac{USL - \bar{X}}{3\sigma} = \frac{25 - 16.78}{3 \times 1.09} = 2.51$$

$$Cpl = \frac{\bar{X} - LSL}{3\sigma} = \frac{16.78 - 11}{3 \times 1.09} = 1.77$$

$$Cpk = \min\{Cpu, Cpl\} = 1.77$$

Conclusion

The preliminary estimates are encouraging. A six sigma quality program requires a Cp of 2 and a Cpk of 1.5, and our results are even higher. If we had evidence that the process was in a state of statistical control, we would include this parameter in the list of "excellent processes." This was based on a histogram. Now let's see what the results will be using a simple manual technique called a "tally sheet."

9.6.3   Performing a TCS Using a Tally Sheet

If we don't have access to a computer, we can perform the TCS with a tally sheet (see Figure 9.9). The procedure is as follows.

Step 1 Set up a convenient arbitrary scale that can be used for preliminary calculations. In our case, we have a scale with midpoints ranging from 13.80 to 19.00 and the interval $(i) = 0.52$ (see column 1 in Figure 9.9).

Step 2 Tally each piece of data into the proper cell (see column 2 in Figure 9.9).

Step 3 Compute the frequency (f) for each cell (see column 3 in Figure 9.9).

Step 4 Locate the center point (C) that has the highest frequency. In our case, 16.40 is the center point.

Step 5 Compute the distance (d) relative to the center point for each cell (see column 4 in Figure 9.9).

Step 6 Fill in the columns "fd" and "$f(d)^2$" as indicated (see columns 5 and 6 in Figure 9.9).

Step 7 Calculate the mean $(\overline{X})$ and the standard deviation (σ) by using the equations given at the bottom of Figure 9.9.

Step 8 Compute the tentative capability indexes using the results from step 7:

$$Cp = \frac{USL - LSL}{6\sigma} = \frac{25 - 11}{6 \times 1.04} = 2.24$$

$$Cpk = \frac{|\overline{X} - \text{nearest spec}|}{3\sigma} = \frac{|16.26 - 11|}{3(1.04)} = 1.68$$

Conclusion

The results from the manual calculations using a tally sheet are very close to the previous results from the computer printout histogram. So the conclusions about the process remain the same.

You can see that when a computer is not available, the tally sheet is a simple technique that can be used to perform a process study. Now let's use another method for the same purpose, using the same data, and see if there will be a difference in the results.

Center Point (mils) (1)	Tally (2)	f (3)	d (4)	fd (5)	$f(d)^2$ (6)
13.80	I	1	+5	+5	25
14.32		0	+4	0	0
14.84	III	3	+3	+9	27
15.36	++++	5	+2	+10	20
15.88	++++ III	8	+1	+8	8
16.40	++++ ++++ I	11	0	0	0
16.92	++++ ++++	10	-1	-10	10
17.44	++++ I	6	-2	-12	24
17.96	I	1	-3	-3	9
18.48	++++	5	-4	-20	80
19.00					

$$\sum f = n = 50$$
$$\sum fd = -13$$
$$\sum f(d)^2 = 203$$
$$C = 16.4$$
$$i = 0.52$$

$$\bar{X} = C + i\left(\frac{\sum fd}{n}\right) = 16.40 + 0.52\left(\frac{-13}{50}\right) = 16.26$$

$$\sigma = i\sqrt{\frac{f(d)^2}{n} - \left(\frac{fd}{n}\right)^2} = 0.52\sqrt{\frac{203}{50} - \left(\frac{-13}{50}\right)^2} = 1.04$$

Figure 9.9   A tally sheet for lead separation data.

9.6.4 Performing a TCS Using a Normal Probability Plot (NPP)

Probability plotting is a simple technique which gives a pictorial representation of the data. By using this technique we can: (1) test the distribution assumption, which helps us select the right method for process capability study, and (2) estimate the distribution parameters needed to determine the process capability.

For a given distribution model, the data are plotted on special graph paper designed for that distribution. For example, if the assumed probability model is a normal distribution, we plot the data on normal probability paper. If the assumed model is correct, the plotted points will tend to fall in a straight line. If the model appears to fit the data reasonably well, distribution parameters such as the average and the standard deviation can be estimated directly from the plot.

TABLE 9.9   Worksheet for Normal Probability Paper

Value (X) (1)	i (2)	$f(i)$ (%) (3)	Value (X) (1)	i (2)	$f(i)$ (%) (3)
13.8	1	—	16.9	26	51
14.9	2	3	16.9	27	53
15.0	3	5	16.9	28	55
15.2	4	7	17.0	29	57
15.4	5	9	17.0	30	59
15.5	6	11	17.0	31	61
15.6	7	13	17.1	32	63
15.7	8	15	17.2	33	65
15.8	9	17	17.2	34	67
15.9	10	19	17.3	35	69
15.9	11	21	17.3	36	71
15.9	12	23	17.3	37	73
15.9	13	25	17.4	38	75
15.9	14	27	17.6	39	77
16.0	15	29	17.6	40	79
16.2	16	31	17.6	41	81
16.2	17	33	17.8	42	83
16.4	18	35	17.9	43	85
16.7	19	37	17.9	44	87
16.7	20	39	18.0	45	89
16.7	21	41	18.5	46	91
16.7	22	43	18.6	47	93
16.7	23	45	18.7	48	95
16.8	24	47	18.9	49	97
16.9	25	49	19.0	50	99

When we have a small number of data, we can plot the individual values directly on the probability paper. But, when there is a large number of data points this becomes tedious, so we need to group the data. The procedures for a normal probability plot for individual data and grouped data are described below.

9.6.4.1   Plotting an NPP Using Individual Data

1. Obtain a worksheet similar to that in Table 9.9.
2. Rank the individual readings (X) from smallest to largest in magnitude. There may be some readings that are repeated [see column (1) in Table 9.9].
3. Assign the individual values with increasing rank order of $i = 1$ for the first smallest value to $i = n$ for the last largest value, where n is the total number of values [see column (2) in Table 9.9].

4. For each reading, calculate the cumulative relative frequency $f_{(i)}$ where $i = 1, 2, \ldots, n$ by using one of the following formulas:

$$f_{(i)} = \frac{\left(i - \dfrac{1}{2}\right)100}{n}$$

or

$$f_{(i)} = \frac{\left(i - \dfrac{1}{2}\right)}{n}$$

depending on whether the marked axis on the probability paper refers to the percentage or the proportion of observations. In our case, we use $f_{(i)}$ in percent [see column (3) in Table 9.9]. To demonstrate the calculation of $f_{(i)}$, let's use the second values from Table 9.9 ($X = 14.9$). The corresponding value for i is 2; therefore,

$$f_{(i)} = \frac{\left(i - \dfrac{1}{2}\right)100}{n} = \frac{\left(2 - \dfrac{1}{2}\right)100}{50} = 3$$

5. Obtain normal probability paper and scale the horizontal axis into divisions that will accommodate all the readings of X (see Figure 9.10).
6. Plot the cumulative relative frequency $f_{(i)}$ against the reading of X for each $(X, f_{(i)})$ pair. *Note*: If there are more than two values of $f_{(i)}$ for the same value of X, plot only the highest and the lowest $f_{(i)}$ values. For example, in our case $X = 15.9$, the highest $f_{(i)} = 27$, the lowest $f_{(i)} = 19$, so we only plotted these two points (see Figure 9.10).
7. If a straight line appears to fit the data, draw a line, (a), that will give the best fit for all the points. When drawing the line, it is best to emphasize the central points on the graph, rather than the extremes.
8. Estimate the process average and the standard deviation from Figure 9.10 as follows.

Determine the Process Average

Based on the assumption of normality, the median can be selected as an estimate of the process average ($\overline{X}$), which is simply the 50% point (see Figure 9.10). To determine the process average, a perpendicular line (c) should be dropped from the intersection of the plotted line (a) and the 50% line (b). This will cut the baseline at the median. In our example,

$$50\% \text{ point} = \overline{X} = 16.80$$

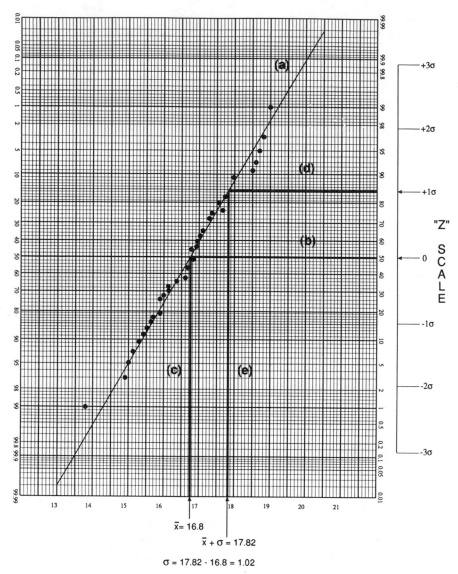

$\bar{x} = 16.8$

$\bar{X} + \sigma = 17.82$

$\sigma = 17.82 - 16.8 = 1.02$

Figure 9.10 Normal probability plot for individual data.

Estimate the Standard Deviation

To estimate the standard deviation, draw a vertical line (e) from the intersection of the plotted line (a) and the 84% line (d) to the horizontal scale. This is the value for $\bar{X} + \sigma$. In our example,

$$84\% \text{ point} = \bar{X} + \sigma = 17.82$$

Thus,

$$\sigma = 84\% \text{ point} - 50\% \text{ point}$$
$$= 17.82 - 16.80$$
$$= 1.02$$

9.6.4.2 Plotting an NPP using Grouped Data

To demonstrate the procedure of plotting a normal probability plot using grouped data, we will use the same figures that were used in demonstrating

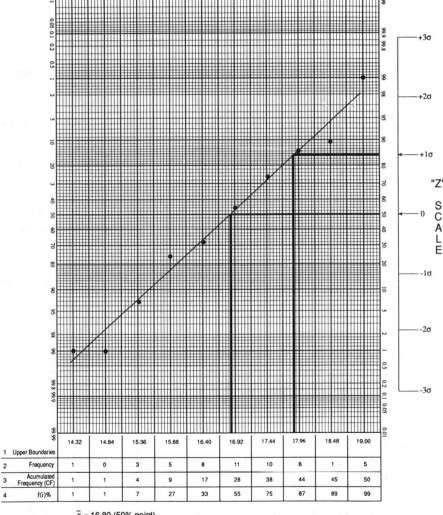

	14.32	14.84	15.36	15.88	16.40	16.92	17.44	17.96	18.48	19.00
1 Upper Boundaries										
2 Frequency	1	0	3	5	8	11	10	6	1	5
3 Acumulated Frequency (CF)	1	1	4	9	17	28	38	44	45	50
4 f(i)%	1	1	7	27	33	55	75	87	89	99

$\bar{x} = 16.80$ (50% point)
$\bar{x} + \sigma = 17.82$ (84% point)

Figure 9.11 Normal probability plot for grouped data.

Upper Boundary (1)	Tally (2)	f (3)
14.32	I	1
14.84		0
15.36	III	3
15.88	++++	5
16.40	++++ III	8
16.92	++++ ++++ I	11
17.44	++++ ++++	10
17.96	++++ I	6
18.48	I	1
19.00	++++	5

Figure 9.12 Worksheet for normal probability paper.

the process of plotting it for individual values. This will enable us to compare the results for the two methods.

1. Obtain probability paper designed for the distribution under examination. We selected the normal probability plot paper (see Figure 9.11).
2. Use the proper cell boundaries for the baseline scale [see column (1) in Figure 9.12].
3. Tally the values on the worksheet [see column (2) in Figure 9.12].
4. Compile the frequency for each cell [see column (3) in Figure 9.12].
5. Transfer the values of the upper boundaries and frequencies onto the bottom of the normal probability paper (see rows 1 and 2 in Figure 9.11).
6. Calculate the cumulative frequency (CF) (see row 3 in Figure 9.11). For example, the CF for the data up to 15.36 is

$$CF = 1 + 0 + 3 = 4$$

   Note: To verify that the CF calculation was correct, the last value of CF must be equal to the total number of samples.
7. For each value, calculate the cumulative relative frequency $f(i)$ (see row 4 in Figure 9.11). *Note*: The formula for deriving $f_{(i)}$ is the same as for individual data.
8. Plot the cumulative relative frequency $f_{(i)}$ against the reading of X for each $(X, f_{(i)})$ pair.

Figure 9.11 shows the results. We can see that by using grouped data, the process average ($\overline{X} = 16.80$) and standard deviation ($\sigma = 1.02$) agree with the results calculated from individual data.

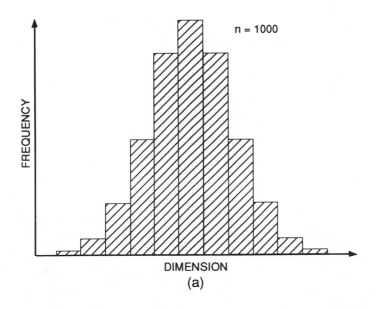

(a)

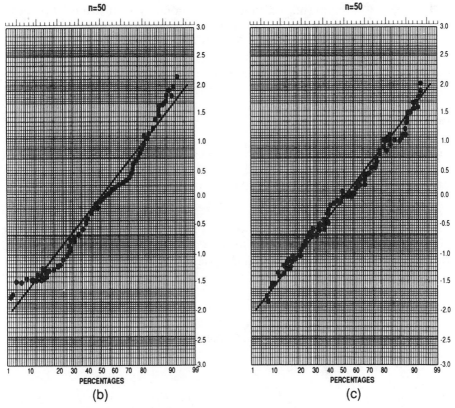

(b)   (c)

Figure 9.13   A histogram made from 1000 measurements and two normal probability plots based on random samples of 50 units from a normal distribution process.

Calculating the Tentative Capability Indexes Using a NPP

Based on the results from the probability plot, we can calculate the tentative capability indexes:

$$Cp = \frac{USL - LSL}{6\sigma} = \frac{25 - 11}{6 \times 1.02} = 2.28$$

$$Cpk = \frac{|\bar{X} - \text{nearest spec}|}{3\sigma} = \frac{|16.80 - 11|}{3(1.02)} = 1.90$$

Interpretation of the Normal Probability Plot

If the normal probability plot is based on data from a normal distribution, the points will cluster around a straight line. However, there can be some deviations because of random sampling fluctuations. And, if the plot deviates from linearity significantly, this means that the data do not represent a normal distribution. The decision of which can be considered a straight line and which cannot is a subjective matter. This is the negative part of this technique. However, the larger the sample size and the greater the divergence from the assumed model, the less likely a wrong decision about normality will be made.

Figure 9.13(a) represents a histogram from a normally distributed process, and Figures 9.13(b) and (c) are normal probability plots from two random samples of 50. As one can see the plots on Figure 9.13(c) are more clustered around the straight line than Figure 9.13(b). This is a result of the random sampling fluctuation effect. All these figures represent the same normally distributed process. So sometimes when the result of a NPP is suspicious, it may be necessary to repeat the plotting by taking another sample to make sure that the decision is right.

It takes a little time to get used to this technique, but after some experience the NPP becomes a very simple and useful approach for studying a process. This technique allows us to not only determine the process parameters, but it is also a test of the normality.

Conclusion

Table 9.10 shows the comparison of the results by applying a histogram, a tally sheet, and a probability plot. One can see that the results are very close. By applying any one of these techniques, we can get a preliminary estimate of the process capability and make plans for improvement based on the information.

TABLE 9.10   A Comparison of the Results Obtained by Applying a Histogram, Tally Sheet, and Probability Plot

Method	$\bar{X}$	σ	Cp	Cpk
Histogram	16.78	1.09	2.14	1.77
Tally Sheet	16.26	1.04	2.24	1.68
Probability Plot	16.80	1.02	2.28	1.90

9.7   FORMAL PROCESS CAPABILITY STUDIES

9.7.1   The Purpose of Process Capability Studies

Before going into a description of the procedure of performing a formal process capability study, it is important to define its purpose. According to the AT&T *Statistical Quality Control Handbook*, process capability study (PCS) is "the systematic study of a process by means of statistical control charts in order to discover whether it is behaving naturally or unnaturally; plus investigation of any unnatural behavior to determine its cause; plus action to eliminate any of the unnatural behavior which it is desirable to eliminate for economic or quality reasons. The natural behavior of the process after unnatural disturbances are eliminated is called the 'process capability'."[12] This means that PCS is a program of process improvement by discovering and eliminating the causes of unnatural variation, and bringing the process into a condition of statistical control.

Once this condition is achieved, the process capability can be determined. The right control charts to control the process can be selected and the process output can be predicted.

9.7.2   Preparation of the Process Capability Study

A formal process capability study (PCS) is the most valuable method of determining the true process potential. The proper way of conducting such a study is under planned operating conditions where a single operator, a single inspector, and a single batch of raw material are used throughout the study. Before starting the PCS, the measurement equipment must be calibrated and, if necessary, checked continuously to assure that the testing or measurement process is correct. In addition to the calibration, which mainly takes care of the instrument accuracy, a repeatability measure should be performed to determine the instrument precision (see Chapter 8 for repeatability study). During the PCS any changes or adjustments of the process should be avoided. If the process in study has a tendency to drift, it is sometimes reasonable to allow the process characteristic to drift and not make numerous adjustments. This will give the true picture of the process stability and capability. However, if the specifics of the process require changes or adjustments, these actions must be properly documented to help interpret the reasons for process changes.

A formal PCS is a study of current production by applying control charts. When dealing with a process that can generate variable data, usually an $\bar{X}$-R chart is used. When variable data are not available or there are other reasons (to be discussed later) to generate only attribute data, p or np control charts are used to demonstrate the process capability. A p or np chart is also used for a process capability study when we want to determine the overall process capability. In this case, the data are collected at the end of line where all characteristics are combined.

Sometimes there are situations where charts other than those mentioned earlier are used. For example, when the sample size is very large and the fraction defective is relatively small, a p chart with control limits based on a moving average is used. In this case, the sample fraction defective is treated as a variable measurement. Another example is when the data are limited. In this case, we can use a control chart for individual measurements.

There are also special situations when the classical statistical control charts are difficult to use. For example, when a process is running with a process average of 1000 or less parts per million (ppm), the p chart is replaced with a cumulative count control chart (CCC chart). Sometimes the capability study requires the use of designed experiments to separate the components of variation. In this case, nested design is an appropriate technique.

As we already know by now, one of the most important requirements before determining the process capability is that the process should be in a state of statistical control. To demonstrate this state, the control chart should consist of 20 to 25 points. As an exception, in the case when a point on the chart represents a long period of time (let's say a month), 10 to 12 consecutive points may be adequate. The control limits on the chart should usually be based on $\pm 3\sigma$ limits. However when working with attribute data, especially in a low ppm environment, $\pm 2\sigma$ limits would be preferable. This will make the chart more sensitive. Now if all the points on the control chart are within the range of control limits, and the pattern is "natural," it is recognized that the process is in control and the capability can be determined.

9.7.3 Determining the Capability of the Process

Having determined that the process is in a state of statistical control, this does not mean that the process is capable of meeting the specification requirements. There are many techniques for assessing the capability of a process. In relation to the specifications, some of them are: capability ratio, capability index, percent defective, parts per million (ppm), process average ($\bar{p}, n\bar{p}, \bar{c}$) and other modifications. Below, by using real case histories, we will demonstrate formal capability studies using variable and attribute data.

Case History 9.4: Process Capability as an Important Component of Supplier Relations

Evaluation of a supplier's process capability is one of the most important elements in a supplier relations program (SRP). We may successfully conduct different activities to improve supplier relations, but if their process is not capable of meeting our requirements, those relations will remain weak.

Process capability is not just an important component of supplier relations, it is the basic foundation of those relations because, by definition, it includes not only machines, materials, and methods, but also the people who work in the process and the management who directs the process. This is why AMD pays serious attention to the supplier's process capability.

In this section we will use actual results from a supplier capability study to illustrate some points related with supplier relations and their capability.

Selecting the Object to Be Studied

The cerdip base and cap are two ceramic parts used to package an electronic device in a hermetically sealed "enclosure." Although this may sound very simple, the technology of producing these parts is complicated and the overall quality of the complete electronic device depends on its quality. This is why all suppliers of this type of product are included in a list of "critical" suppliers.

To perform a special study of base and cap supplier capability, two companies have been selected; let's call them Supplier A and Supplier B. Both suppliers had already developed an SPC program several years ago, and it is known that their processes are in a reasonable state of statistical control. It is also known that Supplier A has a better process than Supplier B. This

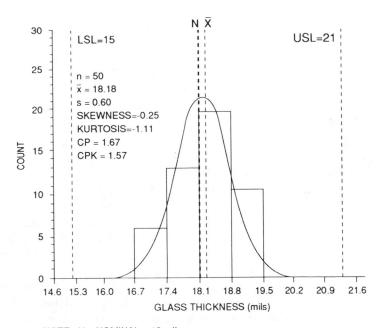

NOTE: N = NOMINAL = 18 mils

Figure 9.14 The results from a tentative estimate for Supplier A.

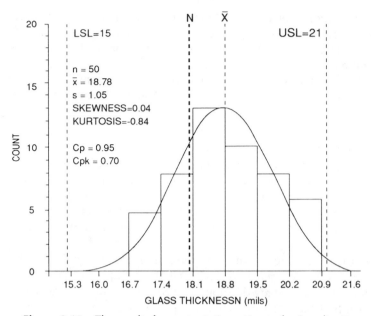

Figure 9.15 The results from a tentative estimate for Supplier B.

was one of the main reasons for selecting them for a study: to see why two plants that have almost the same technology have different capabilities.

AMD, together with the suppliers, approved a list of the most critical parameters to be controlled and agreed that the results were to be reported to AMD on a monthly basis. After analyzing these results it was decided that the sealing glass printing parameter is one of the most important variables which influence the overall quality. This parameter has been selected for the supplier capability study.

A Tentative Estimate of Supplier Capability

To perform a tentative estimate of the process capability (TEPC), a sample of 50 units was measured and a histogram was made for Suppliers A and B separately (see Figures 9.14 and 9.15). The units were taken directly from the process, during which time no adjustments were allowed. The time frame for taking the sample was very short (two hours) because of the high speed of production.

Based on these results, a tentative estimate of the process capability and its indexes was made and the standard deviation was calculated from the individual measurements. Considering that the process was continuously monitored with an $\bar{X}$-R chart and that the measurements were a result of an instantaneous process check, it is possible that this tentative estimate may be very close to the true process capability.

After viewing the histograms and considering the values of the coefficients of skewness and kurtosis (see Sections 10.1 and 10.2), we may conclude that we are dealing with a normal distribution. However, we decided to perform a normal probability plot (NPP) in addition for a "rough" check on whether the samples can reasonably be regarded as having come from a normal population. If so, we may conclude that the estimated $\overline{X}$ and S from the

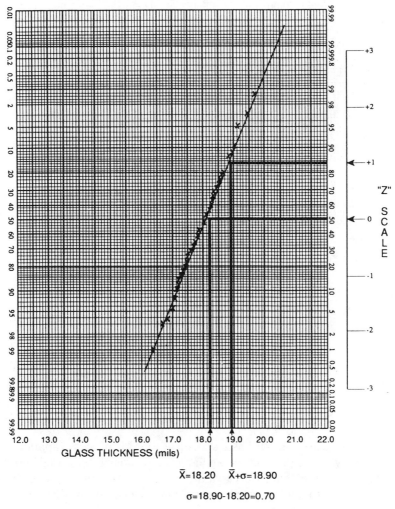

Figure 9.16 A normal probability plot for Supplier A.

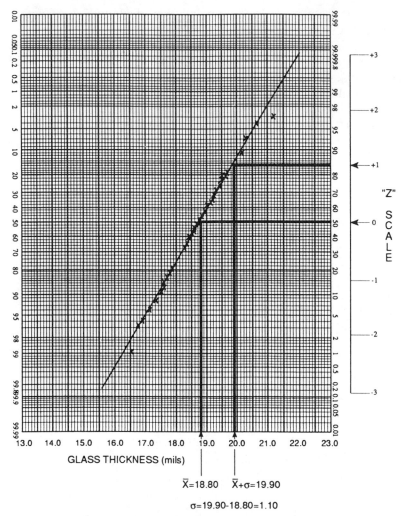

Figure 9.17 A normal probability plot for Supplier B.

sample represent the true process average (μ) and its process variation (σ) (see Figures 9.16 and 9.17).

To demonstrate the NPP as a statistical tool for process capability studies, we also derived the process average and the standard deviation from the plot and calculated the tentative capability indexes. As one can see, the results are very close to the previous conventional calculations when using a histogram (see Table 9.11).

TABLE 9.11   A Comparison of the Values for the Mean ($\bar{X}$), Standard Deviation (S), and Capability Indexes Estimated from a Histogram and a Normal Probability Plot

	From a Histogram	From a Normal Probability Plot
Supplier A	$\bar{X} = 18.18$	$\bar{X} = 18.2$
	$S = 0.6$	$S = 0.7$
	Cp = 1.67	Cp = 1.43
	Cpk = 1.57	Cpk = 1.33
Supplier B	$\bar{X} = 18.78$	$\bar{X} = 18.8$
	$S = 1.05$	$S = 1.10$
	Cp = 0.95	Cp = 0.91
	Cpk = 0.70	Cpk = 0.67

Preliminary Conclusions

The process of Supplier A is set closer to the specification nominal than the process of Supplier B. As we expected, Supplier A has a higher capability (Cp = 1.67 and Cpk = 1.57) than Supplier B. A preliminary "guess" would be that Supplier A is a good candidate to be included in the list of "six sigma quality suppliers."

Based on these preliminary results, Supplier B didn't even achieve the earlier requirements set by AMD. These requirements stipulate that every supplier should achieve a capability index of no less than 1.33 for all critical parameters.

But we should not rush into conclusions based on just a tentative estimate of process capability. Let's perform a formal capability study and see if our guess was right.

Formal (Short-Term) Process Capability Study

In the second phase of process investigation in the two plants, a formal capability study was performed:

1. To see if the process was in a state of statistical control.
2. To determine and introduce measures to stabilize the process if the results show that there is a lack of control.
3. To determine the process capability after a state of statistical control is assessed (or achieved).
4. To determine if the supplier is capable of meeting the specification.
5. To see if the supplier can ship its product directly to our factory stores without our incoming inspection.

The investigation is called a "short-term study" because the procedure requires that the study be conducted in a time frame long enough to collect 20 to 25 consecutive subgroups of 4 to 5 measurements on a control chart and, at the same time, short enough not to introduce excess variations which may be influenced by tool wear, change of the operator or inspector, readjustments, or other factors. Such a study should demonstrate the best result that a process can give under normal conditions without any major expenditures for process improvement.

To do this, we followed the formal procedure for a process capability study as described in Section 9.3.4. The measurement system was calibrated, the process was properly set up, materials were prepared from one batch, etc. Also no adjustments were made when running the process in the period of

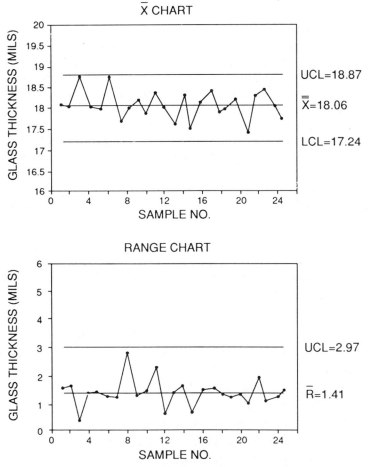

Figure 9.18   An $\bar{X}$-R chart for Supplier A.

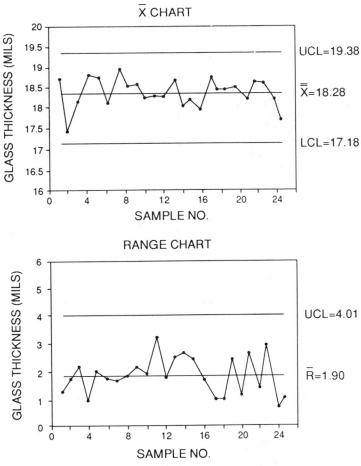

Figure 9.19 An $\bar{X}$-R chart for Supplier B.

the study. The results are shown in two control charts which, again, represent the processes from both plants (see Figures 9.18 and 9.19).

Analyzing these charts, one can see that a reasonable process stability is demonstrated. In addition to the charts, histograms were made to see the shape of the distribution (see Figure 9.20). Both of the histograms show a low coefficient of skewness which can be negligible. Also in both cases, the histograms show that the kurtosis is negative which means the distributions are slightly flat in comparison with a normal distribution. Considering all this together, we can say that we have a reasonably stable process, so we can proceed with the calculation of the capability indexes. The results are shown in Table 9.12. Note that to calculate the standard deviation at this time, we used the average range from the control chart ($\bar{R}$) which only includes the

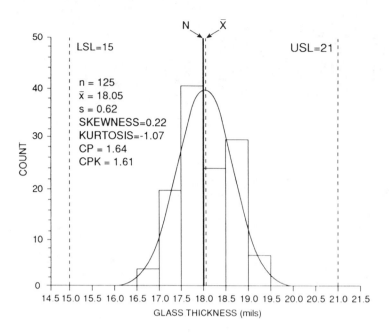

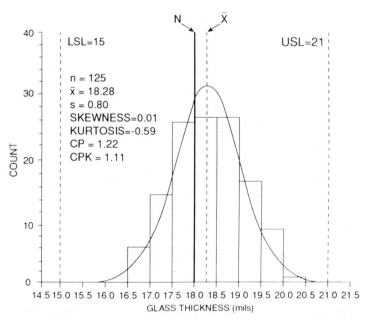

Figure 9.20 (a) Histogram for Supplier A. (b) Histogram for Supplier B.

TABLE 9.12 A Comparison of the Capability Indexes Based on Short-Term Process Capability Studies for Both Suppliers

Supplier A	Supplier B
$\bar{R} = 1.41$	$\bar{R} = 1.90$
$d_2 = 2.326$	$d_2 = 2.326$
$S = \dfrac{\bar{R}}{d_2} = \dfrac{1.41}{2.326} = 0.61$	$S = \dfrac{\bar{R}}{d_2} = \dfrac{1.90}{2.326} = 0.82$
$Cp = \dfrac{USL - LSL}{6S} = \dfrac{21 - 15}{6(0.61)} = 1.64$	$Cp = \dfrac{USL - LSL}{6S} = \dfrac{21 - 15}{6(0.82)} = 1.22$
$Cpk = \dfrac{|\bar{X} - \text{nearest spec}|}{3S_{\text{within}}}$	$Cpk = \dfrac{|\bar{X} - \text{nearest spec}|}{3S_{\text{within}}}$
$= \dfrac{|18.06 - 21|}{3(0.61)} = 1.61$	$= \dfrac{|18.28 - 21|}{3(0.82)} = 1.11$

"within"-subgroup variation. Usually the results from a study are more accurate and differ from a tentative estimate. However, in this particular case, there is a slight difference in the results and the conclusion remains almost the same; both processes run in a state of control, and Process A is significantly better than Process B. This is what we conclude from a short-term study.

But how does the process behave when running in a longer time frame, when the process is influenced by the variations in the material coming from different suppliers, or being run by different operators; when all additional factors mentioned earlier have an influence on the process variability? To answer this question, we should conduct a process performance study.

Process Performance and Long-Term Capability Study
So far, by conducting a short-term capability study, we demonstrated the process potential (its best ability to perform under controlled conditions). In this type of study, because of the short time frame of operation, it is possible to keep a large number of the variables that may influence the process capability almost constant. But when the process is operating over an extended period of time, the influence of variables such as tool changes and adjustments, different operators and inspectors, different lots of material, etc., is unavoidable. How will they influence the process performance and its capability? To find out, we need to continue our study at least long enough to include all the effects mentioned above.

   As we mentioned earlier, one of the reasons for the study is to determine if the supplier's process is capable of meeting the customer's specification requirements. In the short-term capability study, we only used 125 measurements. However, when we want to determine the natural limits of a process, a larger sample is required. According to Dr. Shewhart, "...it is necessary to have a comparatively large sample, usually more than a thousand, as a basis for establishing the tolerance range if one is to keep within practical limits the error in setting such ranges."[13]

   There is another reason why it was decided to conduct the study for a longer period of time. It will allow us to accomplish two things: (1) to determine the *long-term capability* of the process which is based on a measure of the "within"-subgroup variation only and (2) to determine the *process performance* which is based on measuring two types of variations—"within" subgroup and "between" subgroup. These two separate evaluations complement each other and will give a full picture of the process capability and its performance under the influence of a number of time-related variables.

   Normally a long-term capability or performance study can take a month or more, depending on the process production pace. In our case, because of the high-speed process performance, a study conducted for a period of five days would be sufficient to include the influence of all expected process effects. Also, this is the maximum time needed to produce the largest AMD lot size. Figures 9.21 and 9.22 are $\bar{X}$-R charts and Figures 9.23 and 9.24 are histograms which represent the results of the five-day study in both plants. As

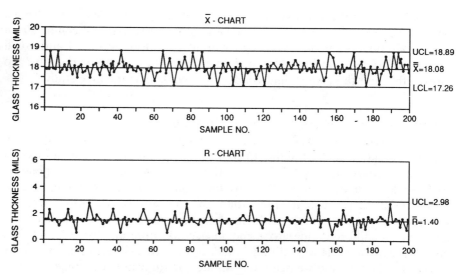

Figure 9.21   A long-term $\bar{X}$-R chart for Supplier A.

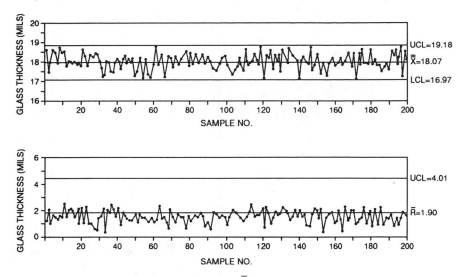

Figure 9.22   A long-term $\bar{X}$-R chart for Supplier B.

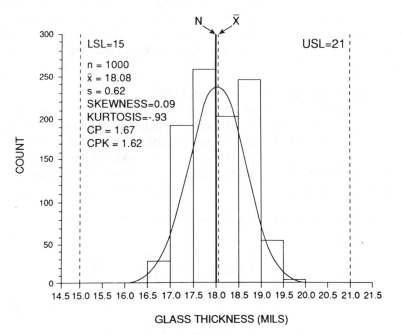

Figure 9.23   A long-term histogram for Supplier A.

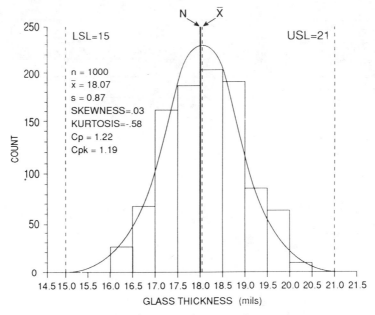

Figure 9.24   A long-term histogram for Supplier B.

one can see, there were some "out-of-control" situations but because of immediate corrective actions, the process was brought back into control quickly. Finally, we can say that both processes exhibit long-term stability, so we can estimate the long-term capability index and the process performance index as follows.

1. Evaluating Long-Term Process Capability

Long-term process capability is evaluated in the same way as short-term capability. Only the "within"-subgroup variation is considered when determining the process standard deviation. The difference is that in the long-term study the process average ($\overline{X}$) and the average range ($\overline{R}$) are influenced by a number of time-related variables and in the short-term study they are practically kept constant. So, the long-term capability is estimated as follows.

   a. Estimating the "within"-subgroup variability:
      for Supplier A,

$$S_{within} = \frac{\overline{R}}{d_2} = \frac{1.40}{2.326} = 0.60$$

      for Supplier B,

$$S_{within} = \frac{\overline{R}}{d_2} = \frac{1.90}{2.326} = 0.82$$

b. Calculating the capability indexes:
   for Supplier A,

$$Cp = \frac{USL - LSL}{6S_{within}} = \frac{21 - 15}{6(0.60)} = 1.67$$

$$Cpk = \frac{|\bar{X} - nearest\ spec|}{3S_{within}} = \frac{|18.08 - 21|}{3(0.60)} = 1.62$$

for Supplier B,

$$Cp = \frac{USL - LSL}{6S_{within}} = \frac{21 - 15}{6(0.82)} = 1.22$$

$$Cpk = \frac{|\bar{X} - nearest\ spec|}{3S_{within}} = \frac{|18.07 - 21|}{3(0.82)} = 1.19$$

2. Evaluating Process Performance

As we mentioned earlier, the difference between long-term capability and process performance is that the latest is a measure of the two types of variation exhibited by a process: "within"-subgroup variation and "between"-subgroup variation. Because of this, we use the following procedure of estimation.

a. Estimate the standard deviation by using

$$S_{total} = \sqrt{\frac{\sum_{i=1}^{n}(X_i - \bar{X})^2}{n - 1}}$$

For Supplier A, $S_{total} = 0.62$, and for Supplier B, $S_{total} = 0.87$.
b. Determine the process performance indexes:
   for Supplier A,

$$Pp = \frac{USL - LSL}{6S_{total}} = \frac{21 - 15}{6(0.62)} = 1.61$$

$$Ppk = \frac{|\bar{X} - nearest\ spec|}{3S_{total}} = \frac{|18.08 - 21|}{3(0.62)} = 1.57$$

for Supplier B,

$$Pp = \frac{USL - LSL}{6S_{total}} = \frac{21 - 15}{6(0.87)} = 1.15$$

$$Ppk = \frac{|\bar{X} - nearest\ spec|}{3S_{total}} = \frac{|18.07 - 21|}{3(0.87)} = 1.12$$

TABLE 9.13   A Comparison of Mean, Standard Deviation, and Capability Indexes from Different Types of Studies

	Supplier A				Supplier B			
	$\bar{X}$	S	Cp	Cpk	$\bar{X}$	S	Cp	Cpk
1. Tentative process performance	18.18	0.60	1.67	1.57	18.78	1.05	0.95	0.70
2. Short-term capability study	18.06	0.61	1.64	1.61	18.28	0.82	1.22	1.11
3. Long-term capability study	18.08	0.60	1.67	1.62	18.07	0.82	1.22	1.19
4. Process performance	18.08	0.62	1.61	1.57	18.07	0.87	1.15	1.12

Now to make a conclusion we can compare the mean, standard deviation, and the capability indexes from different types of studies (see Table 9.13).

By analyzing the results, we can conclude that in this particular case there was no significant difference in the results from different levels of study. This means that we are dealing with a very stable process where the time factor and other process components did not have much influence on the overall process variability. However, this does not always happen. In some cases there will be a significant difference between short-term and long-term capability and between process capability and process performance. Because of this it makes sense to introduce a process stability ratio (PSR) which is the ratio between the standard deviation derived from the short-term process capability (Spc) and the standard deviation derived from process performance (Spp):

$$PSR = \frac{S_{pc}}{S_{pp}}$$

If the process is stable the value of the standard deviation derived from a short-term study should be close to the value of the standard deviation estimated from a process performance study. In an ideal case, when the process average is absolutely stable, the PSR will be equal to 1. Table 9.14 shows the PSR for both suppliers.

The capability indexes reflect the relationship of the process spread to the specification limits. For example, knowing that Supplier A has a Cpk of 1.62, we can predict that in the long run it will submit to the customer product with a 0.608 ppm quality level. As for Supplier B, because its Cpk is 1.19, we can expect a quality level of 179 ppm. To visualize the relationship between the supplier processes and specifications, refer to Figure 9.25.

TABLE 9.14   A Comparison of the Process Stability Ratio for Both Suppliers

	Supplier A	Supplier B
Process stability ratio (PSR)	$S_{pc} = 0.60$ $S_{pp} = 0.62$ $PSR = \dfrac{S_{pc}}{S_{pp}} = \dfrac{0.60}{0.62} = 0.97$	$S_{pc} = 0.82$ $S_{pp} = 0.87$ $PSR = \dfrac{S_{pc}}{S_{pp}} = \dfrac{0.82}{0.87} = 0.94$

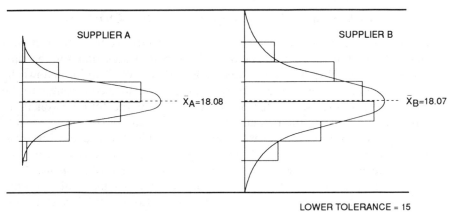

Figure 9.25   A comparison of process spread to tolerance limits for the two suppliers.

This is the answer to our fifth question, "Can we trust the supplier's quality without inspecting the product?"

3. Overall Conclusions

Supplier A has a stable process capable of meeting AMD's requirements of achieving six sigma quality.

Supplier B needs to work on process improvement to reduce the process variability.

AMD's specification requirements are realistic, which is demonstrated by Supplier A.

The product from Supplier A can be shipped directly to stores without inspection at our factory.

To monitor the process improvement, the suppliers should submit the Cp, Cpk, and PSR indexes monthly.

9.8   USING ATTRIBUTE DATA TO DETERMINE THE PROCESS CAPABILITY

Assume a complicated technological process where a unit is going through a large number of operations, control and inspection, and is finally ready to be shipped to the customer. What is the capability of the overall process? In other words, what is the quality level of the product you put in the shipping box? In a low ppm environment, with a long list of quality characteristics, which together represents the overall quality level, there should be a special way to verify the process performance. Below is a hypothetical example of constructing and applying a multi-defective control (MDC) chart that allows us to control the overall quality level at the end of the manufacturing line.

Case History 9.5: An Example of Applying the MDC Chart

To introduce a multi-defective control (MDC) chart, we need to start with the analysis of the historical data. Table 9.15 indicates that for the past period the overall quality level was 0.209% defective. The breakdown by defect categories (see Figure 9.26) shows that the major contributor to the overall quality level was defect code A (62.8% of the total). Even if we consider that most of the defects are not functional, an overall reject rate of 0.2% (2000 ppm) was still too high. This is why the Process Improvement Committee (PIC) decided to investigate this problem to ascertain the major contributors to this defective level. But what kind of statistical tools should be used in this situation? A conventional *p* chart will not work because of the large and mixed sample sizes. Also a regular *p* chart will

TABLE 9.15   Inspection Results from Historical Data

Defect Code	Quantity Rejected	Percentage to Total Defects
A	1,117	62.8
C	226	12.7
B	164	9.2
F	98	5.5
E	75	4.2
D	43	2.4
I	20	1.1
G	13	0.7
J	13	0.7
H	7	0.4
K	7	0.4
Total defects		1,777
Total quantity inspected		850,000
Percent defective		0.209%

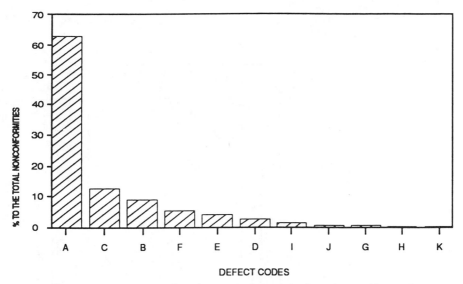

Figure 9.26 A Pareto chart for inspection results (based on Table 9.15).

not give enough information to address the problems. Using, for example, *np* charts to control every defective type separately is too complicated and costly. There are too many categories of defects to be controlled. All these reasons suggest that the best technique is a *p* chart with control limits calculated from the moving range and with an extra attached table that will represent the quality by separate types of nonconformities (see Section 7.7.5 for an explanation of why the control limits are based on the moving average). We will call it the multi-defective control (MDC) chart. Figure 9.27 shows that after applying this technique for a period of six months, the quality level has been improved 50%. The working team accepted this approach and they became confident that the nonconforming data would be gradually reduced to a level of six sigma quality.

To demonstrate how the MDC chart works, we will interpret the results from Figure 9.27. For example, in February a total of 28,000 units were inspected and 16 nonconforming units were found. This gives a reject rate of 0.057% (16/28,000 = 0.00057 = 0.057%). The point on the control chart for this month ($p = 0.057\%$) is within control limits. However, analyzing the results separately by the type of defects, one can see that there are four defectives in the "A" category circles (④). This means that if we maintain a separate chart for this particular parameter, the point for February would be outside the lower control limit, which indicates an improvement. The "H" defect category for this month is marked with a square (②), which means that two defectives are found. If we would

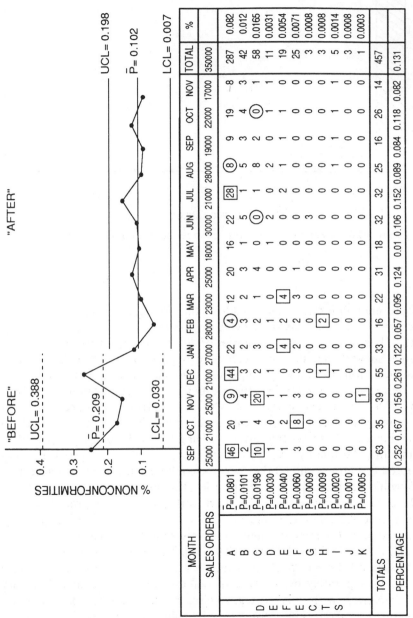

Figure 9.27 *p* chart with moving range "before" and "after" process improvement.

maintain a separate control chart for this type of defect, the point for February would be out of the upper control limit. This indicates a process deterioration has happened on this particular defect category in February. The rest of the defect categories are not marked, which means that they are in a state of statistical control, although process improvements were needed.

This type of chart is very informative because it indicates not only the overall percent defective, but also the condition of control separately by type of defects. It can be applied for any number of defect categories.

Making the MDC Chart

The MDC chart consists of upper and lower portions. The upper portion is a conventional p chart with control limits calculated from the moving range. The lower portion is a table that includes the information about quality levels of all types of defects, which together are the contributors to the overall percent defective. The table is self-explanatory. However, the only column that should be explained is the "estimated average percent defective ($\bar{p}$)" for every type of nonconformity. The MDC chart is used together with a 95% table (see Table 9.16), which gives the upper and lower control limits for a number of defects with a 95% confidence ($\pm 2\sigma$ limits). To use this table, we need to have two numbers: (1) the quantity inspected (n) for a particular period (day, week, month, etc.) and (2) the estimated average percent defective ($\bar{p}$), which is calculated from the historical data or from collecting new data of a sample size to represent the process (see Section 7.6.2). Having these two numbers and using the table, one can find the values of the upper and lower boundaries which are calculated by

$$n\bar{P} \pm 2\sqrt{n\bar{P}}$$

The table can be calculated for a particular group of defective categories and for a particular range of sample sizes depending on the specific situation. Let's illustrate how to use the table. As an example, in February $n = 28,000$ and the process average for defect type "A" is $\bar{p} = 0.080\%$. Using the table, we find that the maximum allowable number of defects is 32, and the minimum allowable defects is 14. Based on this, we consider that if in February only four defective units were found, the point would fall outside the lower control limit, which indicates an improvement and it is valuable to find out from where this improvement came. This is why the number found is circled (④), which indicates that the point is outside the lower control limits.

Similarly, we proceed with the defect type "H." In this case, since the process average is $\bar{p} = 0.0009\%$ and the sample size is $n = 28,000$, we discovered that if two defectives were found the point would be considered as out of upper control limit because the maximum allowable defects is 1.

TABLE 9.16 The 95% Number of Defective Table

<table>
<tr><th rowspan="2">n</th><th colspan="24">p̄ (%)</th></tr>
<tr><th>0.0005</th><th>0.0006</th><th>0.0007</th><th>0.0008</th><th>0.0009</th><th>0.001</th><th>0.002</th><th>0.003</th><th>0.004</th><th>0.005</th><th>0.006</th><th>0.007</th><th>0.008</th><th>0.009</th><th>0.01</th><th>0.02</th><th>0.03</th><th>0.04</th><th>0.05</th><th>0.06</th><th>0.07</th><th>0.08</th><th>0.09</th><th>0.1</th></tr>
<tr><td rowspan="2">15,000</td><td>–</td><td>–</td><td>–</td><td>–</td><td>–</td><td>–</td><td>–</td><td>–</td><td>–</td><td>–</td><td>–</td><td>–</td><td>–</td><td>–</td><td>–</td><td>0</td><td>0</td><td>1</td><td>2</td><td>3</td><td>4</td><td>5</td><td>6</td><td>7</td></tr>
<tr><td>1</td><td>1</td><td>1</td><td>1</td><td>1</td><td>1</td><td>1</td><td>2</td><td>2</td><td>2</td><td>3</td><td>3</td><td>3</td><td>4</td><td>4</td><td>6</td><td>9</td><td>11</td><td>13</td><td>15</td><td>17</td><td>19</td><td>21</td><td>23</td></tr>
<tr><td rowspan="2">16,000</td><td>–</td><td>–</td><td>–</td><td>–</td><td>–</td><td>–</td><td>–</td><td>–</td><td>–</td><td>–</td><td>–</td><td>–</td><td>–</td><td>–</td><td>–</td><td>0</td><td>0</td><td>1</td><td>2</td><td>3</td><td>5</td><td>6</td><td>7</td><td>8</td></tr>
<tr><td>1</td><td>1</td><td>1</td><td>1</td><td>1</td><td>1</td><td>1</td><td>2</td><td>2</td><td>3</td><td>3</td><td>3</td><td>4</td><td>4</td><td>4</td><td>7</td><td>9</td><td>11</td><td>14</td><td>16</td><td>18</td><td>20</td><td>22</td><td>24</td></tr>
<tr><td rowspan="2">17,000</td><td>–</td><td>–</td><td>–</td><td>–</td><td>–</td><td>–</td><td>–</td><td>–</td><td>–</td><td>–</td><td>–</td><td>–</td><td>–</td><td>–</td><td>–</td><td>–</td><td>1</td><td>2</td><td>3</td><td>4</td><td>5</td><td>6</td><td>7</td><td>9</td></tr>
<tr><td>1</td><td>1</td><td>1</td><td>1</td><td>1</td><td>1</td><td>2</td><td>2</td><td>2</td><td>3</td><td>3</td><td>3</td><td>4</td><td>4</td><td>4</td><td>7</td><td>10</td><td>12</td><td>14</td><td>17</td><td>19</td><td>21</td><td>23</td><td>25</td></tr>
<tr><td rowspan="2">18,000</td><td>–</td><td>–</td><td>–</td><td>–</td><td>–</td><td>–</td><td>–</td><td>–</td><td>–</td><td>–</td><td>–</td><td>–</td><td>–</td><td>–</td><td>–</td><td>–</td><td>1</td><td>2</td><td>3</td><td>4</td><td>6</td><td>7</td><td>8</td><td>10</td></tr>
<tr><td>1</td><td>1</td><td>1</td><td>1</td><td>1</td><td>1</td><td>2</td><td>2</td><td>2</td><td>3</td><td>3</td><td>4</td><td>4</td><td>4</td><td>4</td><td>7</td><td>10</td><td>13</td><td>15</td><td>17</td><td>20</td><td>22</td><td>24</td><td>26</td></tr>
<tr><td rowspan="2">19,000</td><td>–</td><td>–</td><td>–</td><td>–</td><td>–</td><td>–</td><td>–</td><td>–</td><td>–</td><td>–</td><td>–</td><td>–</td><td>–</td><td>–</td><td>–</td><td>–</td><td>1</td><td>2</td><td>3</td><td>5</td><td>6</td><td>7</td><td>9</td><td>10</td></tr>
<tr><td>1</td><td>1</td><td>1</td><td>1</td><td>1</td><td>1</td><td>2</td><td>2</td><td>3</td><td>3</td><td>3</td><td>4</td><td>4</td><td>4</td><td>5</td><td>8</td><td>10</td><td>13</td><td>16</td><td>18</td><td>21</td><td>23</td><td>25</td><td>28</td></tr>
<tr><td rowspan="2">20,000</td><td>–</td><td>–</td><td>–</td><td>–</td><td>–</td><td>–</td><td>–</td><td>–</td><td>–</td><td>–</td><td>–</td><td>–</td><td>–</td><td>–</td><td>–</td><td>–</td><td>1</td><td>2</td><td>4</td><td>5</td><td>7</td><td>8</td><td>10</td><td>11</td></tr>
<tr><td>1</td><td>1</td><td>1</td><td>1</td><td>1</td><td>1</td><td>2</td><td>2</td><td>3</td><td>3</td><td>3</td><td>4</td><td>4</td><td>4</td><td>5</td><td>8</td><td>11</td><td>14</td><td>16</td><td>19</td><td>21</td><td>24</td><td>26</td><td>29</td></tr>
<tr><td rowspan="2">21,000</td><td>–</td><td>–</td><td>–</td><td>–</td><td>–</td><td>–</td><td>–</td><td>–</td><td>–</td><td>–</td><td>–</td><td>–</td><td>–</td><td>–</td><td>–</td><td>–</td><td>1</td><td>3</td><td>4</td><td>6</td><td>7</td><td>9</td><td>10</td><td>12</td></tr>
<tr><td>1</td><td>1</td><td>1</td><td>1</td><td>1</td><td>1</td><td>2</td><td>2</td><td>3</td><td>3</td><td>4</td><td>4</td><td>4</td><td>5</td><td>5</td><td>8</td><td>11</td><td>14</td><td>17</td><td>20</td><td>22</td><td>25</td><td>28</td><td>30</td></tr>
<tr><td rowspan="2">22,000</td><td>–</td><td>–</td><td>–</td><td>–</td><td>–</td><td>–</td><td>–</td><td>–</td><td>–</td><td>–</td><td>–</td><td>–</td><td>–</td><td>–</td><td>–</td><td>–</td><td>2</td><td>3</td><td>4</td><td>6</td><td>8</td><td>9</td><td>11</td><td>13</td></tr>
<tr><td>1</td><td>1</td><td>1</td><td>1</td><td>1</td><td>1</td><td>2</td><td>2</td><td>3</td><td>3</td><td>4</td><td>4</td><td>4</td><td>5</td><td>5</td><td>9</td><td>12</td><td>15</td><td>18</td><td>20</td><td>23</td><td>26</td><td>29</td><td>31</td></tr>
<tr><td rowspan="2">23,000</td><td>–</td><td>–</td><td>–</td><td>–</td><td>–</td><td>–</td><td>–</td><td>–</td><td>–</td><td>–</td><td>–</td><td>–</td><td>–</td><td>–</td><td>–</td><td>–</td><td>2</td><td>3</td><td>5</td><td>6</td><td>8</td><td>10</td><td>12</td><td>13</td></tr>
<tr><td>1</td><td>1</td><td>1</td><td>1</td><td>1</td><td>1</td><td>2</td><td>2</td><td>3</td><td>3</td><td>4</td><td>4</td><td>5</td><td>5</td><td>5</td><td>9</td><td>12</td><td>15</td><td>18</td><td>21</td><td>24</td><td>27</td><td>30</td><td>33</td></tr>
</table>

	1	2	3	4	5	6	7	8	9	10	11	12	13	14	15	16	17	18	19	20	21	22
24,000	14	12	10	9	7	5	3	2	0	—	—	—	—	—	—	—	—	—	—	—	—	—
	34	31	28	25	22	19	16	13	9	5	5	5	4	4	3	3	2	2	1	1	1	1
25,000	15	13	11	9	7	5	4	2	1	—	—	—	—	—	—	—	—	—	—	—	—	—
	35	32	29	26	23	20	16	13	9	6	5	5	4	4	3	3	2	2	1	1	1	1
26,000	16	14	12	10	8	6	4	2	1	—	—	—	—	—	—	—	—	—	—	—	—	—
	36	33	30	27	23	20	17	13	10	6	5	5	5	4	4	3	3	2	1	1	1	1
27,000	17	14	12	10	8	6	4	2	1	—	—	—	—	—	—	—	—	—	—	—	—	—
	37	34	31	28	24	21	17	14	10	6	6	5	5	4	4	3	3	2	1	1	1	1
28,000	17	15	13	11	9	7	5	3	1	—	—	—	—	—	—	—	—	—	—	—	—	—
	39	35	32	28	25	21	18	14	10	6	6	5	5	4	4	3	3	2	1	1	1	1
29,000	18	16	14	11	9	7	5	3	1	—	—	—	—	—	—	—	—	—	—	—	—	—
	40	36	33	29	26	22	18	15	11	6	6	5	5	4	4	3	3	2	1	1	1	1
30,000	19	17	14	12	10	7	5	3	1	0	0	—	—	—	—	—	—	—	—	—	—	—
	41	37	34	30	26	23	19	15	11	6	6	6	5	4	4	3	3	2	1	1	1	1
31,000	20	17	15	12	10	8	5	3	1	0	0	—	—	—	—	—	—	—	—	—	—	—
	42	38	35	31	27	23	19	15	11	7	6	6	5	5	4	4	3	2	1	1	1	1
32,000	21	18	15	13	10	8	6	3	1	0	0	—	—	—	—	—	—	—	—	—	—	—
	43	40	36	32	28	24	20	16	11	7	6	6	5	5	4	4	3	2	1	1	1	1
33,000	22	19	16	13	11	8	6	4	2	0	0	—	—	—	—	—	—	—	—	—	—	—
	44	41	37	33	29	25	20	16	12	7	7	6	5	5	4	4	3	2	1	1	1	1
34,000	22	20	17	14	11	9	6	4	2	0	0	—	—	—	—	—	—	—	2	—	—	—
	46	42	38	34	29	25	21	17	12	7	7	6	5	5	4	4	3	2	1	1	1	1
35,000	23	20	17	15	12	9	7	4	2	0	0	—	—	—	—	—	—	—	2	—	—	—
	47	43	39	34	30	26	21	17	12	7	7	6	6	5	4	4	3	2	2	1	1	1

In the MDC chart, the value without any marks indicates that if control charts for that type of defect were maintained the point for this month would be between the 2 sigma control limits.

Constructing the 95% Tables
The idea of constructing tables for this type of chart belongs to Jack S. Gantt, who developed the 95% number defective table. These tables, to the best of our knowledge, have not been published. We would like to express out appreciation to the author for submitting to us a copy of his tables, which are successfully used at AMD. Taking into consideration that for every area of application, only a fraction of the table is used, we developed simple computer software which allowed us to make tables that will reflect the needs of a particular area (see Table 9.16 as an example).

Conclusion
The MDC chart is a simple technique to be used mostly at the end of the product process where large sample sizes of most types of production are used. This chart allows us to establish an effective feedback to the process where the defects were generated. Counting, charting, controlling, and measuring the capability of a process that generates only attribute data makes people aware of what can be done internally and what information can be given externally, as a feedback for improvement.

9.9   THE SEVEN PHASES OF PERFORMING A PROCESS CAPABILITY STUDY

A process capability study consists of seven major steps. Figure 9.28 shows the flowchart of performing the study. Detailed descriptions of each phase are listed below.

Phase 1: Select the Process
As we know, the overall capability of a manufacturing process depends on the capability of the steps that comprise the process. On the other hand, the capability of any process step depends on the capability of the factors (variables) related to this process step. This can be expressed in the following relation:

$$y = f(X_1, \ldots, X_n)$$

where y is the output of a process step, and $X_1, \ldots, X_n$ are the various input and process factors (variables) related to this process step.

For example, in an integrated circuit assembly process die attach is a process step, and the capability of this process step depends on a number of

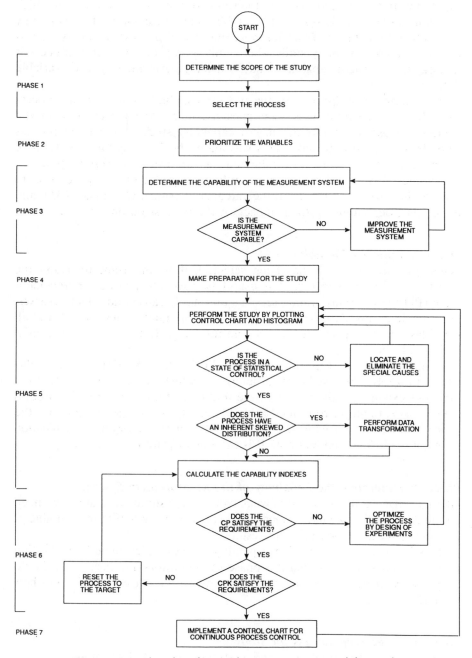

Figure 9.28   Flowchart for conducting a process capability study.

factors: temperature, pressure, paste, etc. During the first phase of a process capability study, we need to define the scope of the area of interest for which we want to perform the study. This can be a macro capability study where by using attribute data we are establishing the overall process capability, or this can be a micro capability study where only a step of the overall process is investigated. A micro capability study is usually performed by using variable data.

To determine the priority of the process capability study for the process steps (or its elements), it is sometimes useful to do an evaluation of the overall process by performing macro capability studies on all the process steps by using attribute data. This will allow the investigator to determine which process step is the most influential on the overall process output.

Another way to select the most influential areas of the process is to classify the process steps into three categories: critical (C), major (M), and minor (m), depending on their impact on the process output.

Phase 2: Prioritize Variables

When the area of study has been selected, it is important to assess the variables to be investigated. It is useful to have the process improvement team (PIT) hold brainstorming sessions and apply cause-and-effect matrixes, fishbone diagrams, Pareto charts, and other techniques to develop a list of all variables which impact the process output and select the most influential variables to be studied.

Tentative process capability studies (TPCS) (see Section 9.6) are also useful to apply when we want to determine which variable needs attention first. For example, if the results of the tentative study show that the process spread of Variable A consumes only 50% of the allowable spread, and the process spread of Variable B consumes 90% of the allowable specification range, it is obvious that Variable B is the higher priority to be studied.

Phase 3: Determine the Capability of the Measurement System

At this stage we need to assess what type of measurement system we need and make sure that we have a measuring/inspection system that is capable of producing accurate measurement results.

This involves test/inspection capability studies which include calibration and precision studies. During this phase the operators and inspectors should be properly instructed on how to use the measurement equipment and how to collect the data.

Phase 4: Make Preparations for Study

Depending on the situation, the investigator should determine if he/she wants to perform a short- or long-term capability study. It is also important

TABLE 9.17   Recommended Minimum Values of the Process Capability Indexes

	Cp	Cpk
Existing processes	1.50	1.33
New processes	1.67	1.50
Critical parameters in existing processes	2.00	1.50
Critical parameters in new processes	2.00	1.67

here to revise the existing manufacturing procedures and specification and determine the frequency of sampling. Special forms for data collection and charting should be prepared which will allow us to keep track of all necessary events and changes made in the period of study.

Phase 5: Perform the Capability Study

The capability study should be performed according to existing procedures. If the control chart and the histogram indicate that the process is in a state of statistical control, the capability index should be determined. If the results show that the process is not in control, engineering efforts should be made to stabilize the process. After eliminating the special causes from the process, the capability study should be repeated. If the results indicate that the process has a skewed distribution, and based on engineering judgment, we can assess that the skewness is part of the nature of the process, then we will need to perform a data transformation before calculating the capability indexes.

Phase 6: Study the Results and Make Calculations

If the Cp and Cpk satisfy the company's and the customer's standards (see Table 9.17), accept the process and establish a control chart to monitor the process. If the Cpk does not satisfy the specification requirements and its value is smaller than the Cp, take actions to reset the process to the target. If the Cp also has not satisfied the specification requirements, a design of experiment technique should be selected and introduced to optimize the process parameters and reduce its variation. This will also let us optimize the existing specification ranges which, together with reduction of process variation, will improve the Cp and Cpk.

Phase 7: Continuous Process Control

After the design of experiment result is introduced, the process capability should be assessed and a control chart for continuous process control should be implemented.

9.10   IF THE PROCESS IS NOT IN CONTROL—SHOULD WE MAKE CAPABILITY PREDICTIONS?

As we have already mentioned, a process capability prediction should not be made until the process is brought into a state of statistical control. However, in practice there are a lot of situations where the process is not in control. In addition, the causes that make the process unstable cannot be economically eliminated or cannot be eliminated in a timely manner. Should we (or can we) make some kind of a process capability prediction?

The statistician might say "No," and he/she would be right. But at the same time, the statistician-practitioner understands, and the manufacturing engineer insists, that some kind of comparison of the capability to the product tolerances *must be made*. The conclusion drawn from this comparison will be a rough one and tentative, but it is more dangerous to delay the analysis and wait until the assignable cause is eliminated from the process. Making tentative predictions from a process that is not even in control is not a rule, but an exception; to be used when it is very difficult, time consuming, and costly to work on stabilizing the system.

In the electronics industry there are some processes where the control chart shows instability and efforts to eliminate the assignable causes are not immediately successful. In this case, experience shows that it is good to start by plotting individual measurements against tolerance limits (see Section 9.5). This will show if the process can meet the product tolerances even with assignable causes present. By using the data from plotting individual measurements, a rough standard deviation can be determined, and a process capability can be computed. We need to bear in mind here that because the process will not be operating at its best, the sigma will be inflated, so the capability index will be reduced. But even if we don't get a true picture because of the out-of-control conditions, the investigation will give us a more intelligent estimate as to where the process lies in comparison with the specifications.

In addition, this investigation will show us how far we are from reaching the desired state of control. To obtain the best estimate of an uncontrolled process, the AT&T *Statistical Quality Control Handbook* gives the following procedures[14]:

1. If the pattern shows a trend, determine the cause of the trend and decide which portion of it represents the way in which the process will be run in the future. Estimate the capability, basing your estimates on the selected portion of the pattern only (see Figure 9.29).

2. If the pattern is interrupted by periodic lack of control, this can sometimes be recognized as indicating the presence of two or more separate patterns in the data. It should be possible to run the process at any one of the indicated levels provided we are able to identify the causes and bring the process, at some later time, into a state of control (see Figure 9.30).

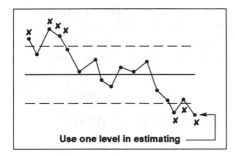

Use one level in estimating

Figure 9.29 Estimating from a trend. Reprinted with permission from Western Electric Co., "Statistical Quality Control Handbook" (1984), AT&T Technologies, Inc., p. 60, Figure 58.

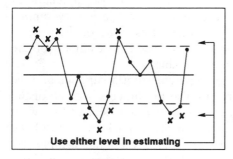

Use either level in estimating

Figure 9.30 Estimating from an interrupted pattern. Reprinted with permission from Western Electric Co., "Statistical Quality Control Handbook" (1984), AT&T Technologies, Inc., p. 60, Figure 59.

Wherever it is possible to pick out such probable levels by eye, this provides a reasonable basis for estimating capability. Use engineering judgment in deciding which points are likely to indicate separate patterns.

3. If the pattern is erratic in such a manner that it is not possible to pick out the separate patterns by eye, it may be that the best available estimate of capability will be the center of the out-of-control pattern. If this estimate is used for want of a better one, keep in mind that it is a very uncertain estimate (see Figure 9.31).

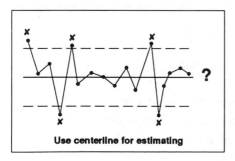

Use centerline for estimating

Figure 9.31 Estimating from an erratic pattern. Reprinted with permission from Western Electric Co., "Statistical Quality Control Handbook" (1984), AT&T Technologies, Inc., p. 60, Figure 60.

When estimates are based on out-of-control patterns, always explain the basis for the estimate and show the pattern of the data from which the estimate came. Also remember that no estimate from uncontrolled data is reliable. Reliability comes from knowledge that the data show control.

9.11   HOW LARGE SHOULD THE SAMPLE SIZE BE TO DETERMINE THE PROCESS AVERAGE?

9.11.1   Optimizing the Sample Size

When we have control charts on the process and they demonstrate a state of statistical control, the determination of the process average is very simple. The center line of the chart ($\bar{X}$, $\bar{p}$, or $\bar{c}$) is the best estimate of the process average. But there are some situations when we need an estimate of the process average and control charts are not available.

For example, when we receive a large shipment from a supplier and we want to estimate not only the lot quality but also the process average of the supplier. Or, let's say we visit a supplier for a source inspection. and we want to pull a random sample from the process to make a quick estimate of the process average. In this case, we need to know what the optimal sample size should be to estimate the process average with the desired precision. Intuitively we know that the larger the sample size, the higher the precision of the estimate is. But when we take a larger sample there is also a larger cost involved. And if we take a smaller sample there is a greater risk of undermining the reliability of the estimate.

There is always a conflict between information needed and the cost of obtaining this information. It is important to note here that the information obtained from a sample is a function of the square root of sample size n. This means that if, for example, we receive 3 units of information from a sample size of 9, then to get 6 units of information, we need a sample size of 36. The point we want to make here is that by increasing the sample size, the cost grows much faster than the information, and it comes to a point when increasing the sample size is not economically reasonable. So a sample is optimal when we obtain the information needed with the lowest cost.

Before going into the calculation of the optimal sample size, the following questions need to be answered:

Is the intent to use variable or attribute data?

Is the test destructive or nondestructive?

How expensive is it to obtain and inspect (or test) a unit?

What accuracy and confidence of an estimate is required?

How much variation among measurements is expected?

TABLE 9.18 The Number of Standard Error Units (Z) Associated with the Level of Confidence Specified

Z	Confidence Level
3	99.7%
2.58	99%
2.33	98%
2	95.5%
1.96	95%
1.64	90%
1.28	80%

Only after answering these questions can we estimate the optimal sample size that will allow us to obtain the most precise information at the lowest cost.

9.11.2 Calculating the Sample Size when Using Variable Data

If the value of the individual items can be assumed to follow a normal distribution, the equation for estimating the sample size (n) is

$$n = \left(\frac{Z\sigma}{E} \right)^2 \tag{9.1}$$

where E is the maximum allowable error between the estimate to be made from the sample and the actual (unknown) process average. Z is the number of standard error units associated with the level of confidence specified (see Table 9.18). The choice of the factor of 3 is recommended for general use. σ is the standard deviation of the process.

From these three elements of the formula, the Z and E values can usually be preassigned. However, the process standard deviation (σ), which is usually unknown, should be at least roughly determined. One way to determine a prior estimate of the standard deviation is by using one of the equations from Figure 9.32.[15] From previous experience, try to decide what the smallest and largest values of the characteristic are likely to be. Also try to picture how the other observations may be distributed. Based on this information, select one of the forms of distribution from Figure 9.32 and calculate the advance estimate standard deviation by using the formula related to the selected shape of distribution (also in Figure 9.32). In case of doubt, or in case the desired precision (E) is a critical matter, the rectangular distribution may be used, which will result in a larger sample.

Another way of estimating the advance standard deviation, if no prior information about the process is available, is to take a pilot sample from the process and compute the S as an estimate of σ. In this case, the process average will be estimated in two stages, by taking two samples.

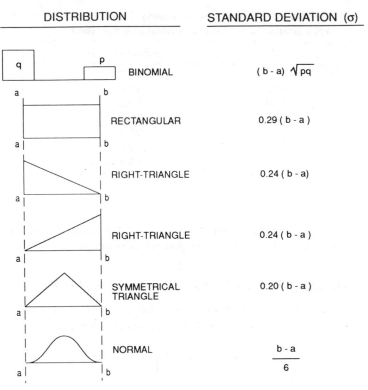

DISTRIBUTION	STANDARD DEVIATION (σ)
BINOMIAL	$(b-a)\sqrt{pq}$
RECTANGULAR	$0.29(b-a)$
RIGHT-TRIANGLE	$0.24(b-a)$
RIGHT-TRIANGLE	$0.24(b-a)$
SYMMETRICAL TRIANGLE	$0.20(b-a)$
NORMAL	$\dfrac{b-a}{6}$

Figure 9.32   Some geometric figures and formulas of estimating deviations. *Note:* The range of variation is in every panel from *a* to *b*. Adapted with permission from W. Edwards Deming, "Sample Design in Business Research" (1960), John Wiley & Sons, Inc., p. 260, Figure 16.

An Example of Estimating the Sample Size for Variable Data   Visiting a potential supplier, the customer wants to make an estimate of the supplier's process average to see if it can meet the customer's requirements. What sample size is needed to obtain an estimate with reasonable precision and confidence?

The test procedure to determine the product quality is nondestructive, and the cost of performing the test is relatively inexpensive, so the customer decides to have a sample size which will give an estimate of the process average with an error (E) of ± 5 measurement units. This is a somewhat high accuracy considering that the specification range is equal to 100 measurement units.

The customer also wants to have 98% confidence in the estimate. By applying such high confidence, this means that the sampling procedure employed may generate estimates that differ from the true (unknown) process average by more than the maximum allowable error (in this case more than ± 5 units) only 2% of the time. And, concerning the estimate of the standard deviation, even when the supplier had some information about the process spread, the customer decided to compute an advance estimate of the standard deviation.

From experience the customer knew that the process may have a distribution skewed slightly to the right-hand side and the spread of the values can be approximately 64 measurement units. With this prior information and applying Figure 9.32, the customer selected the triangular distribution where the formula of the standard deviation is $0.24(b - a)$. Substituting the values in the equation, we have

$$S = 0.24(b - a) = 0.24(64) = 15.36$$

Now all components are available to determine the sample size by using the following formula:

$$n = \left(\frac{Z\sigma}{E} \right)^2$$

Looking it up in Table 9.18, the Z value for 98% confidence is 2.33. Substituting the available figures in the formula above, the sample size is determined:

$$n = \left(\frac{Z\sigma}{E} \right)^2 = \left(\frac{2.33 \times 15.36}{5} \right)^2 = \left(\frac{35.79}{5} \right)^2 = 51.23$$

A sample size of 51 units is adequate to have a rough estimate of the process average with an accuracy of ± 5 measurement units and a confidence of 98%.

Calculating the Sample Size when Using Attribute Data

The equation of estimating the sample size (n) for attribute data is

$$n = \left(\frac{Z}{E} \right)^2 \bar{p}(1 - \bar{p}) \tag{9.2}$$

where Z and E are defined the same as before, and $\bar{p}$ is the estimated average fraction defective rate for the process.

When no data from previous samples are available, proceed as follows. From experience, approximate the band within which the fraction defective is likely to lie. Apply Figure 9.33[16] and locate the value of $\sigma^2 = \bar{p}(1 - \bar{p})$ for the middle of the possible range of $\bar{p}$ and use it in Equation (9.2) to calculate the sample size. In case the desired precision is a critical matter, use the largest value of σ^2 within the possible range of $\bar{p}$ (which is 0.5).

An Example of Estimating the Sample Size for Attribute Data   A customer received the first large shipment of 20,000 parts that will be used in the pilot line of a new product. Assuming that the shipment represents the process, the customer wants to know what sample size needs to be inspected to have an estimate of the supplier's process average. The customer cannot inspect the lot 100%, but he is willing to take a relatively large sample to have a high precision of the estimate because those parts are very influential on the product quality. Looking at Figure 9.33, one can see that the largest value of $\sigma^2 = \bar{p}(1 - \bar{p})$ within the possible range of $\bar{p}$ is 0.5. So, substituting this value in Equation (9.2), we have

$$n_{max} = \left(\frac{Z}{E}\right)^2 \bar{p}(1 - \bar{p}) = \left(\frac{Z}{E}\right)^2 (0.5)(0.5)$$

The customer decided to allow for a maximum sampling error (E) to be 0.02, and he also wants to have a 98% confidence that the sample estimate will not differ from the true process average by more than the allowable sampling error.

Having all those parameters established, we can calculate the maximum sample size needed to determine the process average:

$$n = \left(\frac{Z}{E}\right)^2 \bar{p}(1 - \bar{p})$$

$$= \left(\frac{2.33}{0.02}\right)^2 (0.5)(0.5)$$

$$= (13,572)(0.25)$$

$$= 3393.06$$

$$\approx 3393$$

This means that for a confidence level of 98% and a sampling error of 0.02 (if the process average is 0.5), the maximum sample size would be $n = 3393$.

Now for illustration purposes, let's assume that we receive some information about the process and we estimate the process average ($\bar{p}$) to be 0.35. Then the sample size would be

$$n = \left(\frac{Z}{E}\right)^2 \bar{p}(1 - \bar{p})$$

$$= \left(\frac{2.33}{0.02}\right)^2 (0.35)(0.65)$$

$$= (13,572)(0.2275)$$

$$= 3087$$

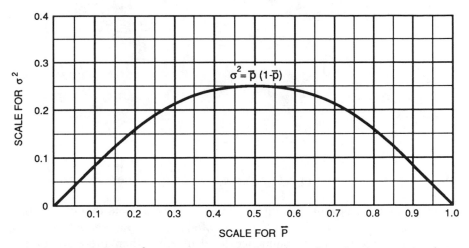

Figure 9.33 Values of σ^2, corresponding to values of $\bar{p}$. Adapted with permission from "ASTM Manual on presentation of Data and Control Chart Analysis" (1976), American Society for Testing and Materials, p. 162.

As we can see, by changing the $\bar{p}$ value from 0.5 to 0.35 (leaving the other conditions the same), the change in the sample size is not significant (from $n_1 = 3393$ to $n_2 = 3087$). To continue the illustration of changing the value of $\bar{p}$, let's assume that the expected process average is 0.20. In this case, the sample size would be

$$n = \left(\frac{Z}{E}\right)^2 \bar{p}(1-\bar{p})$$

$$= \left(\frac{2.33}{0.02}\right)^2 (0.2)(0.8)$$

$$= (13{,}572)(0.16)$$

$$= 2171.52$$

The sample size dropped significantly. This is reflected in Figure 9.33. If $\bar{p}$ is likely to lie between 0.35 and 0.65, the accurate estimation of the sample size does not require a precise estimate of $\bar{p}$. If, however, $\bar{p}$ is near 0% or 100%, the accurate determination of n requires a close guess about the value of $\bar{p}$.

9.11.3 What if We Cannot Afford the Calculated Sample Size?

If the calculated sample size is too large and it is not affordable, there is always a possibility of reducing the sample size by manipulating the probability factor (Z) with the precision factor (E) using the following formulas.

For attribute data,

$$E = \sqrt{\frac{Z^2 \bar{p}(1-\bar{p})}{n}} \qquad Z = \sqrt{\frac{nE^2}{\bar{p}(1-\bar{p})}}$$

For variable data,

$$E = \sqrt{\frac{Z^2 \sigma^2}{n}} \qquad Z = \sqrt{\frac{nE^2}{\sigma^2}}$$

We can determine what precision is possible from an affordable sample size of n.

In an illustrated example using attribute data, we assume that we can only afford a sample size of 1000 and the estimated average fraction defective rate is $\bar{p} = 0.01$. If we want a confidence level of 95%, then the error (E) of estimating the process average by a sample of 1000 units will be

$$E = \sqrt{\frac{Z^2 \bar{p}(1-\bar{p})}{n}} = \sqrt{\frac{(1.96)^2(0.01)(1-0.01)}{1000}} = 0.006$$

Assuming that the process is very critical, this error cannot be accepted and we need to at least have an error of $E = 0.004$, then the factor Z for the confidence level will be

$$Z = \sqrt{\frac{nE^2}{\bar{p}(1-\bar{p})}} = \sqrt{\frac{1000(0.004)^2}{0.01(1-0.01)}} = 1.27$$

From Table 9.18, we can see that the confidence level will be around 80%.

The same analogy can be applied to variable data. Suppose from previous data we estimated that standard deviation of the process as $\sigma = 3.5$, and we can only afford a sample of 50 units. If we want to estimate the process average with a confidence level of 99%, then the error (E) will be

$$E = \sqrt{\frac{Z^2 \sigma^2}{n}} = \sqrt{\frac{(2.58)^2(3.5)^2}{50}} = 1.27$$

Suppose that this error is not acceptable, we'll need the error to be at least 1.0; then the Z factor for the confidence level will be

$$Z = \sqrt{\frac{nE^2}{\sigma^2}} = \sqrt{\frac{50(1.0)^2}{(3.5)^2}} = 2.03$$

Consulting Table 9.18, we find the confidence level is around 95.5%.

REFERENCES

1. J. M. Juran and Frank M. Gryna, "Juran's Quality Control Handbook," 4th ed. (1988), McGraw-Hill Book Co., p. 16.17.

2. J. M. Juran and Frank M. Gryna, "Juran's Quality Control Handbook," 3rd ed. (1980), McGraw-Hill Book Co., p. 9.22.

3. J. M. Juran and Frank M. Gryna, "Quality Planning and Analysis," 2nd ed. (1980), McGraw-Hill, Book Co., p. 285.

4. J. M. Juran and Frank M. Gryna, "Juran's Quality Control Handbook," 3rd ed. (1980), McGraw-Hill Book Co., p. 9.22.

5. Yutaka Osuga, "Process Capability Studies in Cutting Processes," Reports of Statistical Applications Research, Japanese Union on Scientists and Engineers, Vol. 11, No. 1 (1964), pp. 23–35.

6. L. P. Sullivan, "Reducing Variability: A New Approach to Quality," Quality Progress (July 1984), p. 15.

7. Victor E. Kane, "Process Capability Indexes," Journal of Quality Technology, (January 1986), p. 41.

8. J. M. Juran and Frank M. Gryna, "Juran's Quality Control Handbook," 4th ed. (1988), McGraw-Hill Book Co., p. 16.23.

9. J. M. Juran, "Statistical Quality Control Handbook," 4th ed. (1988), McGraw-Hill Book Co., pp. 16–25.

10. C. C. Mentch, "Manufacturing Process Quality Optimization Studies," Journal of Quality Technology, Vol. 12, No. 3 (July 1980), p. 120.

11. Shigeru Mizuno, "Company-Wide Total Quality Control" (1988), Asian Productivity Organization, pp. 218–220.

12. "AT&T Statistical Quality Handbook" (1984), AT&T Technologies, Inc., p. 34.

13. Walter A. Shewhart, "Statistical Methods from the Viewpoint of Quality Control" (1986), Dover Publications, Inc., p. 113.

14. "AT&T Statistical Quality Handbook" (1984), AT&T Technologies, Inc., p. 60.

15. W. Edwards Deming, "Sample Design in Business Research" (1960) John Wiley & Sons, Inc., p. 260, Figure 16.

16. "ASTM Manual on Presentation of Data and Control Chart Analysis" (1976), American Society for Testing and Material, p. 162.

Chapter 10

Working with Skewed Distributions

Most of the statistical methods and techniques applied in the industry are based on the assumption that the characteristics under investigation follow some kind of formal probability distribution such as normal, binomial, or Poisson. The effectiveness and accuracy of the conclusions and decisions made based on the application of statistical principles depend mainly on how true the distribution assumption is.

For example, based on the Cpk value, we predict the process output. This prediction is based on the assumption that the process has a normal distribution. If the assumption is wrong, the prediction is also wrong. To make the right prediction or conclusion from a process under statistical investigation, we need to know the nature of the distribution of this process so we can select the right statistical tools.

Most of the time we use a sample selected from the process to determine the shape of the process distribution. But it is not always possible to determine the form of the process distribution from a sample. This is especially difficult when the sample size is limited. To resolve this problem, a number of techniques are available to test the distribution assumption. One such technique for testing the distribution assumption is the normal probability plot that was described in Section 9.6.4.

Testing the distribution assumption is a very important attribute of the correct decision-making process. However, most of the time in real life, we trust the distribution assumption to our intuition.

Most of the existing statistical methods and techniques used in the industry are based on the assumption that the variables under study follow a known probability distribution. The results and conclusions from the study are, of course, valid *only* to the extent that the distribution assumption is correct. For example, when a process is brought into a state of statistical control (which means that all special causes of variation are eliminated and the remaining process variation is only a result of common causes), we can expect that this process will have practically a normal distribution. So all the statistical tools and procedures to control and study this process are based on the assumption of normality.

However, there are some processes which have a nonnormal or skewed distribution due to their nature. For example, measurements of the thickness from a coating process or moisture content after a sealing process can have a skewed distribution. To apply the methods which are based on the assumption of normality to investigate processes which are skewed due to their nature, we need to apply some techniques that will allow us to transform the data from a skewed distribution to a normal distribution. But, before describing methods of transformation, let's become more familiar with skewed distributions.

10.1 SKEWNESS

Skewness is a measure of asymmetry. It indicates the extent and direction of lopsidedness in a distribution. Let's use Figure 10.1 to visualize the difference between symmetrical and skewed distributions.

1. Figure 10.1(a) is a symmetrical (or nonskewed) distribution.
2. Figure 10.1(b) is a positively skewed distribution. The "tail" of the distribution extends toward the larger values which indicates that there are a few relatively large values compared to the other measurements in the distribution.
3. Figure 10.1(c) represents a negatively skewed distribution. The "tail" of the distribution extends toward the smaller values which indicate that there are a few relatively small values compared to the other measurements in the distribution.

If a distribution is perfectly symmetrical, the measure of skewness will be equal to zero. The mean, mode, and median will also have the same value. For a moderately skewed distribution it has been found that the distance between the mean and the median is one-third of the distance between the mean and the mode.[1]

10.1.1 Two Types of Skewness

Figure 10.2 is a histogram made from a sample of 75 units taken from a gold-plating process. It shows that the distribution is skewed to the right-hand side. Should we conclude that the process under investigation has a skewed distribution by its nature? Certainly not. In many processes the degree of

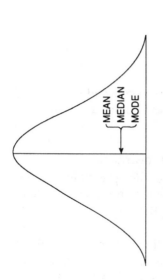

(a) WITH A SYMMETRICAL DISTRIBUTION, THE MODE, MEDIAN, AND MEAN WILL ALL BE THE SAME.

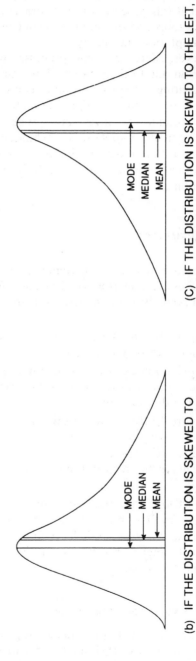

(b) IF THE DISTRIBUTION IS SKEWED TO THE RIGHT, THE MEAN WILL BE TO THE RIGHT OF THE MODE.

(c) IF THE DISTRIBUTION IS SKEWED TO THE LEFT, THE MEAN WILL BE TO THE LEFT OF THE MODE.

Figure 10.1 Shapes of frequency curves.

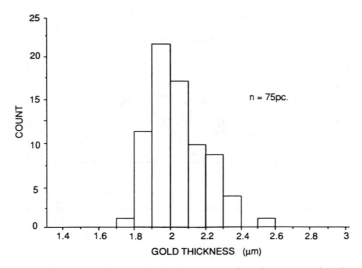

Figure 10.2 A histogram for a gold-plating process based on a sample of 75 units.

skewness is subject to change, and it does not belong to the nature of the process. For example, the skewness may be just a result of sampling variation.

Figure 10.3 is a histogram from the same process. The only difference is that the sample size has been increased to 600 units. Now the histogram shows that the process has a normal distribution. This example demonstrates the importance of the sample size to make the correct conclusion about the process normality.

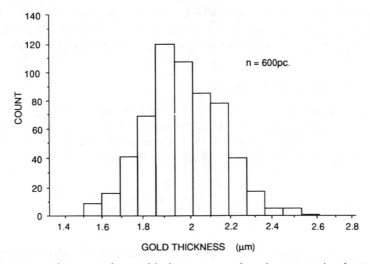

Figure 10.3 A histogram for a gold-plating process based on a sample of 600 units.

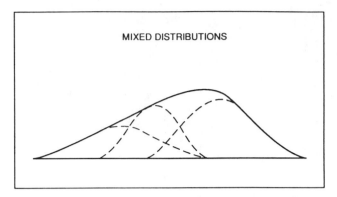

Figure 10.4   An example of a nonpermanent skewed distribution.

When skewness is associated with "out of control" on a control chart, it is likely to be the result of a mixture of two or more distributions (e.g., different suppliers or different measurements). This case belongs to the nonpermanent type of skewness (see Figure 10.4). In this case, engineering should determine the reasons for forming the distribution instead of treating the process as a population with an inherently skewed distribution.

Occasionally skewness may be a result of "freaks" which cause a long tail to spread out on one side. In such cases, before deciding whether to treat the distribution as a skewed distribution, we must decide whether to consider all observations as a regular part of the process.

In this section we are concerned with a type of permanent skewness which is a result of the operation of some strong fundamental factors in the process; factors which are impossible, uneconomical, or very difficult to identify and eliminate. In other words, factors which make the process "naturally" skewed. This may occur, for example, in coating, drilling, molding, and other processes which by their nature will form a nonsymmetrical distribution.

10.1.2   Measuring the Skewness

A measure of the amount of skewness in a population is given by the average value of $(x - \mu)^3$ taken over the population, where μ is the mean of the population and x is the individual measurements. This quantity is called the third moment about the mean (M_3) and can be derived by

$$M_3 = \frac{\Sigma (x - \mu_x)^3}{N}$$

To render this measurement independent of the scale on which the data are recorded, it is divided by σ^3. The resulting coefficient of skewness is denoted by $\sqrt{B_1}$ (or sometimes by γ_1).

The sample estimate of the coefficient is denoted by $\sqrt{b_1}$ (or sometimes by g_1). To determine the skewness from a sample, we compute

$$m_3 = \frac{\Sigma(x - \bar{x})^3}{n}$$

$$m_2 = s^2 = \frac{\Sigma(x - \bar{x})^2}{n}$$

and take

$$\sqrt{b_1} = g_1 = \frac{m_3}{s^3} = \frac{m_3}{m_2\sqrt{m_2}}$$

where $\bar{x}$ is the mean of the sample and s^2 is the variance of the sample. *Note:* To make the calculation of computing the sample variance (s^2 or m_2) slightly easier, we divide by n instead of the customary ($n - 1$).

Let's illustrate the computation with an improvised example. Assume that we are interested in the distribution of a stud pull measurement, typically used to measure bonding strength, and we want to determine its skewness. A random sample of 5 parts (usually a sample of at least 50 measurements is required) is taken from the process and tabulated in Table 10.1(a). The deviations between each measurement and the mean ($x - \bar{x}$) and its square, cube, and fourth root are computed. From Table 10.1 we can conclude that if the positive deviations are larger than the negative then $\Sigma(x - \bar{x})^3$ will be positive, and if the negative deviations are larger than the positive, then the $\Sigma(x - \bar{x})^3$ will be negative. This means that the net sum of the cubed deviations column automatically generates the correct *sign*. At the same time, the more extreme the deviations, the greater the amount by which the positive subtotal exceeds the negative subtotal (or vice versa depending on the direction in which the extreme deviations occur). This means that by evaluating the magnitude of the net total of the $(x - \bar{x})^3$ column, a determination of the *extent of skewness* can be obtained.

To illustrate this point, let's assume that we have selected another sample from a gold-plating process, and the results are as in Table 10.1(b). Observing the results, one can see how the value of m_3 changes when the deviations are more extreme. Because of the change in measurement 5 (from 10 to 20), the m_3 value changed from 12 to 324 [see Table 10.1(b)]. However, m_3 (the average amount of cubed deviation from the mean) has two major disadvantages when expressed in its existing form.

TABLE 10.1 A Comparison of Two Skewed Distributions for Different Processes

(a)				
x (kg)	$x - \bar{x}$	$(x - \bar{x})^2$	$(x - \bar{x})^3$	$(x - \bar{x})^4$
5	-1	$+1$	-1	$+1$
5	-1	$+1$	-1	$+1$
5	-1	$+1$	-1	$+1$
5	-1	$+1$	-1	$+1$
10	$+4$	$+16$	$+64$	$+256$
30	0	$+20$	$+60$	$+260$

$$\bar{x} = \frac{\Sigma x}{n} = \frac{30}{5} = 6 \qquad m_3 = \frac{\Sigma(x - \bar{x})^3}{n} = \frac{60}{5} = 12 \text{ kg}$$

$$m_2 = \frac{\Sigma(x - \bar{x})^2}{n} = \frac{20}{5} = 4 \qquad m_4 = \frac{\Sigma(x - \bar{x})^4}{n} = \frac{260}{5} = 52$$

(b)				
$x(\mu'')$	$x - \bar{x}$	$(x - \bar{x})^2$	$(x - \bar{x})^3$	$(x - \bar{x})^4$
5	-3	$+9$	-27	$+81$
5	-3	$+9$	-27	$+81$
5	-3	$+9$	-27	$+81$
5	-3	$+9$	-27	$+81$
20	$+12$	$+144$	$+1728$	$+20,736$
40	0	$+180$	$+1620$	$+21,060$

$$\bar{x} = \frac{\Sigma x}{n} = \frac{40}{5} = 8 \qquad m_3 = \frac{\Sigma(x - \bar{x})^3}{n} = \frac{1620}{5} = 324 \ \mu''$$

$$m_2 = \frac{\Sigma(x - \bar{x})^2}{n} = \frac{180}{5} = 36 \qquad m_4 = \frac{\Sigma(x - \bar{x})^4}{n} = \frac{21,060}{5} = 4212$$

1. We cannot compare the degree of skewness between two different distributions if they are expressed in different units.
2. If two distributions have the same shape but different values of the variation, the m_3 value will be different (the larger the variance, the larger the m_3 value). To remove the negative effects mentioned above, the third moment is standardized by dividing m_3 by s^3. So the resultant standardized third moment $(\sqrt{b_1})$ is referred to as

$$\sqrt{b_1} = \frac{m_3}{s^3} = \frac{m_3}{m_2\sqrt{m_2}}$$

This $\sqrt{b_1}$ is the average of the cubed standardized deviations from the mean.

Now referring again to Tables 10.1(a) and (b), we calculate the skewness as follows. For example 1 [Table 10.1(a)]:

$$\sqrt{b_1} = \frac{m_3}{m_2\sqrt{m_2}} = \frac{12}{4\sqrt{4}} = 1.5$$

For example 2 [Table 10.1(b)]:

$$\sqrt{b_1} = \frac{m_3}{m_2\sqrt{m_2}} = \frac{324}{36\sqrt{36}} = 1.5$$

From this comparison, one can see that even though the values of m_3 and m_2 for Tables 10.1(a) and (b) are different, the $\sqrt{b_1}$ values are the same because the skewness of the two distributions is the same. The standardized formula allows us to obtain values of $\sqrt{b_1}$ (skewness) that are directly comparable between different distributions regardless of the units of measurement or size of the variations involved.

10.1.3 Interpretation of the Value of Skewness $\sqrt{b_1}$

To interpret the values of skewness ($\sqrt{b_1}$), use Appendix D(a)[2], which gives the one-tailed 5% and 1% significance levels of $\sqrt{b_1}$ for sample sizes between 25 and 500. For example, if the sample size $n = 50$ and skewness $\sqrt{b_1} = 0.9$, in Appendix D(a) we find that the critical value for $n = 50$ and the $\alpha = 1\%$ significance level is 0.787. Since skewness $\sqrt{b_1} = 0.9$ is greater than the critical value, it is significant at the 99% confidence level ($\alpha = 1\%$).

10.2 KURTOSIS AS A DEPARTURE FROM NORMALITY

So far we were concerned with the measure of the distribution symmetry as it is one of the most important parameters of the distribution shape. However, not all symmetrical distributions should be considered as normal distributions. In Figure 10.5 there are two examples of symmetrical distributions that are not normal. Distribution A which is flatter than the normal curve, is called a "platykurtic" distribution. Distribution B which is more peaked than the normal curve, is called a "leptokurtic" distribution. The amount of flatness (or kurtosis) can be measured by a factor known as β_2.

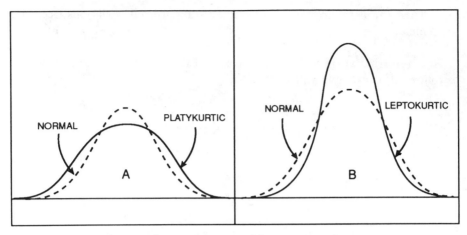

Figure 10.5   Distributions that are symmetrical but not normal.
Adapted with permission from "AT&T Statistical Quality Handbook" (1984), AT&T
Technologies, Inc., p. 134.[3]

10.2.1   Measuring the Kurtosis

A measure of kurtosis in a population is the average value of $(x - u)^4$ divided
by σ^4. For a normal distribution, this ratio has a value of 3. If the ratio
denoted by β_2 exceeds 3, the distribution is longer tailed than a normal
distribution with the same σ. In Figure 10.6 two probability densities are
shown: the uniform or rectangular distribution and the bell-shaped curve that
represents the normal distribution. The values of β_2 for these two distributions are 1.8 and 3.0, respectively. The normal distribution is frequently used
as a standard against which the peakedness of other distributions is compared.

A sample estimate of the amount of kurtosis is given by the following
formula:

$$b_2 = \frac{m_4}{s^4} = \frac{m_4}{m_2^2}$$

where $m_4 = \Sigma(x - \bar{x})^4/n$ is the fourth moment of the sample about the
mean, and $m_2 = s^2$ is the sample variance. Sometimes a value of $g_2 = b_2 - 3$
is used to represent the kurtosis. Notice that the normal distribution value of
3 has been subtracted, with the result that long-tailed distributions show
positive kurtosis and flat-topped distributions show negative kurtosis.

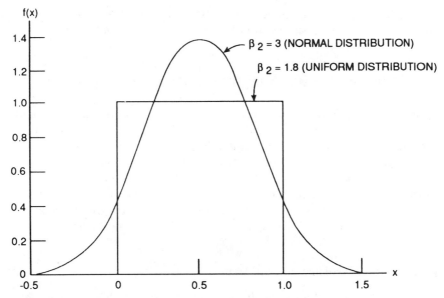

Figure 10.6 Comparison of normal and uniform probability density functions. Adapted with permission from Gerald J. Hahn and Samuel J. Shapiro, "Statistical Models in Engineering" (1967), John Wiley & Sons, Inc., p. 46.[4]

10.2.2   Interpretation of the Value of Kurtosis (b_2)

To interpret the value (b_2), we need to apply Appendix D(b). This appendix gives the critical values of b_2 at the $\alpha = 1\%$ and 5% significance levels for sample sizes between 50 and 2000. For $b_2 > 3$, the critical values under the column "Upper" should be used. If b_2 is greater than the critical value, it is significant. For $b_2 < 3$, the critical values under the column "Lower" should be used. If b_2 is less than the critical value, it is considered significant. As an illustration example, suppose from a sample of 50 units we obtain $m_4 = 52$, $m_2 = 4$, then

$$b_2 = \frac{m_4}{m_2^2} = \frac{52}{(4)^2} = 3.25$$

Because $b_2 = 3.25 > 3$, we compare it with the critical values in Appendix D(b) under the column "Upper." For $n = 50$, $\alpha = 5\%$, the critical value is 3.99. Since $b_2 = 3.25 < 3.99$, we can conclude that it is insignificant.

Now that we know how to measure the shape of the distribution, and we can recognize the difference between a permanent and nonpermanent skewness, we can continue our discussion of how to convert an inherent skewed distribution to a normal distribution by applying transformation principles.

10.3   ACHIEVING SYMMETRY BY TRANSFORMATION

In the preceding section, we discussed situations where the probability distribution of a variable is asymmetrical (skewed). And, at the same time, we know that most of the statistical methodology used in the industry is based on the assumption of normality, which requires the distribution to be symmetric.

To be able to apply statistical methods which are based on the assumption of normality, we can transform the original observations of a skewed distribution so that it will make the new measurements more symmetrically distributed. Described below are some methods of data transformation.

10.3.1   Before Transforming the Data

Don't rush to transform the data! In real life, few processes will strictly satisfy the assumption of normality. The practitioner often uses the term "approximately normal" because this is the best we can expect from a manufacturing process. But what is approximately normal? When we take a random sample, even from a "normal" process, some moderate deviations are to be expected because of sampling variation. To confirm that the sample represents the theoretical population assumption, we can use a normal probability plot (NPP), which is a good approximation for a test of normality. From this, we would calculate the average and standard deviation. Having these distribution parameters, we can calculate the process capability and its indexes. To decide if the data are approximately normally distributed, we can also use the results of the tests of skewness and kurtosis.

And, finally, before deciding to transform the original data, we need to make sure that the skewness is not a result of different kinds of mixture, freaks, unpredictable trends, and tool wear. We need to be sure that the skewness belongs to the nature of the process and can be explained by engineering reasons and accepted as an inherent, constant component of the process. The transformation procedure will make sense only after considering all the elements mentioned above.

10.3.2   Forms of Transformation

To get a better feeling for how transformation works, let's consider some contrived examples (see Figure 10.7). In each of the contrived examples, the original data are so absolutely skewed that there is no need to prove the skewness. However, when we transform the data by using the log and

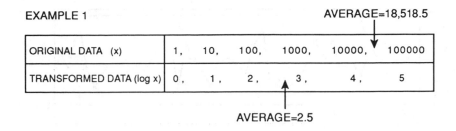

EXAMPLE 1   AVERAGE=18,518.5

| ORIGINAL DATA (x) | 1, | 10, | 100, | 1000, | 10000, | 100000 |
| TRANSFORMED DATA (log x) | 0, | 1, | 2, | 3, | 4, | 5 |

AVERAGE=2.5

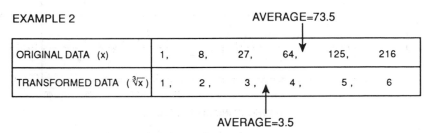

EXAMPLE 2   AVERAGE=73.5

| ORIGINAL DATA (x) | 1, | 8, | 27, | 64, | 125, | 216 |
| TRANSFORMED DATA ($\sqrt[3]{x}$) | 1, | 2, | 3, | 4, | 5, | 6 |

AVERAGE=3.5

Figure 10.7   Examples of skewed data.

cube-root transformations, the data become normalized and the average appears in the center position of the data.

It is understandable that these improvised examples do not reflect real-life situations, but they do illustrate the importance and power of transformation. So, how do we make a skewed distribution look normal in practice? It is difficult to give the exact principles of when to use what kind of transformation; however, some general guidelines for selecting an appropriate transformation can be offered.

First of all, a large amount of data is required before selecting the appropriate transformation. With a small sample of 20 or 30, it is difficult to determine the degree of asymmetry and to decide what form of transformation is needed. But if we know the true distribution of the data, we may utilize this information in choosing the correct form of transformation. If the data are positive and skewed to the right, the appropriate transformation would be $x^{1/2}$, $x^{1/3}$, ln x, or in general x^p, with $p < 1$. This will make the distribution more symmetrical. Note that the square-root transformation should be used with data that are not as skewed as data for which $x^{1/3}$ or ln x are used. When we have a situation in which the data are skewed to the left and positive, the appropriate transformations would be x^2, x^3, or x^p with $p > 1$. If the data follow the log-normal distribution, then the logarithmic transformation $x = \log x$ is appropriate. Below are some examples to illustrate the importance of selecting the right transformation.

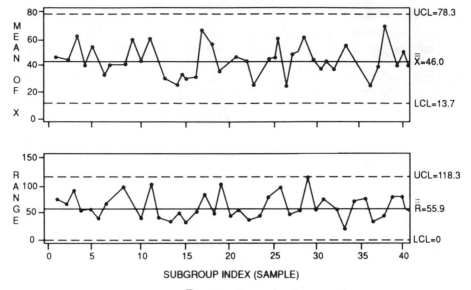

Figure 10.8 $\bar{X}$-R chart for stud pull test results.

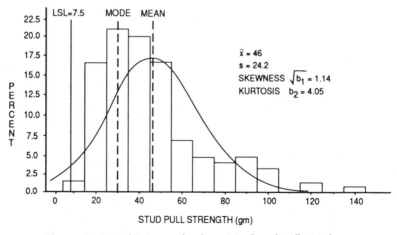

Figure 10.9   A histogram for the original stud pull test data.

Example 1   The process improvement team (PIT) wanted to evaluate the process capability of a die attach process based on the stud pull test. A random sample of 200 units was selected and tested. Using the data, an $\bar{X}$-R chart was plotted which showed that the process is in a state of statistical control (see Figure 10.8). A histogram was also completed (see Figure 10.9) but the results showed that the distribution is skewed with a skewness

Lower Class Limits (lb)	f	U	fU	U²	fU²	U³	fU³	U⁴	fU⁴
5	2	-2	-4	4	8	-8	-16	16	32
15	33	-1	-33	1	33	-1	-33	1	33
25	42	0	0	0	0	0	0	0	0
35	39	1	39	1	39	1	39	1	39
45	32	2	64	4	128	8	256	16	512
55	14	3	42	9	126	27	378	81	1134
65	10	4	40	16	160	64	640	256	2560
75	8	5	40	25	200	125	1000	625	5000
85	10	6	60	36	360	216	2160	1296	12960
95	7	7	49	49	343	343	2401	2401	16807
105	0	8	0	64	0	512	0	4096	0
115	2	9	18	81	162	729	1458	6561	13122
125	0	10	0	100	0	1000	0	10000	0
135	1	11	11	121	121	1331	1331	14641	14641

$n = 200$ $\sum fU = 326$ $\sum fU^2 = 1680$ $\sum fU^3 = 9614$ $\sum fU^4 = 66840$

Test of Skewness:

$$h_1 = \frac{\sum fU}{n} = \frac{326}{200} = 1.63$$

$$h_2 = \frac{\sum fU^2}{n} = \frac{1680}{200} = 8.4$$

$$h_3 = \frac{\sum fU^3}{n} = \frac{9614}{200} = 48.07$$

$$m_2 = h_2 - h_1^2 = 8.4 - (1.63)^2 = 5.74$$

$$m_3 = h_3 - 3h_1 h_2 + 2h_1^3 = 48.07 - 3(1.63)(8.4) + 2(1.63)^3 = 15.65$$

$$\sqrt{b_1} = \frac{m_3}{m_2\sqrt{m_2}} = \frac{15.65}{5.74\sqrt{5.74}} = 1.14$$

Test of Kurtosis:

$$h_4 = \frac{\sum fU^4}{n} = \frac{66840}{200} = 334.2$$

$$m_4 = h_4 - 4h_1 h_3 + 6h_1^2 h_2 - 3h_1^4 = 334.2 - 4(1.63)(48.07) + 6(1.63)^2(8.4) - 3(1.63)^4 = 133.51$$

$$b_2 = \frac{m_4}{m_2^2} = \frac{133.51}{(5.74)^2} = 4.05$$

Figure 10.10 Computation for the skewness and kurtosis of the original data of the stud pull test.

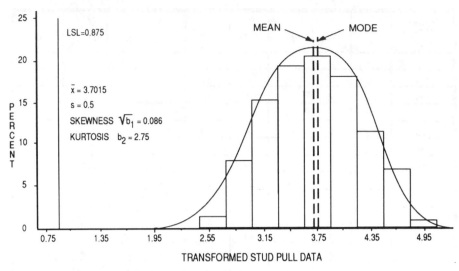

Figure 10.11   The histogram for the log-transformed stud pull test data.

$\sqrt{b_1} = 1.14$ and a kurtosis $b_2 = 4.05$. Figure 10.10^5 is the computation of the coefficient of skewness $(\sqrt{b_1})$ and kurtosis (b_2) based on grouped data. All the calculations are self-explanatory.

As seen in Appendix D(a), it was found that with a sample size $n = 200$ the critical value at the $\alpha = 1\%$ level for skewness is 0.403 and for kurtosis is 3.98. Since $\sqrt{b_1} = 1.14 > 0.403$ and $b_2 = 4.05 > 3.98$, the PIT concluded that the distribution is significantly skewed and not normal. Based on engineering experience, the skewness is considered inherent. In order to calculate the capability index, the data must be transformed to a normal distribution.

Since the distribution is skewed to the right-hand side and follows a log-normal distribution, it was decided to use a log transformation to normalize the data. Figure 10.11 shows the histogram after transformation. Applying the same procedure before transformation (see Figure 10.10), the coefficient of skewness $(\sqrt{b_1})$ for the transformed data is equal to 0.086. Since this value is much less than the critical value of 0.403, the skewness is considered insignificant. The kurtosis after transformation is 2.75, which is greater than the critical value 2.51 at the $\alpha = 5\%$ level, so it is also considered insignificant. In addition, a normal probability plot (NPP) was completed (see Figure 10.12) to test the normality of the data. Before transformation the NPP of the data shows a curve, whereas the NPP of the transformed data shows a reasonably straight line. Based on all of these results, the PIT concluded that the data after log transformation represents a normal distribution. So the calculation of the capability indexes from transformed data would be legitimate.

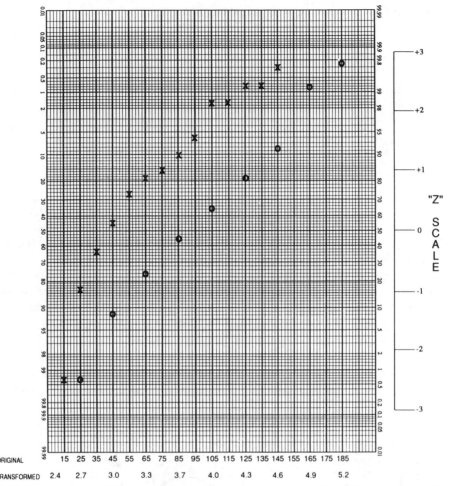

Figure 10.12 The normal probability plots for the original and transformed stud pull test results ($\times$ =original, $\bigcirc$ =transformed).

Table 10.2 shows a comparison of the capability index (Cpk) before and after the transformation. The low Cpk value before transformation is due to the skewness. The Cpk after transformation represents the true capability of the process.

Example 2 The die shear test data of a sample of 150 units from the die attach process was analyzed. The histogram shows a skewness to the left-hand side (see Figure 10.13). Applying the procedure to calculate the skewness and kurtosis, we derived $\sqrt{b_1} = -0.69$ and $b_2 = 3.05$. The kurtosis is very close to

TABLE 10.2   A Comparison of the Capability Index (Cpk) Before and After the Transformation for Stud Pull Test Results

Before	After
LSL = 7.5 lb	LSL = 0.875
$\bar{X} = 46$	$\bar{X} = 3.7$
$s = 24.2$	$s = 0.5$
$\text{Cpk} = \dfrac{\bar{X}-\text{LSL}}{3s} = \dfrac{46-7.5}{3(24.2)} = 0.53$	$\text{Cpk} = \dfrac{\bar{X}-\text{LSL}}{3s} = \dfrac{3.7-0.875}{3(0.5)} = 1.88$

the normal distribution value of 3, but for the skewness [from Appendix D(a)] we find for $n = 150$ the critical value = 0.464 at the $\alpha = 1\%$ significance level. Since $|\sqrt{b_1}| = 0.69 > 0.464$, the skewness is significant. Based on engineering judgement, the skewness is considered inherent.

Considering that the distribution is skewed to the left-hand side, it is decided to perform an x^2 transformation. The histogram for the transformed data is shown in Figure 10.14. The skewness $(\sqrt{b_1}) = -0.09$, and the kurtosis $(b_2) = 2.98$. Since $|\sqrt{b_1}|$ is smaller than the critical value (0.464), the skewness is insignificant at the $\alpha = 1\%$ level. Also the kurtosis (b_2) is very close to normal distribution. Based on this information, the transformed data can be considered to represent a normal distribution. The capability indexes are calculated for the original data and transformed data (see Table 10.3). Before

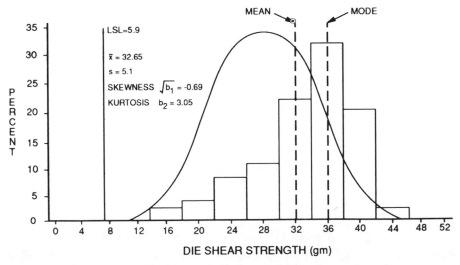

Figure 10.13   The histogram for the original die shear test results.

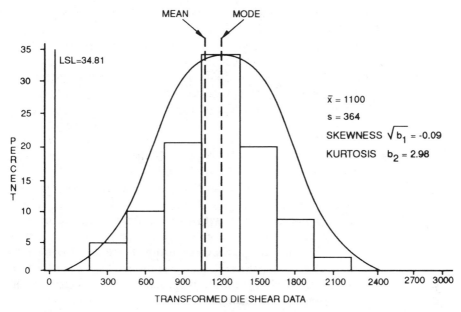

Figure 10.14   The histogram for the transformed die shear test results.

transformation the Cpk is better, but this is due to skewness. After transformation the Cpk represents the true process capability.

Case History 10.1: Process Performance Study—Water Vapor Content in a Device

In the electronics industry, the water vapor content of the atmosphere inside a hermetically sealed device is a very important element of quality and reliability. According to the customer specification requirements, a device may contain moisture after sealing no more than 5000 ppmv (parts per million in volume).

TABLE 10.3   A Comparison of the Capability Index (Cpk) Before and After Transformation for Die Shear Test Results

Before	After
LSL = 5.9	LSL = 34.81
$\overline{X} = 32.65$	$\overline{X} = 1100$
$s = 5.1$	$s = 364$
$\text{Cpk} = \dfrac{\overline{X} - \text{LSL}}{3s} = \dfrac{32.65 - 5.9}{3(5.1)} = 1.75$	$\text{Cpk} = \dfrac{\overline{X} - \text{LSL}}{3s} = \dfrac{1100 - 34.81}{3(364)} = 0.98$

Water vapor content depends on a number of variables, such as piece parts quality, baking and sealing processes, etc. All these variables are continuously controlled at various stages in the assembly process. However, in addition to this, a periodical final check of the finished product quality is required to the customer. This is done by taking three to five devices from every package type, cavity size, lead count, type of material, etc., and performing a destructive test on a periodic basis.

Since the testing method is destructive, complicated, and expensive, it was decided to perform a process performance study to see if there is any possibility of minimizing the testing program. If the results of the study show that the process is capable of meeting the specifications (meaning that any shift of the process can produce nonconforming products), it is out of question to reduce the testing program. However, if the results show that we have a stable process with low variability and the process average is far away (let's say 5 to 6 sigma) from the specification, it is possible to discuss reducing the amount of product testing. Taking into consideration that the sealing process is based on two different methods, it was decided to conduct the study separately for each method. Here we will describe only the results of the glass sealing study.

Historical data for a six-month period were analyzed to see if the data could be used as a substitute for a capability study. This can be a good

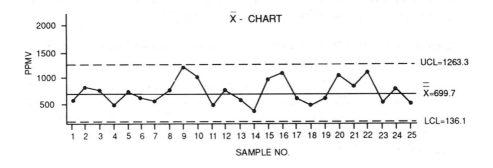

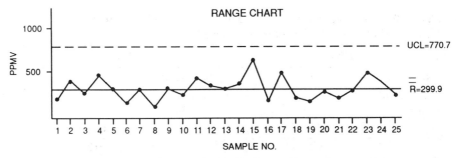

Figure 10.15 $\bar{X}$-R chart for the water vapor content process.

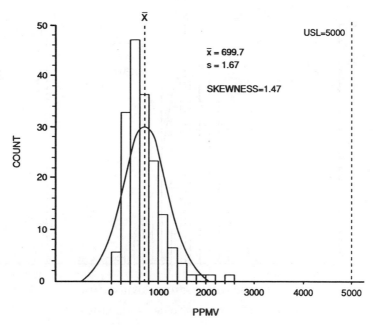

Figure 10.16 A histogram for the original data.

reason to revise the testing procedure. The $\bar{X}$-R control charts for a six-month period show that the water vapor content in the process is reasonably stable (see Figure 10.15). To determine the process capability, in addition to the control chart a histogram was made (see Figure 10.16). The results show that for the six-month period no parts were produced outside the specification. However, the measurements have a skewness equal to 1.47.

Based on engineering experience, it was concluded that the skewness could be considered an inherent characteristic of the process, and that there was no technical or economical reason to try to determine and eliminate the causes of skewness.

However, considering that the method of calculating capability indexes is based on the assumption of normality, it was decided to transform the original data before calculating the indexes to receive a normal distribution. In this case, the appropriate transformation would be $\sqrt[3]{x}$. Figure 10.17 represents the results after transformation. This shows that the distribution is now reasonably normal. The coefficient of skewness $\sqrt{b_1}$ is 0.11. In Appendix D(a) we find for $n = 175$, the critical value is 0.430 at the $\alpha = 1\%$ level. Since $\sqrt{b_1} < 0.430$, we can conclude that the skewness is not significant.

Table 10.4 is a comparison of the capability indexes before and after transformation. You can see that if we ignore the skewness and calculate

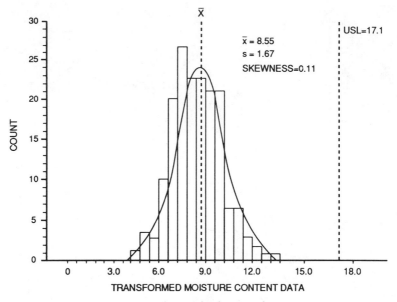

Figure 10.17 The result after transformation.

the capability index without transformation, the capability would be even higher than after transformation, but this would not be a valid conclusion about the process. The capability index after transformation is lower but the calculations are taken from an approximately normal distribution, so the conclusion about the process is more valid.

A Cpk of 1.7 indicates the process can be considered very close to six sigma quality (see Chapter 12), and because of this a proposal to reduce

TABLE 10.4 A Comparison of Capability Indexes Before and After Transformation

Before	After
USL = 5000	USL = 17.1
$\bar{X} = 699.7$	$\bar{X} = 8.55$
$s = 454.7$	$s = 1.67$
$\text{Cpk} = \dfrac{\text{USL} - \bar{X}}{3s}$	$\text{Cpk} = \dfrac{\text{USL} - \bar{X}}{3s}$
$= \dfrac{5000 - 699.7}{3(454.7)} = 3.15$	$= \dfrac{17.1 - 8.55}{3(1.67)} = 1.71$

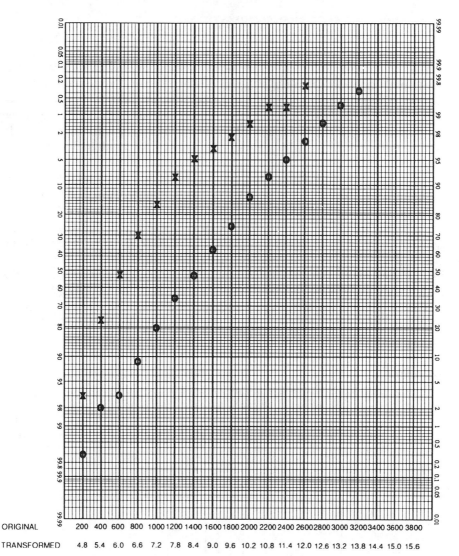

Figure 10.18 Normal probability plots for the data before and after transformation (X =original, O =transformed).

the destructive testing to control the moisture content can be submitted. To demonstrate the effect of transformation and perform a test for normality, in addition to this study a normal probability plot (NPP) was plotted to see how the distributions before and after transformation fit a straight line (see Figure 10.18). The NPP shows that after transformation the data lie in a reasonably straight line. In contrast, before transformation the line gave strong evidence of asymmetry.

10.4   MIRROR IMAGE TRANSFORMATION

General Motors[6] recommends a very interesting form of transformation called the "mirror image method (MIM)." This method is widely used by AMD practitioners because of its simplicity and because after transformation the measurement units remain the same. To perform the mirror image transformation, software has been developed so that all the calculations can be performed by the computer. However, to understand the principles behind this method, we will demonstrate the manual version of the calculation.

To illustrate how the mirror image transformation works, we will use the stud pull test results from a die attach process as an example. (*Note:* The data for this example are coded.)

From the histogram (Figure 10.19), one can see that the distribution is skewed to the right side. This is a "normal" situation for the existing die attach process quality. The process has an inherent skewed distribution.

The die attach process has a unilateral tolerance—lower specification limit, which means that the capability index is Cpl. If we would calculate the Cpl without using mirror image transformation the Cpl would be 1.00 (see Figure 10.19). This is a misleading result because of the skewness of the distribution.

Procedure
As stated in the introduction, the mirror image method transforms a nonsymmetrical into a symmetrical distribution. This is accomplished by generating a new distribution about the zero original class interval (mode) and then calculating the mean and standard deviation in the usual manner. Having a

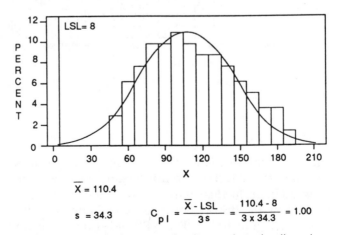

$$\overline{X} = 110.4$$

$$s = 34.3 \qquad C_{pl} = \frac{\overline{X} - LSL}{3s} = \frac{110.4 - 8}{3 \times 34.3} = 1.00$$

Figure 10.19   The histogram for die attach stud pull results.

transformed symmetrical distribution, we calculate a new Cpl (or Cpu) by using the new mean and standard deviation.

Following are the steps of the mirror image transformation (MIT) procedure.

Step 1 Determine the sample size and the number of classes needed to construct a histogram.

A sample of no less than $n = 50$ is recommended and the number of classes should be no less than 8.

A good rule of thumb is to let the number of classes be no less than $\sqrt{n}$ where n is the sample size.

Step 2 Select the class interval with the greatest frequency (mode class) and establish it as the zero origin class. In our example,

$$\text{Class interval } (i) = 10$$

$$\text{Mode } (A) = 100$$

(see Figure 10.20).

Original Data					
Midpoint Value	Tally	f	d	fd	$f(d^2)$
50	11	2	-5	-10	50
60	1111	4	-4	-16	64
70	11111	5	-3	-15	45
80	11111 11	7	-2	-14	28
90	11111 11	7	-1	-7	7
Mode (A) — 100	11111 111	8	0	0	0
110	11111 11	7	1	7	7
120	11111 1	6	2	12	24
130	11111 1	6	3	18	54
140	11111	5	4	20	80
150	1111	4	5	20	100
160	111	3	6	18	108
170	11	2	7	14	98
180	11	2	8	16	128
190	1	1	9	9	81
TOTAL		69		72	874

$$\overline{X} = A + \frac{i \Sigma fd}{n}$$

$$= 100 + \frac{(10)(72)}{69} = 110.4$$

$$Cpl = \frac{\overline{X} - LSL}{3s} = \frac{110.4 - 8}{3 \times 34.3} = 1.00$$

$$s = i \sqrt{\frac{\Sigma f (d^2) - (\Sigma fd)^2 / n}{n - 1}}$$

$$= 10 \sqrt{\frac{874 - (72)^2 / 69}{69 - 1}}$$

$$= 34.3$$

Figure 10.20 Original data.

Transformed Data After Mirror Image of the Lower Side Distribution

Midpoint Value	Tally	f	d	fd	$f(d^2)$
50	11	2	-5	-10	50
60	1111	4	-4	-16	64
70	11111	5	-3	-15	45
80	11111 11	7	-2	-14	28
90	11111 11	7	-1	-7	7
Mode (A) —— 100	11111 111	8	0	0	0
110	11111 11	7	1	7	7
120	11111 11	7	2	14	28
130	11111	5	3	15	45
140	1111	4	4	16	64
150	11	2	5	10	50
	TOTAL	58		0	388

$$\overline{X} = A + \frac{i \, \Sigma \, fd}{n}$$

$$= 100 + \frac{(10)\,(0)}{58} = 100$$

$$Cpl = \frac{\overline{X} - LSL}{3s} = \frac{100 - 8}{3 \times 26.09} = 1.18$$

$$s = i \sqrt{\frac{\Sigma f\,(d^2) - (\Sigma\,fd)^2 \, / n}{n - 1}}$$

$$= 10 \sqrt{\frac{388 - (0) \, / \, 58}{58 - 1}}$$

$$= 26.09$$

Figure 10.21   Transformed data after mirror image of the left-side distribution.

Step 3 Duplicate the cell frequencies, from the zero origin class in the skewness direction, on the nonskewed side of the zero class.

In our example, since we have a lower specification limit, we are interested in the left side of the distribution. Therefore, we duplicate the left side distribution. This results in a symmetrical frequency distribution. The right side of the distribution is replaced by the mirror image of the left side (see Figure 10.21).

Step 4 Calculate the mean ($\overline{X}$) and standard deviation for the new distribution(s) by the following formula:

$$\overline{X} = A + \frac{i\Sigma fd}{n}$$

$$s = i\sqrt{\frac{\Sigma f(d^2) - (\Sigma fd)^2 / n}{n - 1}}$$

where i is the distance between each midpoint and A is the mode.

In our example (see Figure 10.21), we have

$$\overline{X} = 100 + \frac{(10)(0)}{58} = 100$$

$$s_1 = 10\sqrt{\frac{388 - (0)^2/58}{58 - 1}} = 26.09$$

Step 5 Determine the capability index using the new standard deviation (s_1). In our example (see Figure 10.21), we have

$$\mathrm{Cpl} = \frac{\overline{X} - \mathrm{LSL}}{3s_1} = \frac{100 - 8}{3 \times 26.09} = 1.18$$

The capability index based on transformed data (Cpl = 1.18) represents a quality level of 200 ppm. On the other hand, the capability index based on the original skewed data (Cpl = 1.0) represents a quality level of 1350 ppm. As one can see, it is a big difference. Therefore, before calculation and interpretation of capability indexes, we need to be sure that the data are from a normal distribution.

10.5   APPLICATION OF THE SECOND APPLICATION[6]
TO THE NORMAL CURVE

One of the common theoretical distributions involving skewness is known as the "second approximation to the normal curve." Figure 10.22[7] shows distributions that are "second approximation" curves with $\sqrt{b_1} = +1$ and $\sqrt{b_1} = -1$, respectively. Tables 10.5 and 10.6 show the percentages of areas associ-

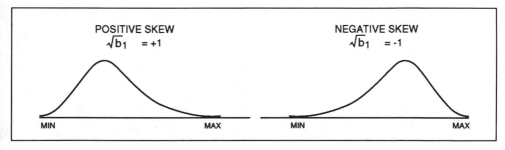

Figure 10.22   Second approximation curves.
Reprinted with permission from Western Electric Co., "Statistical Quality Control Handbook" (1984), AT&T Technologies, Inc., p. 134, Figure 26.[7]

TABLE 10.5 Second Approximation with $\sqrt{b_1} = +1$

	Percentage Outside of Maximum $t = \dfrac{\bar{X} - \text{USL}}{\sigma}$			Percentage Outside of Minimum $t = \dfrac{\text{LSL} - \bar{X}}{\sigma}$	
	(1)	(2)		(3)	(4)
t	If t Is Negative	If t Is Positive	t	If t Is Negative	If t Is Positive
0.0	43.3%	43.3%	0.0	56.7%	56.7%
0.1	39.4%	47.4%	0.1	52.6%	60.6%
0.2	35.8%	51.6%	0.2	48.4%	64.2%
0.3	32.5%	55.0%	0.3	44.6%	67.5%
0.4	29.3%	60.4%	0.4	39.6%	70.7%
0.5	26.4%	64.7%	0.5	35.3%	73.6%
0.6	23.8%	69.0%	0.6	31.0%	76.2%
0.7	21.5%	73.1%	0.7	26.9%	78.5%
0.8	19.4%	77.0%	0.8	23.0%	80.6%
0.9	17.4%	80.7%	0.9	19.3%	82.6%
1.0	15.8%	84.1%	1.0	15.0%	84.2%
1.1	14.3%	87.2%	1.1	12.8%	85.7%
1.2	12.9%	89.9%	1.2	10.1%	87.1%
1.3	11.6%	92.3%	1.3	7.7%	88.4%
1.4	10.4%	95.4%	1.4	5.6%	89.6%
1.5	9.3%	96.0%	1.5	4.0%	90.7%
1.6	8.3%	97.1%	1.6	2.9%	91.7%
1.7	7.4%	98.5%	1.7	1.5%	92.6%
1.8	6.5%	99.3%	1.8	0.7%	93.5%
1.9	5.7%	99.9%	1.9	0.5%	94.3%
2.0	4.9%	100.0%	2.0	0.0%	95.1%
2.1	4.2%	—	2.1	—	95.8%
2.2	3.6%	—	2.2	—	96.4%
2.3	3.1%	—	2.3	—	96.9%
2.4	2.6%	—	2.4	—	97.4%
2.5	2.1%	—	2.5	—	97.9%
2.6	1.7%	—	2.6	—	98.3%
2.7	1.4%	—	2.7	—	98.6%
2.8	1.1%	—	2.8	—	98.9%
2.9	0.9%	—	2.9	—	99.1%
3.0	0.7%	—	3.0	—	99.3%
3.1	0.5%	—	3.1	—	99.5%
3.2	0.4%	—	3.2	—	99.6%
3.3	0.3%	—	3.3	—	99.7%
3.4	0.2%	—	3.4	—	99.8%
3.5	0.2%	—	3.5	—	99.8%
3.6	0.1%	—	3.6	—	99.9%
3.7	0.07%	—	3.7	—	99.9%
3.8	0.03%	—	3.8	—	99.9%
3.9	0.01%	—	3.9	—	99.99%
4.0	0.00%	—	4.0	—	99.99%

Reprinted with permission from Western Electric Co., "Statistical Quality Control Handbook" (1984), AT&T Technologies, Inc., p. 135.[8]

TABLE 10.6   Second Approximation with $\sqrt{b_1} = -1$

	Percentage Outside of Maximum $t = \dfrac{\bar{X} - \text{USL}}{\sigma}$			Percentage Outside of Minimum $t = \dfrac{\text{LSL} - \bar{X}}{\sigma}$	
	(1)	(2)		(3)	(4)
t	If t Is Negative	If t Is Positive	t	If t Is Negative	If t Is Positive
0.0	56.7%	56.7%	0.0	43.3%	43.3%
0.1	52.6%	60.6%	0.1	39.4%	47.4%
0.2	48.4%	64.2%	0.2	35.8%	51.6%
0.3	44.6%	67.5%	0.3	32.5%	55.0%
0.4	39.6%	70.7%	0.4	29.3%	60.4%
0.5	35.3%	73.6%	0.5	26.4%	64.7%
0.6	31.0%	76.2%	0.6	23.8%	69.0%
0.7	26.9%	78.5%	0.7	21.5%	73.1%
0.8	23.0%	80.6%	0.8	19.4%	77.0%
0.9	19.3%	82.6%	0.9	17.4%	80.7%
1.0	15.9%	84.2%	1.0	15.8%	84.1%
1.1	12.8%	85.7%	1.1	14.3%	87.2%
1.2	10.1%	87.1%	1.2	12.9%	89.9%
1.3	7.7%	88.4%	1.3	11.6%	92.3%
1.4	5.6%	89.6%	1.4	10.4%	95.4%
1.5	4.0%	90.7%	1.5	9.3%	96.0%
1.6	2.9%	91.7%	1.6	8.3%	97.1%
1.7	1.5%	92.6%	1.7	7.4%	98.5%
1.8	0.7%	93.5%	1.8	6.5%	99.3%
1.9	0.5%	94.3%	1.9	5.7%	99.9%
2.0	0.0%	95.1%	2.0	4.9%	100.0%
2.1	—	95.8%	2.1	4.2%	—
2.2	—	96.4%	2.2	3.6%	—
2.3	—	96.9%	2.3	3.1%	—
2.4	—	97.4%	2.4	2.6%	—
2.5	—	97.9%	2.5	2.1%	—
2.6	—	98.3%	2.6	1.7%	—
2.7	—	98.6%	2.7	1.4%	—
2.8	—	98.9%	2.8	1.1%	—
2.9	—	99.1%	2.9	0.9%	—
3.0	—	99.3%	3.0	0.7%	—
3.1	—	99.5%	3.1	0.5%	—
3.2	—	99.6%	3.2	0.4%	—
3.3	—	99.7%	3.3	0.3%	—
3.4	—	99.8%	3.4	0.2%	—
3.5	—	99.9%	3.5	0.2%	—
3.6	—	99.9%	3.6	0.1%	—
3.7	—	99.9%	3.7	0.07%	—
3.8	—	99.9%	3.8	0.03%	—
3.9	—	99.99%	3.9	0.01%	—
4.0	—	99.99%	4.0	0.00%	—

Reprinted with permission from Western Electric Co., "Statistical Quality Control Handbook" (1984), AT&T Technologies, Inc., p. 136.[8]

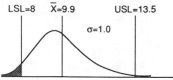

Figure 10.23   A distribution with skewness $\sqrt{b_1}$ =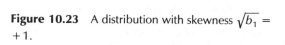
+ 1.

ated with these two curves. As in the case of normal distribution, these tables can be used for estimating the percentage outside the specification limits. The following example demonstrates the application of these tables. Given a process with a skewness of $\sqrt{b_1} = +1$, $\bar{X} = 9.9$, and $\sigma = 1.0$; LSL = 8 and USL = 13.5 (see Figure 10.23). Determine the ppm level for the process.

Step 1 For the right-hand side distribution:

$$t_U = \frac{\bar{X} - \text{USL}}{\sigma} = \frac{9.9 - 13.5}{1.0} = -3.6$$

For the left-hand side distribution:

$$t_L = \frac{\text{LSL} - \bar{X}}{\sigma} = \frac{8 - 9.9}{1.0} = -1.9$$

Step 2 Apply Table 10.5 to determine the percentage of nonconforming parts outside the specifications. The percentage is given opposite the value of t in the appropriate column, depending on whether the value of t is found to be negative or positive. In our particular case, for the right-hand side of the distribution $t_U = -3.6$; therefore, we use column (1), and find the percentage outside the USL (P_U) = 0.1%. For the left-hand side distribution, $t_L = -1.9$; therefore, we use column (3) and find the percentage outside the LSL (P_L) = 0.5%.

Step 3 Determine the overall percentage defective (P_T) outside the specification limits:

$$P_T = P_U + P_L = 0.1\% + 0.5\% = 0.6\%$$

Step 4 Transform the percentages to a ppm scale:

$$P_T = 0.6\% = \left(\frac{0.6}{100}\right) \times 1,000,000 = 6000 \text{ ppm}$$

When we have distribution skewed on the left-hand side ($\sqrt{b_1} = -1$), we proceed in the same way; the only difference is that we use Table 10.6.

REFERENCES

1. Herbert Arkin and Raymond R. Colton, "Statistical Methods," 5th ed. (1970), Harper & Row Publishing, Inc., p. 28.
2. E. S. Pearson and H. O. Hartley, "Biometrika Tables for Statisticians," Vol. 1 (1976), Biometrika Trust, pp. 207–208, Tables 34B and C.
3. Western Electric Co., "Statistical Quality Control Handbook" (1984), AT&T Technologies, Inc., p. 134, Figure 125.
4. Gerald J. Hahn and Samuel S. Shapiro, "Statistical Models in Engineering," 7th ed. (1967), John Wiley & Sons, Inc., p. 46.
5. George W. Snedecor and William G. Cochran, "Statistical Methods" (1980), The Iowa State University Press, pp. 78–79.
6. Harvey C. Charbonneau and Gordon L. Webster, "Industrial Quality Control" (1978), Prentice-Hall, Inc., pp. 45–48.
7. Western Electric Co., "Statistical Quality Control Handbook" (1984), AT&T Technologies, Inc., p. 134, Figure 126.
8. Western Electric Co., "Statistical Quality Control Handbook" (1984), AT&T Technologies, Inc., pp. 135–136.

Chapter **11**

Engineering Specifications

To determine the capability of a manufacturing process by using the formula mentioned earlier:

$$\text{Process capability} = 6\sigma$$

we don't need to know what the specification of the product is. This formula makes no reference to the specification parameters. Still, the major reason for quantifying process capability is not just to know the process spread, but also to be able to compute the ability of the process to maintain product tolerances. To do this, we compute the process capability ratio or other related indexes (Cp and Cpk) as described in Chapter 9.

A capability ratio (or index) would compare the distribution width with the engineering specification width:

$$\text{Cp} = \left(\frac{6\sigma \text{ variation}}{\text{total tolerance}} \right)$$

thereby providing us with a unitless number of comparative purposes. This means that a capability ratio (or index) has a value which is free of measurement scale units. Therefore, we can use these indexes for direct comparisons of the quality of one product characteristic to another, even though the characteristics may be completely different.

As we have seen in Chapter 9, a process capability ratio (or index) is always related to a design specification and a process standard deviation. Given this, the capability of a particular manufacturing process might be high when referenced to a certain tolerance, say T_1, but very low when compared to some other tolerance, let's say T_2. This means that just by opening the tolerance widths we can "improve" the capability index, which has nothing to do with process improvement.

So to make the capability index more informative, it is essential to better understand where the tolerance limits are coming from. This will be the discussion in this chapter.

11.1   THE ARBITRARY APPROACH OF TOLERANCING

In a meeting related to process improvement, the presenter was asked what was the base of establishing the specification limits. The answer was that the tolerances were developed a long time ago and they "stood the test of time." In his report, a number of capability indexes were less than one, and the reason could be due to unnecessarily tight tolerances, which is often the case in practice. In real life, tolerances are often established just by bargaining between the customer and supplier; without really knowing the customer's need and the supplier's capability.

In statistical literature it is often stated that the base for establishing tolerances should be the customer requirements, and this is absolutely right. But first of all what is the base for establishing the customer requirements? Second, if the supplier's process is not capable of meeting the customer's requirements, it simply will not meet them. So establishing tight tolerances by intuition will not increase the quality, it will only increase the cost. At the same time, setting loose tolerances can jeopardize product quality and decrease stability. So before calculating capability indexes, it is important to introduce scientific methods of tolerance design and establish the proper balance between the value of precision and the cost of that precision.

11.2   THE SCIENTIFIC APPROACH OF TOLERANCING

A number of effective statistical methods exist to establish the optimal tolerance limits. Some of the methods are more precise than others but most of them have their applicability in different situations and requirements. Taking into consideration that there are too many quality characteristics to be studied, and in real life most of the time we are unable to apply scientific methods of tolerancing for all quality characteristics, it is very important before deciding what method of tolerancing is to be used, to divide all parameters which need to be specified into three or four categories such as:

Critical ($\boxtimes$)
Major A ($\boxdot$)
Major B ($\circ$)
Minor (not marked)

Based on this classification, a plan of establishing or revising the specifications can be developed where one of the elements of the plan will be methods of statistical tools to be applied in specification development.

In general, the most important characteristic also calls for the greatest precision of tolerance development. In this section we will demonstrate the application of some of the methods and forms of determining statistical tolerance design.

11.2.1   Using the Normal Probability Plot (NPP) for Setting Specification Limits

At the process development stage when only laboratory data are available and we cannot afford to select large samples, but we want to design preliminary specification limits, a technique which is based on the application of the normal probability plot (the NPP method) is very useful (see Section 9.6.4 for more information on the procedure of applying NPP). Let's explain this technique by using an example.

Suppose you develop a new tin-plating process and you are interested in determining the lower specification limit of the tin thickness required to achieve the best quality and the highest yield. In the laboratory environment you receive only 10 measurements with the following results in microinches: 758, 680, 668, 732, 790, 774, 627, 728, 715, and 732. Based on this result, it was thought reasonable to plan a lower specification limit of 620 μ''. It was then decided to check this by performing a probability plot (see Figure 11.1).

Extrapolating the line to 620, the plot predicts that about 25% of the plated parts will fall below 620, even though all of the sample data exceed 620 μ''. Of course, if a larger sample size was available, more fundamental investigations could be made. But with such a small sample size the NPP method is a good tool for preliminary specification settings.

11.2.2   Determining Tolerance Limits Using the Z Table

When a process is in the stage of development and has not been introduced to the manufacturing floor, we don't really know its capability. To develop specifications, the designer would obtain a sample of data from the laboratory or pilot line to calculate the process limits and compare them with the specifications he has in mind. If he has "no idea" what the specifications should be, he would take the sample results as a base to calculate the specifications and compare them with the factual needs of the product or with the customer requirements. Statistically speaking, the problem is to predict the limits of variation of individual items in the population (process) based on sample data. This can be done in a number of ways, but we will

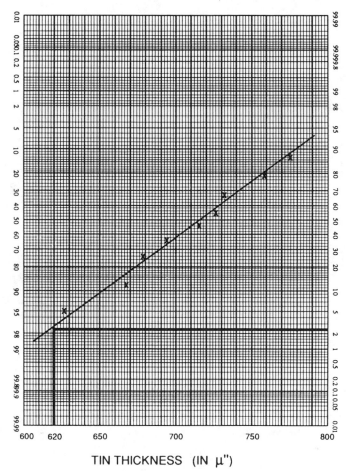

TIN THICKNESS (IN μ'')

Figure 11.1   Normal probability plot for a sample of 10 measurements from tin-plating process.

illustrate the simplest method based on the normal distribution table (Z table).

An Illustrative Example   From the pilot line of a new bonding process, an engineer selected a random sample of 50 units and measured the bond pull strength. using the data, he made a histogram and a normal probability plot (NPP) (see Figures 11.2 and 11.3). The distribution fell into a bell-shaped curve and the NPP fits reasonably into a straight line. Because of the product limits, the engineer couldn't make a control chart but from the results he assumes that he is dealing with a "normal" process.

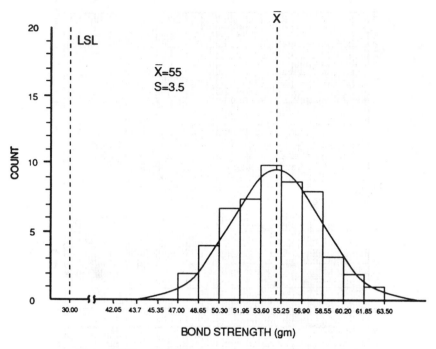

Figure 11.2   A histogram for the bond pull test results.

Based on the sample data, the process average ($\overline{X}$) and standard deviation (s) have been estimated as

$$\overline{X} = 55 \text{ gms}$$

$$s = 3.5 \text{ gms}$$

Using these results as estimates of the (unknown) population process average (μ) and standard deviation (σ) and using the normal distribution table (see Appendix C), we can calculate limits which will include any given percentage of the population. Expressed in an equation, the limit will be set at

$$\overline{X} \pm Z\left(\frac{\alpha}{2}\right)s$$

For example, if we wish to include $p = 99\%$ of the population, then we first locate $Z(\alpha/2)$:

$$\alpha = 1 - p = 1 - 0.99 = 0.01$$

$$\left(\frac{\alpha}{2}\right) = \left(\frac{0.01}{2}\right) = 5.000E - 03$$

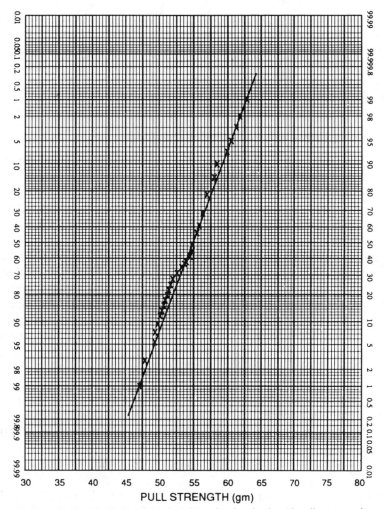

Figure 11.3   The normal probability plot for the bond pull test results.

From Appendix C, we find that

$$Z = 2.575$$

In our case, the limits should be set at

$$\overline{X} \pm 2.575s = 55 \pm 2.575(3.5) = \begin{cases} 64 \\ 46 \end{cases}$$

This means that 99% of the individual measurements in the population will have values between 64 and 46. But this conclusion is made on a number of assumptions that may or may not be true: First, the process is in a state of statistical control; second, the average and standard deviation of the sample will be equal to the population values; and third, the process is set on the nominal and no shifts are considered. If the assumptions are correct, we have developed realistic limits. (These limits are frequently called "natural tolerance limits.")

11.2.3 Using Linear Regression to Set Tolerances

"Regression analysis" is a statistical technique for estimating the parameters of an equation relating a particular variable to a set of one or more variables. The resulting equation is called a "regression equation." In this particular application (setting tolerances) we will use "simple linear regression" because it is the simplest equation to resolve a two-variable problem. To describe the procedure, let's use an improvised example.

Suppose a customer requires a solder-plating thickness in the range of 300 μ'' (microinches) to 1000 μ''. let's call them the "low" and "high" values and make a profile to satisfy the customer. The manufacturer wants to establish realistic tolerances on the current density in amperes per square foot (ASF) which he believes is the most influential variable on the solder-plating thickness. To do this, we use a simple regression analysis and proceed as follows.

Step 1 We select a range of values for the current density, x (predictor variable), run the process, and note the corresponding solder thickness value, y (predictand variable). Based on the results, we plot a graph of the data on a scatter diagram (see Figure 11.4). The convention is to plot the predictand variable (plating thickness) on the vertical axis and the predictor variable (current density) on the horizontal axis. Figure 11.4 suggests that plating thickness is linearly related to current density over the range of this experiment.

Step 2 At this stage, a regression line is plotted over the data. Roughly half the data points should be above the line and half below it. In addition, the line should pass exactly through the point $(\bar{x}, \bar{y})$.

Step 3 Now when the regression line is determined and plotted, we also locate the customer-specified "low" and "high" values for plating thickness on the vertical axis which are 300 μ'' and 1000 μ'', respectively, and extend them horizontally until the regression line is intersected (see Figure 11.4). Dropping two lines from these intersection points, the tolerance limits for the critical variable can then be read on the horizontal axis. In our case, we found the tolerance limits for current density are 50 and 155 ASF.

The horizontal intercept on the x axis between the two vertical lines determines the maximum tolerances permitted for the predictor variable in order to assure conformance to customer requirements. Assuming normality, this would assure roughly a $Cp = 1$ (2700 ppm). If we want to guarantee the customer a close to perfect product (3.4 ppm), the horizontal intercept should be divided into four equal parts and only the middle half should be

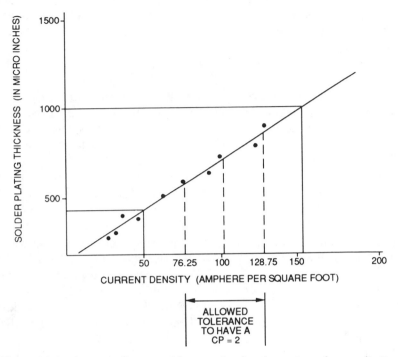

Figure 11.4   A scatter diagram with regression line for setting tolerance limits.

the preferred tolerance for the x parameter. In our case, $(155 - 50)/4 = 26.25$, then the middle half will be from $50 + 26.25 = 76.25$ to $155 - 26.25 = 128.75$. If we control the current density in this range, the solder-plating process would have a capability index (Cp) roughly equal to 2. Furthermore, the target value of the variable x should be at the center of the horizontal intercept; that is, the target value for the current density should be set at $(128.75 + 76.25)/2 = 102.5$ ASF. These new target values and tolerances calculated for the current density should be compared against the existing values and tolerances, and the necessary changes made to assure zero defects and 100% yields.

11.3   ESTABLISHING STATISTICAL TOLERANCE LIMITS

In the previous example, we used the formula:

$$\bar{X} \pm Z\left(\frac{\alpha}{2}\right)s$$

where $\bar{X}$ and s are estimates of the unknown population average (μ) and standard deviation (σ), and the Z value is taken from the Appendix C. Since

$\overline{X}$ and s are only estimates and not the true parameter values, we cannot say that the above interval always contains $100(1 - \alpha)\%$ of the distribution.

For example, if we use the formula:

$$\overline{X} \pm 3s$$

to determine the natural tolerance limits for a normal distribution variable, we cannot be certain that exactly 99.73% of the product from a process which is in a state of statistical control will fall within the limits calculated above. This approximation ignores the possible error in both the average and the standard deviation as estimated from the sample.

There is another method of establishing engineering tolerance limits by using *statistical tolerance limits*. The formula for this method is

$$\overline{X} \pm ks$$

where k is a factor obtained from a standard table [see Appendixes M(a) and M(b)][1]. This method is more precise because the numerical value of k is governed by the sample size employed, the confidence level chosen, and the proportion of the population to be included within the limits.

For example, suppose that a sample of 10 wires from a process yielded an average and standard deviation of 76.12 mils and 1.58 mils, respectively. The tolerance limits are to include $p = 99\%$ of the population, and the tolerance statement is to have a confidence level of $\gamma = 95\%$. Referring to Appendix M(b), the value of k is 4.43, and the tolerance limits are then calculated as

$$76.12 \pm 4.43(1.58) = \begin{cases} 83.12 \\ 69.12 \end{cases}$$

We can interpret the results as following: We are 95% confident that at least 99% of the wires in the population will have a length between 69.12 and 83.12 mils.

These type of limits are often called statistical tolerance limits which indicates that the estimation procedure is involved, utilizing sample data. The formal procedure to determine statistical tolerance interval can be simply summarized as follows.

Procedure

1. Obtain a sample of n observations.
2. Perform a rough test of normality by applying some simple statistical checks (e.g., histogram or normal probability plot), and adjust the sample data as necessary (e.g., remove outliner or perform a normalizing transformation).
3. From the data which show normality, obtain parameter estimates of $\overline{X}$ and s.
4. Select the desired confidence level (γ) and the desired population proportion (p) to be covered by the limits.

5. Obtain the appropriate k value from Appendix M(a) if the process has a one-sided specification limit, or from Appendix M(b) if there are two-sided specification limits.
6. Obtain the upper limit as $\overline{X} + ks$ and/or lower limit $\overline{X} - ks$.

Example 1

1. A sample of 60 bumps was taken from the bump fab process and the degree of hardness was measured for each bump.
2. A histogram and a normal probability plot were made to test for normality (see Figures 11.5 and 11.6). The results showed that the data represent a normal distribution.
3. The average ($\overline{X}$) and the standard deviation (s) were calculated:

$$\overline{X} = 61.1$$

$$s = 5.1$$

4. The confidence level was selected as $\gamma = 0.95$ and the desired population portion to be covered by the limits (p) was determined to be 99%.
5. In this example we have two-sided limits. Therefore, we use Appendix M(b) to obtain the k value. For $\gamma = 0.95$, $p = 0.99$, and $n = 60$, we get $k = 3.07$.

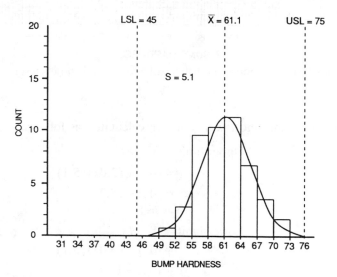

Figure 11.5 A histogram for the bump hardness test results.

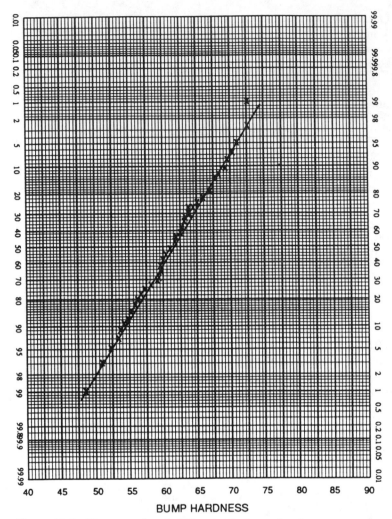

Figure 11.6   The normal probability plot for bump hardness test results.

6. The upper limit and lower limit are calculated as follows:

$$\text{Upper limit} = \overline{X} + ks$$
$$= 61.1 + (3.07)(5.1)$$
$$= 61.1 + 15.7$$
$$= 76.8$$
$$\text{Lower limit} = \overline{X} - ks$$
$$= 61.1 - (3.07)(5.1)$$
$$= 61.1 - 15.7$$
$$= 45.4$$

Working on developing new and revising existing specification requirements and their tolerance limits is one of the important problems in process improvement. These activities should be based on the application of statistical principles that will allow us to derive specification limits based on the process capability and also identify areas where the process capability should be improved to satisfy customer specification requirements. Considering that process improvement is based on continuously reducing the variation around the target, specification and tolerance improvements should also be a continuous process.

REFERENCES

1. ISO, "Statistical Interpretation of Data—Determination of a Statistical Tolerance Interval," Report ISO 3207, 1975, International Organization for Standardization, Geneva.

Chapter 12

Zero Defects Process Capability

Two years ago in the electronics industry a process with a quality level in the range of 500 to 1000 parts per million (ppm) was acceptable. Now a lot of processes approach a quality level of 50 ppm or less, which is close to perfection. But how about products that are totally defect free? Can we achieve this? The only way to do it is to have a process capable of working on a zero defects quality level.

The concept of zero defects has a long history. Philip Crosby, the quality guru who takes credit for this concept said, "...Remember that Zero Defects is not a motivation method, it is a performance standard. And it is not just for production people, it is for everyone...."[1]

In this chapter we give an overview of two aspects of zero defects. One is a motivational aspect which was used in the 1960s, and the second is a preventional aspect called "six sigma quality" which was developed in 1987 by Motorola and adapted by a number of other companies such as IBM, AMD, etc.[2] In this chapter the reader will see that just by introducing psychologically oriented motivational programs such as "zero defects," high quality cannot be achieved. In contrast, we will describe Motorola's six sigma quality approach to show that zero defects can be achieved when we motivate people and, at the same time, take care of the process to enable it to continually improve, thereby leading to perfect products.

12.1 WHAT WENT WRONG WITH THE ZERO DEFECTS PROGRAM IN THE 1960s?

In 1961, the Orlando Division of the Martin Company, which at the time had been producing missiles for the military, came up with an initiative where it exhorted its inspectors to find all discrepancies so that perfect missiles could be delivered. When this succeeded, the company carried this initiative to the entire workforce. This illustrates the inception of the zero defects (ZD) program. This initiative has been adapted from the military by civilian industry in the United States, and in other countries where the ZD program has different names. For example in the Soviet Union, this program is called the "Sarator system (SS)" because Sarator was the first region in the Soviet Union where the zero defects program was adapted. In Poland ZD has been renamed "Do Ro" which is an acronym for "dobra robota" ("good work" in Polish). In Germany the system is called "Ohne Fehler" which means "without defects." Under different names the zero defects program has been introduced throughout the world, but the lifetime of this program was very short in many countries (including the United States). Although "quality" remains a buzzword in all industries, the slogan "zero defects" became less popular. Why?

The zero defects campaigns that swept a lot of countries in the late 1960s were based on kickoff campaigns, including rallies, posters, pins, and slogans. After a kickoff of the program, instead of giving the employees the tools that would make zero defects possible, employees were invited to sign the zero defects pledge, exhorting them to eliminate all defects. The operators were told to do their best, but they were not taught how to identify and eliminate special causes and situations that cannot be corrected without changing the process. The late 1960s was a period when awards were given to contractors merely for creating ZD programs without waiting for them to improve the quality. Undertaking a zero defects program was a sign of good customer relations.[3] Such an approach to quality improvement could not be maintained and the campaigns soon faded away.

Today making defect-free products is still a major long-term objective for any company that wants to survive and succeed in the marketplace. However, companies that develop new quality programs to achieve zero defects are naming those programs differently. Xerox, for example, has its leadership through quality program. At Intel, it's PDQ[2] (perfect design quality, pretty darn quick). Hewlett-Packard refers to its program as TQC (total quality control). IBM has a defect reduction program which is named "excellence in execution."[4] When Motorola developed its long-term program, it was looking for a "novel name to shake up the company,"[5] and Motorola named it "six sigma quality." This name reflects Motorola's philosophy and its approach to perfection. In the remaining part of this chapter, we will describe this approach.

12.2 CAN WE HAVE AN EXCELLENT PROCESS?

Historically a process was considered capable when at least 99.73% of the parts produced by this process fall within the engineering specification limits

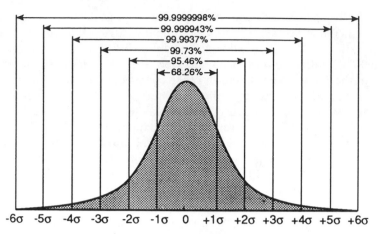

Figure 12.1 Typical areas under the normal curve.

set at ± 3 sigma from the center or "nominal" value of the specification (see Figure 12.1). Expressed differently, a process is considered capable if it will produce only 0.27% (or 2700 ppm) defective (or nonconforming) parts.

However, the times have changed. The customer now demands higher levels of quality at a lower cost. Can we say that 2700 ppm is really good enough? Well, let's answer this question by giving an improvised example. Suppose your company is producing a product which contains 1500 components. If the supplier(s) submits to your company components each from a "capable" process with a quality level of 2700 ppm, the yield of the product (before testing) in your company would be 0.01733 or 1.733% ($0.9973^{1500} = 0.01733$). This means that, on the average, only about two out of every hundred units of product will go through the entire manufacturing process without a defect. Expressing this quality level in defects per unit, we can say that on the average a unit will contain 4.05 defects ($|\ln 0.01733| = 4.05$).

This example was illustrated to make the point that in the electronics industry where a product may contain thousands of components, a ± 3 sigma process (2700 ppm) is not good enough. So what is a good or even an excellent process? We will try to give an answer in this chapter.

12.3 MOTOROLA'S SIX SIGMA QUALITY PROGRAM

In January 1987, Motorola launched a long-term program called "six sigma quality." According to this program, Motorola is planning to reach quality standards approaching zero defects throughout the company in five years. Speaking more precisely, in January 1992, Motorola is planning to achieve a

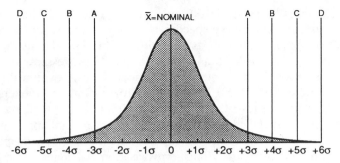

Figure 12.2 A process with four different customer specification requirements.

quality level of 3.4 nonconformities per million opportunities. What does this mean?

This means that Motorola will need to improve its quality and service 100 times by the end of 1991.[6] To complete such a task, there has to be something special about the concept of six sigma quality. To introduce the concept of six sigma quality to its employees, Motorola spent 25 million dollars, just in 1987.[7] What is the philosophy behind this concept? For discussion purposes, let's assume that we have a manufacturing process running in a state of statistical control, and Figure 12.2 represents the distribution of this process. Let's also assume that the product from this process is to be sent to four customers (A, B, C, and D) who have different specification requirements. In this improvised situation the customers will receive a different quality product from the same process, and the same process will have different capability indexes for different customers (see Table 12.1).

For example, Customer A can expect 0.27% (100−99.73) of the parts to fall outside the engineering specification range because the upper and lower

TABLE 12.1 The Same Process with Different Specification Requirements which Cause Different PPM Levels and Capability Indexes

Customer	± Sigma	PPM	Percentage Defective	Cp	Cpk
A	3	2700	99.73	1.0	1.0
B	4	63	99.9937	1.33	1.33
C	5	0.57	99.999943	1.67	1.67
D	6	0.002	99.9999998	2.0	2.0

Note: For simplicity we are assuming that there is one nominal for all customers and the process is constantly set exactly on the nominal, so Cp = Cpk.

specification limits (USL and LSL, respectively) are ±3 sigma from the center (or nominal) value of the specification. In other words, the process will produce 2700 nonconforming parts per million (ppm) on the average. In this situation, we say that we have a ±3 sigma quality process capability which has a Cp index of 1. In contrast, Customer D can expect 0.0000002% $(100 - 0.0000002 = 99.9999998)$ of the parts to fall outside the engineering specification range because the specification limits are ±6 sigma from the center value of the specifications. This is equivalent to 0.002 ppm (practically zero defects). In this situation, we say that we have a ±6 sigma process capability with a Cp index of 2. This improvisation was made to demonstrate the relationship between the process capability and the specification require-ments, and their influence on the capability indexes. When working in a six sigma quality program, there is some room for negotiation between the supplier and customer to relax the specification requirements to improve the Cp or Cpk value, but this has nothing to do with quality improvement.

Quality improvement is related to the reduction of the process variation around the target. This will bring you the same results as relaxing the specification limits, but it is really quality improvement.

In the beginning where we had a ±3 sigma quality process with a Cp index of 1 and, by reducing the process spread in half, we achieved a 6 sigma quality process which is practically a perfect process (see Figure 12.3). This is really an achievement in quality improvement and can only be done by

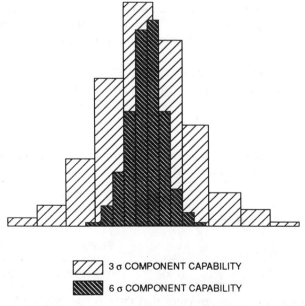

3 σ COMPONENT CAPABILITY

6 σ COMPONENT CAPABILITY

Figure 12.3   Reducing the process spread by half to achieve six sigma quality.

a) A PROCESS WITH ±3σ SPECIFICATION

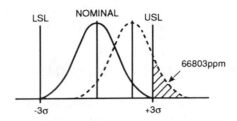

b) A PROCESS WITH ±6σ SPECIFICATION

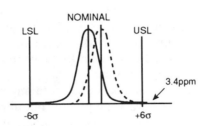

Figure 12.4   The ppm levels for a process assuming the process average will be shifted 1.5 sigma from the nominal of the specifications.

applying design of experiments to improve the process. SPC would not help here because we assumed that the distribution came from a process that was in a state of statistical control. To reduce the remaining variation that is related to common causes, we need to change the system. This is the responsibility of management. The supervisors and workers on the line cannot do it by themselves.

So far the description has been based on the assumption that the process average is constantly adjusted to the specification nominal. But in reality the process is dynamic. In addition, if we run the process for a longer time, we will observe changes in the average. Let's take a look at how these changes will impact the quality.

Motorola's six sigma program is based on the assumption that a maximum factor of 1.5 sigma is sufficient to account for shifts and drifts. Its assumption is supported by a number of publications on this subject by different authors.[8] From Figure 12.4 one can see that if a process with ±3 sigma limits (Cp = 1) shifts 1.5 sigma from the nominal the nonconforming level will change from 1350 ppm to 66,803 ppm. But if a process with ±6 sigma limits (Cp = 2) shifts 1.5 sigma, the ppm level will change from 0.002 ppm to 3.4 ppm which is still very low. So if a company is planning to achieve a close to perfect quality level (3.4 ppm), it needs to have a normal process with a spread that will consume no more than 50% of the customer specifications. Assuming that the process average is set on the nominal and will not make shifts or drifts more than 1.5 sigma. This is not an easy task, but Motorola and other companies that adapted this approach proved that this can be done.

12.4 SIX SIGMA QUALITY: ATTRIBUTE DATA

In this chapter we described the application of the six sigma concept in relation to the manufacturing process where variable data are available. But how about the manufacturing process where variable data are not available? As we mentioned earlier, in the electronics industry there are many processes that are measured only by attribute data. Also, how do we measure six sigma capability of the nonmanufacturing or service processes such as shipping, purchasing, sales, engineering, marketing, accounting, etc.? To become a perfect company, we need to be able to measure and improve all processes throughout the entire organization.

Motorola developed a method to measure any meaningful activity. Scientifically the method may have some pitfalls, but in practice it works very well. Applying six sigma as a measure of manufacturing and nonmanufacturing processes makes it possible to discuss and compare quality levels between plants, departments, suppliers, and individuals, even if the functions and activities are absolutely different.

12.4.1 Estimating Six Sigma Capability

For some people, especially for those who are not familiar with statistical principles, the term "six sigma quality" sounds a little bit unusual, especially when you want to apply it to measure the performance of a salesperson, accountant, or executive. This is why Motorola developed some examples related to real life to give some feeling for what "six sigma quality" means. For example, Scott Shamlin,[9] Director of Manufacturing Operations at Motorola, uses the following analogy when describing the concept. "The odds of a person getting on an airplane and reaching his or her destination alive is about 6.2 sigma; the odds of the same person and his or her bags arriving at the same place at the same time is about 4.1 sigma."

In the same way we could, for example, estimate the six sigma quality of the Persian Gulf War. Knowing the number of sorties accomplished and the number of casualties, we can calculate the six sigma military capability.

These examples illustrate that six sigma is a universal measure of quality. Now let's go back to the manufacturing application and illustrate the methodology of calculating six sigma capability by using an example.

Example 1 In the semiconductor industry before the finished devices are shipped they go through a final operation called "visual/mechanical inspec-

**TABLE 12.2   PPM in Relation to the Number
of Sigmas Between the Process Average
and the Specification Limit**

PPM Level	Sigma
100 ppm	3.72
200 ppm	3.54
300 ppm	3.44
400 ppm	3.35
500 ppm	3.29
600 ppm	3.24
700 ppm	3.19
800 ppm	3.16
900 ppm	3.12
1000 ppm	3.09

tion." We would like to determine the six sigma quality of this operation. Let's assume that the results for a particular period show that a quantity of 10,000 devices has been inspected and only 3 devices are found to be rejects. This would give a reject rate of (3/10,000 = 0.0003). Translating the result in parts per million (ppm), we would have 300 ppm ($10^6 \times 0.0003 = 300$ ppm). We can also say that the yield is 0.9997 or 99.97%. Now we want to know how many standard deviation units would need to be taken into consideration in order to encompass 99.97% of the area under the normal curve? Consulting Table 12.2, we find that the answer is 3.44 sigma (assuming that the 0.0003 area is only on the right side of the normal distribution). (*Note:* Motorola has established a convention of distributing the total proportion nonconforming to the right side of the normal curve.) Finally, to calculate the process capability, we need to adjust the value for "equivalent shift and drifts" by adding the correction factor of 1.5 sigma to our 3.44 sigma estimate. Doing this, we obtain a result of 4.94 sigma. Now we can make a conclusion that the process has a 4.94 sigma quality. From these results we also can determine the Cp value as follows:

$$\frac{4.94\sigma}{3\sigma} = 1.65$$

So the Cp is 1.65.

Example 2   Now let's use a second example from the nonmanufacturing area. For example, the payroll department processed 20,000 paychecks for the previous two weeks' period and it found that 4 checks had mistakes. So the reject rate is $4/20{,}000 = 0.0002$. Translating in ppm, we have

$$10^6 \times 0.0002 = 200 \text{ ppm}$$

The yield will be 0.9998 or 99.98%. From Table 12.2 we find the sigma value is 3.54. Adding the correction factor of 1.5 sigma, we have a process of $3.54 + 1.5 = 5.04$ sigma. The Cp value is

$$\frac{5.04\sigma}{3\sigma} = 1.68$$

The six sigma quality and the capability index (Cp) can be determined for other areas in the same way. To summarize the discussion, we should say that to determine the six sigma capability we need to go through three major steps.

In the first step we must determine how many opportunities there are for a nonconformity or a defect to occur. For example, in a shipping department the number of opportunities to have a nonconformity (late shipment) is the total shipments made. For a typist, the number of opportunities to have a defect (misspelled word) is the total number of words typed. The number of opportunities for a tester to make an error is the number of parts tested. A note of caution should be made when determining the number of opportunities in complex processes. For example, if there are 1,000,000 opportunities for a defect to occur in a computer terminal and by testing 100 terminals we find 200 defects (on the average 2 defects per terminal), the number of opportunities should be 100 million, so the ppm level in this case would be

$$\frac{200}{100{,}000{,}000} = 0.000002 = 2 \text{ ppm}$$

The second step is to have a system of counting the actual number of nonconformities or defects associated with the particular product or activity during the process for a period of time. Having the number of opportunities and the actual number of defects or nonconformities, we can calculate the ppm level.

And the final, third, step is the actual conversion of the ppm into sigma units using the table of normal distribution. It takes a while to get used to measuring quality this way, but by applying this on a regular basis, one can see how easy it is to measure the level of excellence.

"Six sigma quality" is a program to achieve the "best in class" product from an excellent process which is quantitatively defined as Cp = 2.0, Cpk = 1.5, and ppm = 2.0. This approach is not just a motivational technique, but a complex, intellectualized concept based on continuous feedback and application of statistical principles such as process capability studies, design of experiment, process optimization, tolerance design, pre-control, etc.

It can be applied to achieve and measure the quality levels not only for an individual product parameter, but also for an entire manufacturing process. It can also be applied to evaluate the performance of an individual or a group of people who perform different functions. Six sigma quality as an approach is usually used in connection with other programs such as[10]

Short cycle manufacturing (SCM)

Design for producibility (DFP)

Statistical process control (SPC)

Supplier SPC (SSPC)

Participative management practices (PMP)

Part standardization and supplier qualification

Computer simulation

Quality function deployment (QFD)

Total quality control (TQC)

Profit improvements

Product and manufacturing leadership

This does not mean that to achieve six sigma quality a company should introduce all those programs. The point we want to make is that every program separately takes care of one aspect of customer satisfaction. Using a number of programs in parallel as subsystems of a complex system of quality improvement will achieve total customer satisfaction. For example, the introduction of the quality function deployment (QFD) subsystem will cause constant feedback of the continuously changing customer demands and needs. By knowing this, the supplier will know how to develop the capability to satisfy these needs. Applying total quality control (TQC) will allow the manufacturer to fulfill the supplier's demands and needs. But even if we achieve the best product quality, we need to have a system to deliver this product quality within the right time. So we need a subsystem such as short cycle manufacturing (SCM) that will take care of cycle time reduction and on-time delivery. These examples demonstrate the relationship between different types of functional subsystems, which together work as an overall system of continuous process improvement.

There was a time when people thought that having a vision of a perfect company was more a dream than a reality. For example, zero defects can be just a dream if you are only using slogans and asking the workers to do it "right the first time." But showing the workers how to "make it right," developing an environment where everything can be made "right the first time," and developing a system of achieving zero defects can make the dream about a perfect company become a reality. Motorola developed such a system, and this is one of the reasons why it received the first Malcolm Baldrige Award for quality.

REFERENCES

1. Philip B. Crosby, "Quality Is Free" (1979), McGraw-Hill Book Company, p. 172.
2. Bruce C. P. Rayner, "Commitment to Quality," Electronic Business, October 15, 1990, p. 68.
3. J. M. Juran, "Quality Problems, Remedies, and Nostrums," Industrial Quality Control, June 1966, pp. 647–653.
4. Bruce C. P. Rayner, "Commitment to Quality," Electronic Business, October 15, 1990, p. 70.
5. Mark Stuart Gill, "Stalking Six Sigma," Business Month, January 1990, p. 45.
6. Jerry Wolak, "Motorola Revisited," Quality, May 1987, p. 19.
7. Mark Stuart Gill, "Stalking Six Sigma," Business Month, January 1990, p. 45.
8. David H. Evans, "Statistical Tolerancing: The State of the Art, Part III: Shifts and Drifts," Journal of Quality and Technology, Vol. 7, No. 2 (April 1975), pp. 72–76.
9. Tom Inglesby, "How They Brought Home the Prize," Manufacturing Systems, April 1989, p. 30.
10. Mike L. J. Harry, "The Nature of Six Sigma Quality," Motorola, Inc., p. 4.

Off-Line and On-Line Design of Experiments

So far we have described statistical methods related to process control. We have learned how to bring a process into a state of statistical control by performing process capability studies, and how to monitor the process by applying Shewhart control charts and other techniques. By using these techniques we can achieve the first step of process improvement. But this improvement has its own limits, which are related to the inherent process capability. It is important to note that as soon as we bring the process into a state of control, further improvement is not possible until we change the system of causes, which means working on the process to reduce its inherent variability (see Figure III.1). This cannot be done by simply applying SPC (in its narrow definition). This can only be done by applying on-line and/or off-line design of experiments.

The purpose of this part of the book is to demonstrate some principles of design of experiments by applying actual examples of application. For those who are not familiar with design of experiments, these examples can be a good starting point in the study of powerful techniques for continuous process improvement.

By going through the pages of this part of the book, the reader will become familiar with some off-line design of experiments such as the simplest one-factor-at-a-time experiment, factorial, Taguchi, and nested design of experiments. Also the reader will be introduced to some of the on-line design of experiments such as evolutionary operations (EVOP), simplex, rotating square evolutionary operation (ROVOP), and plant experimentation (PLEX). All these will be a good first step in the world of designed experiments.

Before starting this part, we would recommend the reader take a brief look at Table III.1 to become familiar with some of the frequently used terms related to design of experiments.

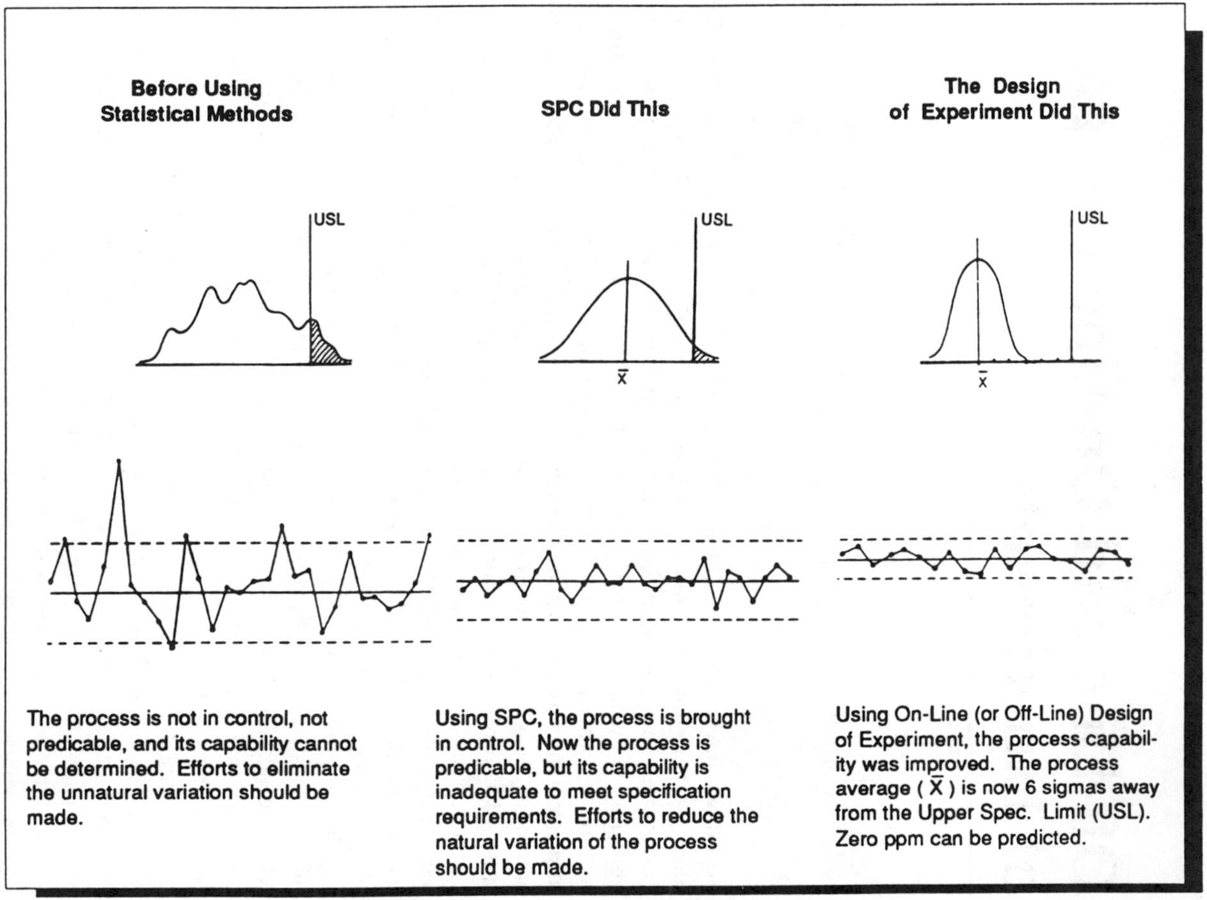

Before Using Statistical Methods

SPC Did This

The Design of Experiment Did This

The process is not in control, not predicable, and its capability cannot be determined. Efforts to eliminate the unnatural variation should be made.

Using SPC, the process is brought in control. Now the process is predicable, but its capability is inadequate to meet specification requirements. Efforts to reduce the natural variation of the process should be made.

Using On-Line (or Off-Line) Design of Experiment, the process capability was improved. The process average ($\bar{\bar{x}}$) is now 6 sigmas away from the Upper Spec. Limit (USL). Zero ppm can be predicted.

Figure III.1 Process improvement using statistical tools.

TABLE III.1   General Terminology to Be Used in Design of Experiments

Factor	A factor is the element of a designed experiment to which we make deliberate changes to observe the impact of these changes on the response.
Level	A value at which each factor in the experiment is set.
Response	A value we receive by performing an experiment.
Treatment combination	The test that we run with each level.
Replicate	A repeat of a set of experimental conditions. Not a re-reading of a value, but an entirely new run.

Chapter **13**

Off-Line Design of Experiments

Through the years, scientists have accumulated a wide variety of methods for off-line design of experiments. For an engineer, it is very important to know how to select the proper method to conduct a meaningful and economic design of experiment. Section 26 of Juran's *Quality Control Handbook* [1] gives a classification of some common statistical experimental designs that can be helpful in selecting the right method for engineering experiments. Table 13.1 is an extraction of this classification.

In this chapter the reader will be introduced to some techniques related to off-line design of experiments.

13.1 THE CLASSICAL ONE-FACTOR-AT-A-TIME EXPERIMENT

This technique is one of the oldest approaches to experimentation. The basic concept of this method is that the experimental factors are varied one at a time, with the remaining factors holding constant. As a result, we receive an estimate of the effect of a single variable at selected fixed conditions of the other variables.

The one-factor-at-a-time technique is appropriate to use when we can assume (from past experience) that the variables in the study act independently on the response (i.e., there are no interactions). This is one of the major weaknesses of the technique, but considering its simplicity in application, the classical strategy of the one-factor-at-a-time technique should not be ignored, even when other designs have an advantage over this method.

The easiest way to describe the application of the one-factor-at-a-time method is to use an example.

TABLE 13.1 Classification of Designs

Design	Type of Application	Structure	Information Sought
Completely randomized	Appropriate when only one experimental factor is being investigated	Basic: One factor is investigated by allocating experimental units at random to treatments (levels of the factor) Blocking: None	1. Estimate and compare treatment effects 2. Estimate variance
Factorial	Appropriate when several factors are to be investigated at two or more levels and interaction of factors may be important	Basic: Several factors are investigated at several levels by running all combinations of factors and levels Blocking: None	1. Estimate and compare effects of several factors 2. Estimate possible interaction effects 3. Estimate variance
Fractional factorial	Appropriate when there are many factors and levels and it is impractical to run all combinations.	Basic: Several factors are investigated at several levels but only a subset of the full factorial is run Blocking: sometimes possible	1. Estimate and compare effects of several factors 2. Estimate certain interaction effects (some may not be estimable) 3. Certain small fractional factorial designs may not provide sufficient information for estimating the variance
Randomized block	Appropriate when one factor is being investigated and experimental material or environment can be divided into blocks or homogeneous groups	Basic: Each treatment or level of factor is run in each block Block: Usually with respect to only one variable	1. Estimate and compare effects of treatments free of block effects 2. Estimate block effects 3. Estimate variance

Type	When appropriate	Basic/Blocking	Results
Latin square	Appropriate when one primary factor is under investigation and results may be affected by two other experimental variables or by two sources of nonhomogeneity. It is assumed that no interaction exists	Basic: Two cross groupings of the experimental units are made corresponding to the columns and rows of a square. Each treatment occurs once in every row and once in every column. Number of treatments must equal number of rows and number of columns Blocking: with respect to two other variables in a two-way layout	1. Estimate and compare treatment effects, free of effects of the two blocked variables 2. Estimate and compare effects of the two blocked variables 3. Estimate variance
Nested	Appropriate when objective is to study relative variability instead of mean effect of sources of variation (e.g., variance of tests on the same sample and variance of difference samples)	Basic: Factors are strata in some hierarchical structure; units are tested from each stratum	Relative variation in various strata, components of variance
Response surface	Objective is to provide empirical maps (contour diagrams) illustrative of how factors under the experimenter's control influence the response	Factor settings are viewed as defining points in the factor space (may be multidimensional) at which the response will be recorded	Maps illustrating the nature of the response surface

Reprinted with permission from J. M. Juran and Frank M. Gryna, "Juran's Quality Control Handbook," 4th ed. (1988), McGraw-Hill Book Company, pp. 26.7–26.10.

TABLE 13.2   Setting Conditions for the Experiment

Factor	Low Level	High Level
1: Pressure (P)	100 psi (P_1)	200 psi (P_2)
2: Temperature (T)	170°C (T_1)	190°C (T_2)
3: Time (t)	20 minutes (t_1)	40 minutes (t_1)

13.1.1   An Example of Applying the One-Factor-at-a-Time Design of Experiment

Suppose we have a semiconductor manufacturing process that needs to be studied to find out how to reduce the number of nonconforming devices. After a brainstorming session, it was decided that the three factors most likely to be the main contributors to the problem were pressure, temperature, and time. The objective of the study was to prove that these selected factors really have an impact on the nonconforming rate of the product.

To achieve the objective, it was decided to apply the one-factor-at-a-time design of experiment. The settings of the experiment have been determined as shown in Table 13.2.

According to the one-factor-at-a-time technique, we should start by establishing a baseline on the low values of each of the three factors, and then make systematic changes to each of the factors, one at a time. By taking the difference between the baseline and each change, we can determine the *effect* of each factor on the response. Having established the baseline and the changes (see Table 13.2), we can proceed with our experiment and observe the results.

On the first test, we run all three factors at their low levels. The next step is to run a treatment combination by changing only one of the factors (e.g., pressure) to its high level, leaving all other factors at their low levels. Making similar changes (one at a time) to the remaining two factors and observing the nonconforming rate (the response), we come up with a four-treatment combination that can be analyzed to obtain the answer to the question that was raised by our objective.

But, as we know, making decisions based on one single isolated point can bring us to a misleading conclusion. Therefore, this type of experiment is usually repeated a number of times to obtain the average results. In this example, it was decided to repeat each treatment four times. The results of the experiment are shown in Table 13.3.

According to the results, one can conclude that the best process settings would be to have all factors set at the low levels. This would give a nonconforming rate of 1.55. Any other combination of process settings would increase the percentage of nonconforming parts.

TABLE 13.3 The Experimental Results

Combination	Treatment Number	Trial 1	Trial 2	Trial 3	Trial 4	Average
$P_1T_1t_1$	1	1.3	1.6	1.5	1.8	1.55
$P_2T_1t_1$	2	2.4	2.6	2.3	2.7	2.50
$P_1T_2t_1$	3	7.2	8.2	7.8	7.2	7.60
$P_1T_1t_2$	4	4.2	4.6	4.1	4.5	4.35

The top spanning header: Percentage Nonconforming Parts (%)

TABLE 13.4 The Experimental Factor Effects

Factor	Effect	Treatment Number
1: Pressure (P)	$2.50 - 1.55 = 0.95$	$2-1$
2: Temperature (T)	$7.60 - 1.55 = 6.05$	$3-1$
3: Time (t)	$4.35 - 1.55 = 2.80$	$4-1$

To evaluate the effect of each factor, we compute the average difference between each of the changed treatments (treatments 2, 3, and 4) and the base case (treatment 1). The results are shown in Table 13.4. The temperature has the highest effect value of 6.05, which means that this factor has more impact on the process than the other two factors.

As one can see, the one-factor-at-a-time technique is very simple. The questions, then, are: Is this experiment efficient? Did we obtain the correct information? How confidently can we make a conclusion based on these results? These and other similar questions will be answered by learning about more complicated, efficient, and accurate techniques. For now we can say that there are conditions where this technique can be used successfully.

13.2 INTRODUCTION TO ANALYSIS OF VARIANCE

Analysis of variance (ANOVA) is a statistical technique that breaks down the total process variation into its components, and it is used to interpret experimental data to make the necessary decisions. ANOVA is based on the following assumptions.

1. Assumption of Normality
This requires that the underlying distribution of the response variable is normally distributed. However, since ANOVA is considered a "robust" test, the normality of the response variable is not a strict requirement.

2. Assumption of Additivity

This means that each response variable is comprised of the sum of an overall mean plus the sum of all main and interaction effects among the factors and the effect due to the experimental error.

3. Assumption of Homogeneity

This requires that the underlying distribution of replicate observations each have the same variance. This is a very strict requirement, and ANOVA does not apply if this assumption is not met.

4. Assumption of Statistical Independence

This means that each response value is independent of all previous values. This is guaranteed by randomization of the treatment combinations to be used as test specimens.

When these four assumptions are met, the results of a matrix of experimental runs can be economically analyzed together by using a statistical test known as the F ratio. This test compares the resulting variation caused by a matrix variable with normal "noise."

Also, it is important to note that the ANOVA technique is fairly reliable even when the assumptions are not 100% valid. These assumptions will become more understandable when the reader becomes more familiar with the ANOVA principles.

Following is the step-by-step procedure for performing ANOVA. The procedure is demonstrated by using an actual example where two controllable parameters are involved. Because of this, a two-way ANOVA is applied.

13.2.1   Example of Applying a Two-Way ANOVA

An integrated circuit manufacturer decided to perform an experiment to determine what causes the increase in the reject rate from an invalid open/short circuit test when using an "X-Y-Z" handler to test the devices. It was decided to perform a two-factor, two-level full factorial design of experiment.

From the list of factors that may impact the reject rate, the process improvement team selected the "contact pressure" and "distance between the contacts" as the two controllable variables that may be the largest contributors to the problem. They decided to start the experiment by setting these two variables as in Table 13.5. As for the response variable, it was

TABLE 13.5   The Experimental Settings for Open/Short Circuit Test

Factor	Level 1	Level 2
A: Pressure	50 psi	70 psi
B: Distance	0.453"	0.457"

TABLE 13.6 The Results from a Two-Factor, Two-Level Experiment

| | | Factor A: Pressure | | | |
		50 psi (A_1)		70 psi (A_2)	
Factor B: Distance	0.453" (B_1)	1.0	1.5	2.0	2.0
	0.457" (B_2)	9.7	9.3	8.0	4.0

Note: The response is the percentage relative invalid opens and shorts.

decided to use the relative percentage of invalid open/short rejects. Table 13.6 shows the results of running the experiment.

Before describing the procedure for ANOVA, we will define the notation to be used as follows:

$Y_{i..} = \Sigma A_i$ is the sum of observations under the A_i level
$Y_{.j.} = \Sigma B_j$ is the sum of observations under the B_j level
$\bar{Y}_{i..} = \bar{A}_i$ is the average of observations under the A_i level
$\bar{Y}_{.j.} = \bar{B}_j$ is the average of observations under the B_j level
$Y_{...}$ is the sum of all observations
$\bar{Y}_{...}$ is the average of all observations
a is the number of levels for factor A
b is the number of levels for factor B
r is the number of replicates for each observation
N is the total number of observations

With this notation we will describe the procedure of ANOVA as follows.
Step 1 Compute the sum of squares (SS).
The SS is an estimation of the variation. The total variation (total sum of squares, SS_T) can be expressed as

$$SS_T = SS_M + SS_A + SS_B + SS_{A \times B} + SS_E$$

where SS_M is the variation due to the overall mean, SS_A is the variation due to factor A, SS_B is the variation due to factor B, $SS_{A \times B}$ is the variation due to the interaction between factors A and B, and SS_E is the variation due to random errors.

The computations for all these sums of squares are as follows.
A. *Sum of Squares due to the Mean (Correction Factor).* This is an estimate of the variation due to the mean, and is computed as follows:

$$SS_M = \frac{(\text{grand total of all the observations})^2}{\text{total number of the observations}} = \frac{Y_{...}^2}{N}$$

In our example, $a = 2$, $b = 2$, $r = 2$, then

$$N = a \times b \times r$$
$$= 2 \times 2 \times 2$$
$$= 8$$

and

$$SS_M = \frac{(1.0 + 1.5 + 9.7 + 9.3 + 2 + 2 + 8 + 4)^2}{8}$$

$$= \frac{37.5^2}{8}$$

$$= 175.8$$

The variation due to the mean (SS_M) does not affect the calculations for the variation due to the factor effects or errors. In most experimental situations, the variation due to the mean has no practical value; therefore, it is subtracted from the total variation (SS_T), and is referred to as the correction factor (CF).

B. *Total Sum of Squares* (SS_T).   The total variation can be estimated by summing the squares of each observation and subtracting the correction factor as follows:

$$SS_T = \sum_{i=1}^{a} \sum_{j=1}^{b} \sum_{k=1}^{r} Y_{ijk}^2 - \frac{Y_{...}^2}{N}$$

$$= \sum (\text{each datum})^2 - CF$$

$$= (1.0^2 + 1.5^2 + 9.7^2 + 9.3^2 + 2^2 + 2^2 + 8.0^2 + 4.0^2) - 175.8$$

$$= 96.05$$

C. *Sum of Squares due to Each Factor*.   The sum of squares due to factor A is the average variation due to each level of factor A relative to the total average variation. Mathematically, this can be expressed as follows:

$$SS_A = \sum_{i=1}^{a} \frac{Y_{i..}^2}{b \times r} - \frac{Y_{...}^2}{N}$$

$$= \frac{(\Sigma A_1)^2}{b \times r} + \frac{(\Sigma A_2)^2}{b \times r} - CF$$

$$= \frac{(1.0 + 1.5 + 9.7 + 9.3)^2}{2 \times 2} + \frac{(2 + 2 + 8 + 4)^2}{2 \times 2} - 175.8$$

$$= 3.78$$

Similarly,

$$SS_B = \sum_{j=1}^{b} \frac{Y_{\cdot j \cdot}^2}{a \times r} - \frac{Y_{\cdots}^2}{N}$$

$$= \frac{(\Sigma B_1)^2}{a \times r} + \frac{(\Sigma B_2)^2}{a \times r} - CF$$

$$= \frac{(1.0+1.5+2+2)^2}{2 \times 2} + \frac{(9.7+9.3+8.0+4.0)^2}{2 \times 2} - 175.8$$

$$= 75.03$$

D. *Sum of Squares due to Interactions.* To calculate the variation due to the interaction of factors A and B ($SS_{A \times B}$), we first compute the sum of squares between the cell subtotals ($SS_{subtotal}$). Table 13.7 is a response table showing the cell subtotals. Expressing this mathematically, we have

$$SS_{subtotal} = \sum_{i=1}^{a} \sum_{j=1}^{b} \frac{Y_{ij\cdot}^2}{r} - \frac{Y_{\cdots}^2}{N}$$

$$= \frac{(\Sigma A_1 B_1)^2}{r} + \frac{(\Sigma A_2 B_1)^2}{r} + \frac{(\Sigma A_1 B_2)^2}{r} + \frac{(\Sigma A_2 B_2)^2}{r} - CF$$

$$= \frac{2.5^2}{2} + \frac{4^2}{2} + \frac{19^2}{2} + \frac{12^2}{2} - 175.8$$

$$= 87.84$$

The $SS_{subtotal}$ also contains the variation due to factor A (SS_A) and variation due to factor B (SS_B). Therefore, the second step is to compute the $SS_{A \times B}$ as

$$SS_{A \times B} = SS_{subtotal} - SS_A - SS_B$$

$$= 87.84 - 3.78 - 75.03 = 9.03$$

TABLE 13.7   A Response Table with Cell Subtotals

Factor	A_1	A_2
B_1	$1.0+1.5 = 2.5$	$2+2 = 4$
B_2	$9.7+9.3 = 19.0$	$8+4 = 12$

E. *Sum of Squares due to Error* (SS_E). The variation due to error is a measure of the variation around the average value. The easiest way to compute the error sum of squares is by using the following formula:

$$SS_E = SS_T - SS_A - SS_B - SS_{A \times B}$$

$$= 96.05 - 3.78 - 75.03 - 9.03 = 8.21$$

Step 2 Determine the degrees of freedom (df). This is computed as follows:

$$df_T = (\text{total number of observations}) - 1 = 8 - 1 = 7$$

$$df_A = (\text{number of levels for factor } A) - 1 = 2 - 1 = 1$$

$$df_B = (\text{number of levels for factor } B) - 1 = 2 - 1 = 1$$

$$df_{A \times B} = df_A \times df_B = 1 \times 1 = 1$$

$$df_E = df_T - df_A - df_B - df_{A \times B} = 7 - 1 - 1 - 1 = 4$$

Step 3 Determine the variance (V). This is sometimes called the mean square (MS) and is determined as follows:

$$V_A = \frac{SS_A}{df_A} = \frac{3.78}{1} = 3.78$$

$$V_B = \frac{SS_B}{df_B} = \frac{75.03}{1} = 75.03$$

$$V_{A \times B} = \frac{SS_{A \times B}}{df_{A \times B}} = \frac{9.03}{1} = 9.03$$

$$V_E = \frac{SS_E}{df_E} = \frac{8.21}{4} = 2.05$$

Step 4 Determine the F ratio (F). The F test is used here to determine whether the factors are statistically significant with respect to the components in the error variance. Thus, the F ratios are determined as follows:

$$F_A = \frac{V_A}{V_E} = \frac{3.78}{2.05} = 1.84$$

$$F_B = \frac{V_B}{V_E} = \frac{75.03}{2.05} = 36.58$$

$$F_{A \times B} = \frac{V_{A \times B}}{V_E} = \frac{9.03}{2.05} = 4.40$$

Step 5 Determine whether the F ratios are significant. To determine whether an F ratio is statistically significant, three pieces of information are considered:[2]

1. The confidence necessary.
2. The degrees of freedom associated with the sample variance in the numerator.
3. The degrees of freedom associated with the sample variance in the denominator.

$F_{\alpha, V1, V2}$ is the format for determining an explicit F value, where α is the risk, the confidence is equal to $1 - \alpha$, v_1 is the degrees of freedom for the numerator, and v_2 is the degrees of freedom for the denominator.

In our example, the α value has been determined to be 0.01, which means the confidence level required is 99%. The numerators in the F ratios are factors A and B and the $A \times B$ interaction. Their degrees of freedom are all equal to 1 (see Step 2); therefore, $v_1 = 1$. The denominator in the ratios is the error variance, and its degrees of freedom is 4 (see Step 2); therefore $v_2 = 4$. From Appendix E we find

$$F_{0.01, 1, 4} = 21.20$$

And, comparing the F ratios calculated from the data (see Step 4) with this critical value, we find

$$F_A = 1.84 < F_{0.01, 1, 4} \qquad \text{Not significant}$$

$$F_B = 36.58 > F_{0.01, 1, 4} \qquad \text{Significant}$$

$$F_{A \times B} = 4.40 < F_{0.01, 1, 4} \qquad \text{Not significant}$$

Step 6 Complete the ANOVA table. The ANOVA summary table is completed as shown in Table 13.8.

TABLE 13.8   Two-Way ANOVA Summary Table

Source of Variation	SS	df	V	F
A	3.78	1	3.78	1.84
B	75.03	1	75.03	36.58*
$A \times B$	9.03	1	9.03	4.40
Error	8.21	4	2.05	
Total	96.05	7		

*Significant at the 99% confidence level.

Conclusion

The ANOVA results indicated that factor A (pressure) has no significant effect on the invalid open/short reject rate, factor B (distance between contacts) had a substantial effect on the reject rate, and that there is no significant interaction effect between the pressure and distance. We can conclude from this result that by changing the levels of distance between contacts, the percentage of invalid open/short rejects will be changed significantly.

To determine which level we should set factor B at, we first find the average response at each level of factor B by using the data from the experiment (see Table 13.6) as follows:

$$\bar{B}_1 = \frac{(1.0 + 1.5 + 2.0 + 2.0)}{4} = 1.625$$

$$\bar{B}_2 = \frac{(9.7 + 9.3 + 8.0 + 4.0)}{4} = 7.75$$

It is clear that at level 1 (distance at 0.453″), the percentage of invalid open/short rejects is lower; therefore, we should set the distance between contact at this level.

This example demonstrates the effectiveness of applying ANOVA to identify which parameters significantly impact the process. Based on the results of the analysis, we can properly adjust the parameters to improve the performance of the process. We can also obtain information about how to continue to improve the handler's response by investigating the contactor design more thoroughly.

13.3   INTRODUCTION TO FACTORIAL EXPERIMENTS

Factorial experiments are employed to simultaneously study the effects of two or more factors, and all possible combinations of the levels of factors are

TABLE 13.9 Two-Way Experimental Data

Level	A_1	A_2
B_1	50	100
B_2	150	0

investigated in the experiment. Factorial designs have several advantages. They are more efficient than one-factor-at-a-time experiments. Furthermore, a factorial design is necessary when interactions may be present, to avoid misleading conclusions. Finally, factorial designs allow the effects of a factor to be estimated at several levels of the other factors, yielding conclusions that are valid over a range of experimental conditions.

13.3.1 The Concept of Main Effect and Interaction Effect

The effect of a factor is defined as the change in response produced by a change in the level of the factor.[3] This is called a "main effect" because it refers to the primary factors in the study. For example, consider the data in Table 13.9.

The main effect of factor A is the difference between the average response at A_1 and the average response at A_2, which is

$$A = \frac{(50+150)}{2} - \frac{(100+0)}{2} = 50$$

That is, changing factor A from level 1 to level 2 causes an average response decrease of 50 units. Similarly, the main effect of factor B is

$$B = \frac{(50+100)}{2} - \frac{(150+0)}{2} = 0$$

However, we cannot conclude that there is no B effect. This is because when we examined the effects of B at different levels of factor A, we saw that the difference in response between the levels of factor B is not the same at A_1 and A_2. At A_1, the B effect is

$$B = 50 - 150 = -100$$

and at A_2, the B effect is

$$B = 100 - 0 = 100$$

Since the effect of B depends on the level chosen for factor A, we can conclude that there is an interaction between A and B. In the case when an

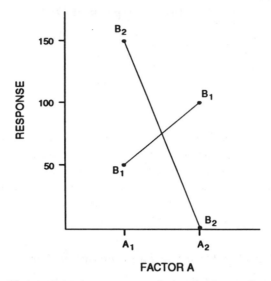

Figure 13.1   A response graph showing interaction.

interaction exists, it is meaningless to interpret the main effects, because a significant interaction can mask the significance of the main effects.

The concept of interaction can be illustrated graphically. Figure 13.1 is a plot of the response based on the levels of A for both levels of B. The B_1 and B_2 lines are not parallel, which indicates that factors A and B interact. If there is no interaction between the two factors, the two lines will be parallel.

In the presence of significant interaction, the experimenter must examine the levels of one factor, say A, with the levels of the other factors, to draw conclusions about the main effect of A.

13.3.2 2^k and 3^k Factorial Designs

Factorial designs are widely used in experiments involving several factors where it is necessary to study the joint effect of these factors on a response.[4] There are several special cases of the general factorial design that are important because they are widely used in research work, and also because they form a basis of other designs of considerable practical value. The first of these important designs is the 2^k factorial design, which involves k factors, each at only two levels. The second of these important designs is the 3^k factorial design, which involves k factors, each at three levels.

This section introduces these two useful series of designs. Throughout this section we assume that (1) the factors are fixed, (2) the designs are completely randomized, and (3) the usual normality assumptions are satisfied.

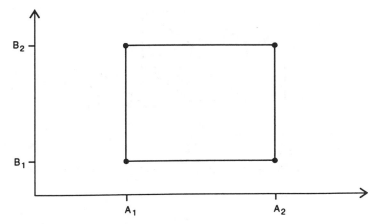

Figure 13.2   A graphical representation of the 2^2 factorial design.

13.3.2.1   Analysis of the 2^k Factorial Design

The 2^k design is particularly useful in the early stages of experimental work, when many factors are likely to be investigated. It provides the smallest number of treatment combinations with which k factors can be studied in a complete factorial arrangement. Because there are only two levels for each factor, we must assume that the response is approximately linear over the range of the factor levels chosen.

The experiment we used to illustrate the procedure of ANOVA involves two factors, each run at two levels. This type of design is called a 2^2 factorial design. The treatment combinations in this design can be graphically represented as a square (as shown in Figure 13.2). If there are three factors, say A, B, and C, under study, each at two levels, the design is called a 2^3 factorial design. The eight treatment combinations can be displayed graphically as a cube (as shown in Figure 13.3). We will use another actual example to demonstrate the design.

An Example of Applying the 2^3 Factorial Design

A fine leak test is performed after an integrated circuit is sealed to determine its hermeticity. To perform the test, the unit is first subjected to high pressure in a gaseous medium that acts as a tracer; then the unit is dried and placed into the test port where the tracer that is emitted from the leaking units is detected and measured.

A process engineer at the fine leak test area observed that units with a protective tape has a higher fine leak failure rate. It was suspected that this was due to the excess amount of tracer that was entrapped in the tape. It was then theorized that drying the units at 25°C might not have been sufficient to remove the excess tracer. So, it was decided to conduct an experiment to

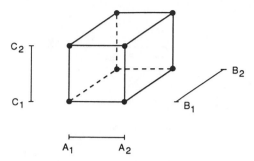

Figure 13.3 A graphical representation of the 2^3 factorial design.

evaluate the effect of increasing the drying temperature to 75°C to see if it was enough to remove the excess tracer, but not enough to use up all the tracer in the leakers.

Taking all the conditions into consideration, it was decided to perform a three-factor, two-level factorial design with two replicates. The setting conditions for the experiment are as in Table 13.10. The relative rate of fine leak was chosen as the response variable. The results of the experiments are shown in Table 13.11. To analyze the experimental results, the ANOVA procedure was performed as follows.

Step 1 Make a response table as shown in Table 13.12.
Step 2 Determine the correction factor (CF):

$$CF = \frac{(\text{grand total of all the observations})^2}{\text{total number of the observations}}$$

$$= \frac{Y_{....}^2}{a \times b \times c \times r}$$

TABLE 13.10   Setting Conditions for Fine Leak Test Experiment

Factor	Level	
	1	2
A: Dry temperature	25°C	75°C
B: Dry time	1.0 minutes	2.5 minutes
C: Protective tape	With	Without

TABLE 13.11 Results from the Experiment for Fine Leak Test

Dry temperature	25° C (A_1)		75° C (A_2)	
Dry time	1.0 minutes (B_1)	2.5 minutes (B_2)	1.0 minutes (B_1)	2.5 minutes (B_2)
Without tape (C_1)	81.92 146.50 (228.42)	39.83 60.50 (100.33)	58.25 102.67 (160.92)	40.58 63.13 (103.71)
With tape (C_2)	279.83 450.42 (730.25)	145.33 161.00 (306.33)	238.86 354.92 (593.78)	101.75 115.08 (216.83)

Values in parentheses indicate cell totals.

TABLE 13.12 Response Table for the Fine Leak Experiment

Level	A	B	C
1	$\Sigma A_1 = 1365.33$ $\bar{A}_1 = 170.67$	$\Sigma B_1 = 1713.17$ $\bar{B}_1 = 214.15$	$\Sigma C_1 = 593.38$ $\bar{C}_1 = 74.18$
2	$\Sigma A_2 = 1075.04$ $\bar{A}_2 = 134.43$	$\Sigma B_2 = 727.20$ $\bar{B}_2 = 90.90$	$\Sigma C_2 = 1846.99$ $\bar{C}_2 = 203.87$

Note: ΣA_1 is obtained by summing all the responses for factor A at level 1. By using the cell totals from Table 13.11, we can compute the value for ΣA_1 as follows:

$$\Sigma A_1 = 228.42 + 100.33 + 730.25 + 306.33$$

$$= 1365.33$$

Then

$$\bar{A}_1 = \frac{\Sigma A_1}{\text{number of responses in } A_1}$$

$$= \frac{1365.33}{8}$$

$$= 170.67$$

Using the same method, one can obtain all the values in the response table as above.

In this example, $a = 2$, $b = 2$, $c = 2$, and $r = 2$, so

$$N = a \times b \times c \times r$$

$$= 2 \times 2 \times 2 \times 2$$

$$= 16$$

Therefore,

$$CF = \frac{(81.92 + 146.50 + \cdots + 115.08)^2}{16}$$

$$= 372{,}212.86$$

Step 3 Determine the sum of squares (SS). The total correct sum of squares is found as follows:

$$SS_T = \sum_{i=1}^{a} \sum_{j=1}^{b} \sum_{k=1}^{c} \sum_{l=1}^{r} Y_{ijkl}^2 - \frac{Y_{\cdots}^2}{N}$$

$$= \sum (\text{each datum})^2 - CF$$

$$= 81.92^2 + 146.50^2 + \cdots + 115.08^2 - 372{,}212.86$$

$$= 215{,}521.37$$

The sum of squares for the main effects are calculated as follows:

$$SS_A = \sum_{i=1}^{a} \frac{Y_{i\cdots}^2}{b \times c \times r} - \frac{Y_{\cdots}^2}{N}$$

$$= \frac{(\Sigma A_1)^2 + (\Sigma A_2)^2}{b \times c \times r} - CF$$

$$= \frac{(1365.33)^2 + (1075.04)^2}{2 \times 2 \times 2} - 372{,}212.86$$

$$= 5266.77$$

In the same way, we can obtain

$$SS_B = 60{,}758.55$$

$$SS_C = 98{,}221.13$$

TABLE 13.13   Response Table for Two-Factor Interactions

	A_1	A_2
B_1	$\Sigma A_1 B_1 = 958.67$	$\Sigma A_2 B_1 = 754.50$
B_2	$\Sigma A_1 B_2 = 406.66$	$\Sigma A_2 B_2 = 320.54$
C_1	$\Sigma A_1 C_1 = 328.75$	$\Sigma A_2 C_1 = 264.63$
C_2	$\Sigma A_1 C_2 = 1036.58$	$\Sigma A_2 C_2 = 810.41$
	B_1	B_2
C_1	$\Sigma B_1 C_1 = 325.32$	$\Sigma B_2 C_1 = 204.04$
C_2	$\Sigma B_1 C_2 = 1323.83$	$\Sigma B_2 C_2 = 522.16$

Note: $\Sigma A_1 B_1$ is obtained by summing all the responses when factor A is at level 1 and factor B is at level 1. By using the cell totals from Table 13.11, we can compute the value for $\Sigma A_1 B_1$ as follows:

$$\Sigma A_1 B_1 = 228.42 + 730.25 = 958.67$$

To calculate the sums of squares for the two-factor interactions, response tables are constructed as shown in Table 13.13. Based on the results of Table 13.13:

$$SS_{A\,B\ \text{subtotal}} = \sum_{i=1}^{a} \sum_{j=1}^{b} \frac{Y_{ij..}^2}{c \times r} - \frac{Y_{....}^2}{N}$$

$$= \frac{(\Sigma A_1 B_1)^2 + (\Sigma A_1 B_2)^2 + (\Sigma A_2 B_1)^2 + (\Sigma A_2 B_2)^2}{c \times r} - \text{CF}$$

$$= \frac{(958.67)^2 + (406.66)^2 + (754.50)^2 + (320.54)^2}{2 \times 2} - 372{,}212.86$$

$$= 66{,}896.31$$

$$SS_{A \times B} = SS_{A\,B\ \text{subtotal}} - SS_A - SS_B$$

$$= 66{,}896.31 - 5266.77 - 60{,}758.55$$

$$= 870.99$$

By using the results in Table 13.13, in the same way we can obtain

$$SS_{A \times C} = 1641.26$$

$$SS_{B \times C} = 23{,}667.51$$

The three-factor interaction sum of squares is computed by using the $A \times B \times C$ cell totals (which are in parentheses in Table 13.11) as follows:

$$SS_{A B C \text{ subtotal}} = \sum_{i=1}^{a} \sum_{j=1}^{b} \sum_{k=1}^{c} \frac{Y_{ijk.}^2}{r} - \frac{Y_{....}^2}{N}$$

$$= \frac{(\Sigma A_1 B_1 C_1)^2 + (\Sigma A_1 B_1 C_2)^2 + \cdots + (\Sigma A_2 B_2 C_2)^2}{r} - CF$$

$$= \frac{(228.42)^2 + (730.25)^2 + \cdots + (216.83)^2}{2} - 372{,}212.86$$

$$= 190{,}461.34$$

$$SS_{A \times B \times C} = SS_{A B C \text{ subtotal}} - SS_A - SS_B - SS_C - SS_{A \times B} - SS_{A \times C} - SS_{B \times C}$$

$$= 190{,}461.34 - 5266.77 - 60{,}758.55 - 98{,}221.13$$

$$\quad - 870.99 - 1641.26 - 23{,}667.51$$

$$= 35.14$$

$$SS_E = SS_T - SS_{A B C \text{ subtotal}}$$

$$= 215{,}521.37 - 190{,}461.34$$

$$= 25{,}060.03$$

Step 4 Determine the degrees of freedom (df). The degrees of freedom are computed as follows:

$$df_T = (\text{total number of responses}) - 1 = 16 - 1 = 15$$

$$df_A = (\text{number of levels for factor } A) - 1 = 2 - 1 = 1$$

$$df_B = (\text{number of levels for factor } B) - 1 = 2 - 1 = 1$$

$$df_C = (\text{number of levels for factor } C) - 1 = 2 - 1 = 1$$

$$df_{A \times B} = df_A \times df_B = 1 \times 1 = 1$$

$$df_{A \times C} = df_A \times df_C = 1 \times 1 = 1$$

$$df_{B \times C} = df_B \times df_C = 1 \times 1 = 1$$

$$df_{A \times B \times C} = df_A \times df_B \times df_C = 1 \times 1 \times 1 = 1$$

$$df_E = df_T - df_A - df_B - df_C - df_{A \times B} - df_{A \times C} - df_{B \times C} - df_{A \times B \times C}$$

$$= 15 - 1 - 1 - 1 - 1 - 1 - 1 - 1 = 8$$

TABLE 13.14 ANOVA Summary Table of 2^3 Factorial Design for Fine Leak

(1) Source of Variation	(2) Sum of Squares (SS)	(3) Degrees of Freedom (df)	(4) Variance (V)	(5) F Ratio (F)
A	5,266.77	1	5,266.77	1.68
B	60,758.55	1	60,758.55	19.40*
C	98,221.13	1	98,221.13	31.36*
$A \times B$	870.99	1	870.99	0.28
$A \times C$	1,641.26	1	1,641.26	0.52
$B \times C$	23,667.51	1	23,667.51	7.56*
$A \times B \times C$	35.14	1	35.14	0.01
Error	25,060.03	8	3,132.50	
Total	215,521.37	15		

*Significant at the 95% confidence level.
$$F_{(0.05, 1, 8)} = 5.32.$$

Step 5 Determine the variance (V) by the following formula:

$$V = \frac{\text{sum of squares}}{\text{degree of freedom}}$$

For example,

$$V_A = \frac{SS_A}{df_A}$$

$$= \frac{5266.77}{1}$$

$$= 5266.77$$

The results of variance calculations are shown in Table 13.14, column (4).
Step 6 Determine the F ratios by the following formula:

$$F = \frac{\text{variance of factor}}{\text{variance of error}}$$

For example,

$$F_A = \frac{V_A}{V_E}$$

$$= \frac{5266.77}{3132.50}$$

$$= 1.68$$

The results of the F ratio calculations are shown in Table 13.14, column (5).

Step 7 Determine whether the F ratios are significant. In our example, it was decided to use a significance level of $\alpha = 0.05$, v_1 (the degrees of freedom for the numerator) $= 1$, and v_2 (the degrees of freedom for the denominator) $= 8$, from Appendix E, $F_{(0.05, 1, 8)} = 5.32$. Comparing F ratios with this critical value, we found

$$F_A = 1.68 < 5.32 \qquad \text{Not significant}$$

$$F_B = 19.40 > 5.32 \qquad \text{Significant}$$

$$F_C = 31.36 > 5.32 \qquad \text{Significant}$$

$$F_{A \times B} = 0.28 < 5.32 \qquad \text{Not significant}$$

$$A_{A \times C} = 0.52 < 5.32 \qquad \text{Not significant}$$

$$F_{B \times C} = 7.56 > 5.32 \qquad \text{Significant}$$

$$F_{A \times B \times C} = 0.01 < 5.32 \qquad \text{Not significant}$$

Conclusion

From the ANOVA table, we can conclude that the main effects of factor B (dry time) and factor C (with/without tape) are significant at the $\alpha = 0.05$ level. Also, interaction exists between factors B and C.

To interpret the results of this experiment, a graph for the average responses at each combination of factors B and C has been constructed as shown in Figure 13.4. From the graph, we can conclude that with dry time $= 2.5$ minutes (B_2) and without tape (C_1), the amount of fine leak is the lowest. Therefore, a conclusion can also be made that this particular tape impacts negatively on the response.

The engineer conducting this experiment needs to conduct further investigations to find a brand of bumper tape that would not entrap the tracer, and determine a suitable duration of drying time to avoid over rejection, or resolve the root causes of the reasons for using protective tape.

Since factor A (dry temperature) does not have a significant effect on the response (fine leak test readout), blowing hot air to dry the units is not the solution to the potential false leaker problem. This experiment helped the engineer make the correct decision not to install a costly air heater that would not change the effect of the protective tape to reduce the open/short circuit reject rate.

13.3.2.2 3^k Factorial Design

In industry there are some relationships that cannot be assumed to be linear, and experience shows that much more information can be obtained by testing (1) below the current level, (2) at the current level, and (3) above the current level. Therefore, the 3^k design of experiments is frequently used.

The 3^k factorial designs are a factorial arrangement with k factors, each at three levels. The three levels of the factors may be referred to as low,

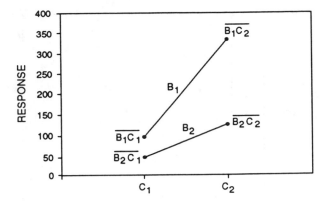

Figure 13.4 Response graph for factor B and C interaction.
Note: $\overline{B_1C_1}$ is obtained by using Table 13.13 and computing as follows:

$$\overline{B_1C_1} = \frac{\Sigma B_1C_1}{\text{number of responses in this combination}}$$

$$= \frac{325.32}{4}$$

$$= 81.33$$

In the same way, we obtain

$$\overline{B_1C_2} = 330.96$$

$$\overline{B_2C_1} = 51.01$$

$$\overline{B_2C_2} = 130.54$$

medium, and high. These levels will be denoted by the digits 1 (low), 2 (medium), and 3 (high).

The simplest design in the 3^k system is the 3^2 design, that is, two factors each at three levels. The treatment combinations can be graphically represented as in Figure 13.5.

Now suppose there are three factors (A, B, and C) under study, and each factor is arranged in three levels in an experiment. This is a 3^3 factorial design, and the experimental layout is shown in Figure 13.6.

The elements of the ANOVA for 3^k designs can be computed by the usual methods as described above. Due to space limitations, we will not describe the actual calculations here. The reader will find an example of the three-level design using Taguchi's method in the following section.

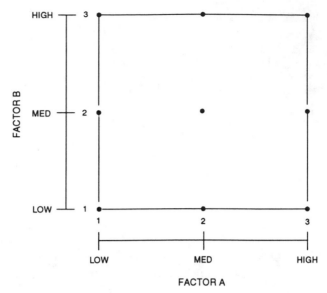

Figure 13.5   A graphical representation of a 3^2 factorial design.

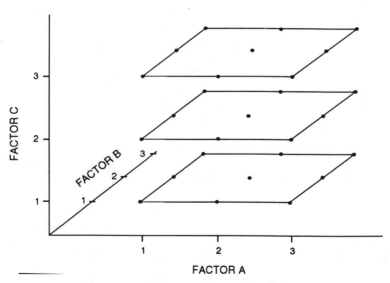

Figure 13.6   A graphical representation of a 3^3 factorial design.

13.4   THE TAGUCHI APPROACH TO QUALITY IMPROVEMENT

Genichi Taguchi developed a philosophy and methodology of quality improvement that introduced a significant change in the way of thinking about quality. The Taguchi approach is so different and versatile that it is very difficult to describe concisely. However, since AMD adopted Taguchi's concept of quality improvement, and because of the increasing interest in his approach, we will try to give at least some overview of Taguchi's concept, and demonstrate only one aspect of his methodology by applying design of experiments, using orthogonal arrays.

13.4.1   What Difference Does Taguchi's Approach Make?

Taguchi has developed new techniques and modified a number of existing statistical techniques, giving them a second life. Recognizing the importance of his statistical methods, which have caused a significant change in engineering technology, we would say that the most important part of Taguchi's philosophy is his concept, which can be reflected in two elements:[5]

1. Quality losses must be defined as the deviation from the target.
2. Quality should be designed in the product.

These are the most essential elements that influence a change in the definition of "quality" and how it can be achieved.

We have been taught for many years that "quality" is making parts to the specification requirements. This definition has been implanted in people's mentality. We know that if all parts are made within the specification limits, no further improvement is necessary because there will be no losses to the producer and the customer will be happy. Concern arises only when the parts fall outside the limits, because of extra scrap, rework, retest, and possible returns from the customer (see Figure 13.7). In reality this is the wrong interpretation of quality and losses.

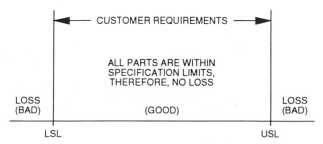

Figure 13.7   Traditional view of loss.

First of all, we know from experience that not all product within the specification has the same quality. Parts that are closer to the upper or lower specification have a different quality than the parts distributed closer around the target (nominal). Second, parts that are just outside the specification limits exhibit no significant difference in quality compared to the parts just within the specification limits.

This superstitious definition of quality is not good enough anymore, especially when we need to compete in international markets. It is important to remember that in nature there is not normally such an abrupt change from "perfect" to "useless" when some arbitrary boundary is crossed. What really happens is that performance gradually deteriorates as the quality measure deviates further and further from the intended target. This is reflected in Taguchi's concept.

13.4.2 Taguchi's Theory of Loss Function

Taguchi[6] defines quality as "the losses a product imparts to the society from the time the product is shipped" (see Figure 13.8). Taguchi describes this concept in his theory of loss function, where the customer's desire to have more consistent products is recognized. The loss function can be illustrated by a quadratic function as shown in Figure 13.8. The more the product deviates from the target value, the more loss is incurred.[7] Figure 13.9 shows two processes with different capabilities. Process A has a capability index of Cp = 1 and Process B has a capability index of Cp = 2. As one can see, the higher the capability index, the lower the loss is to the customer (see shaded area). In other words, the tighter the process spread is (the more parts that

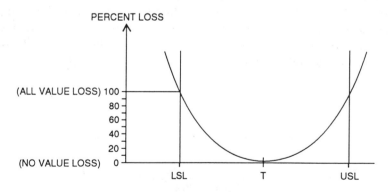

ON THE VERTICAL AXIS IS A MEASURE OF "VALUE LOSS" DUE TO DEVIATION OF THE CHARACTERISTIC FROM THE TARGET (T). FOR SIMPLICITY A PERCENTAGE VALUE VARYING FROM 0 TO 100% IS USED.

Figure 13.8 Taguchi's view of loss.

(A)
LOSSES FROM A PROCESS OF Cp=1

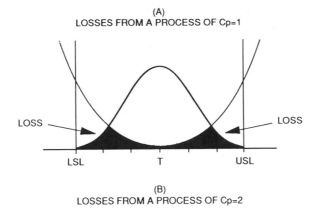

(B)
LOSSES FROM A PROCESS OF Cp=2

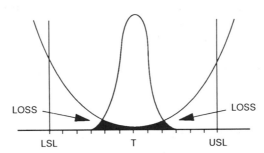

Figure 13.9   Losses from two different distributions.

are made closer to the target value), the higher the quality and the lower the losses.

From this we can conclude that reduction of the process variation around the target is the only way to ensure continuous process improvement and customer satisfaction. Regardless of the existing statistical controversy related to Taguchi's methods, we think that his concept of quality improvement is very progressive and reflects the current trend toward continuous improvements and customer satisfaction. This is why AMD adopted Taguchi's concept and some of his methodology. AMD's statistical "toolbox" consists of tools developed by various authors, who may have different approaches or points of view on quality improvement, but by using these tools properly, they work together very well.

As we demonstrated earlier, for on-line design of experiment AMD is using Box's EVOP, and for off-line design of experiment AMD is using the classic design of experiments and Taguchi's methods. In the following subsection a case history of applying Taguchi's method will be illustrated.

Taking into consideration that the Taguchi approach is new for some readers, before introducing the case history, we will give an overview of the new terms related with Taguchi's experiments. We will start with an introduction to orthogonal arrays.

13.4.3   Introduction to Orthogonal Arrays

In the previous subsection, we described a technique that belongs to the full factorial type of design of experiment. The examples we used were based on two factors each at two levels (2^2), and three factors each at two levels (2^3). So, to conduct a full factorial, we need to perform only four and eight runs (or treatments), respectively. But wnen we need to conduct an experiment that involves a larger number of factors, the number of runs required to perform a full factorial experiment will increase geometrically as the number of factors (k) increases. For example, if we plan to perform a two-level experiment involving seven factors, we would need to perform $2^7 = 128$ runs. However, it is important to note that when k becomes larger, the desired information can often be obtained by performing only a fraction of the full factorial design.

G. E Box[8] developed a concept called "fractional factorial design," which allows us to use a significantly lower number of runs and still obtain enough information to make relatively accurate decisions. For example, the full 2^5 factorial requires 32 runs, but when applying a half-factorial (2^{5-1}), we would employ only 16 runs and obtain essentially the same information. It is obvious that the fractional factorial concept will be more attractive for the experimenter since only a fraction of the number of runs need be used and the same information can be obtained as if a full factorial experiment were run. Continuing the effort of reducing the numbers of required runs, Taguchi[9] has developed a family of fractional factorial experiment matrixes that can be utilized in different situations. These matrixes are called "orthogonal arrays" (OAs). The OAs are labeled as

$$L_a(b)^c$$

where a is the number of tests required, b is the number of levels, and c is the maximum number of factors. For example, $L_9(3^4)$ means the design of the experiment will suit a maximum number of factors $c = 4$, each at $b = 3$ levels, and the number of tests required is $a = 9$.

Table 13.15 shows the orthogonal array $L_9(3^4)$. The numbers in the left-hand column are called the "experiment numbers" and they range from 1 to 9. The vertical alignment arrangements are called the "columns of the orthogonal array" and every column contains the values 1, 2, and 3, which represent the factor levels. Since the combinations of numbers in any two columns are combinations of 1, 2, and 3, there are nine ordered pairs. The nine combinations of the two columns: $(1,1)$, $(1,2)$, $(1,3)$, $(2,1)$, $(2,2)$, $(2,3)$,

TABLE 13.15   Orthogonal Array L_9

Experiment	Column			
Number	1	2	3	4
1	1	1	1	1
2	1	2	2	2
3	1	3	3	3
4	2	1	2	3
5	2	2	3	1
6	2	3	1	2
7	3	1	3	2
8	3	2	1	3
9	3	3	2	1

$(3,1)$, $(3,2)$, and $(3,3)$ appear with the same frequency (once each in this case).

Two such columns are said to be balanced or orthogonal. Since there are four columns in the orthogonal array $L_9(3^4)$, we can experiment on as many as four factors of three levels each.

The assignment of the factors by orthogonal array and other details of application and interpretation will be described later when an actual example will be used to demonstrate the application of the Taguchi method.

13.4.3.1   Selecting the Appropriate Orthogonal Array

We need to know two things to be able to select the appropriate orthogonal array (OA): the minimum required degrees of freedom in the experiment and the total degrees of freedom available in an OA. To determine the minimum required degrees of freedom in an experiment, we need to know:

1. The number of factors and interactions of interest.
2. The number of levels for the factor of interest.

These two elements determine the total number of degrees of freedom required for the entire design of experiment. As we mentioned earlier, the degrees of freedom for each factor is the number of levels minus one; for example,

$$\mathrm{df}_A = (\text{number of levels for factor } A) - 1$$

and the degrees of freedom for an interaction is calculated by multiplying all the degrees of freedom of the interaction factors together; for example,

$$\mathrm{df}_{A \times B} = \mathrm{df}_A \times \mathrm{df}_B$$

Note that the minimum required degrees of freedom in the experiment is the sum of all the factors and interaction degrees of freedom.

The determination of the degrees of freedom available in an OA is very simple. The number in the array designation indicates the number of trials in the array. For example, L_9 has nine trials. The total degrees of freedom available in an OA (df_{Ln}) is equal to the number of trials (n_t) minus one. Mathematically, this can be expressed as

$$df_{Ln} = n_t - 1$$

Having all this information, we can select the appropriate OA for an experiment. The following inequality may be satisfied to have the right OA:

$$df_{Ln} \geq df \text{ required for factors and interactions}$$

One more element should be taken into consideration when determining the appropriate OA. This is the number of levels. If the factors are two-levels, the basic kinds of OA are: L_4, L_8, L_{12}, L_{16}, and L_{32}. If the factors are three-levels, the basic OAs are: L_9, L_{18}, and L_{27} (see Appendix F).[9]

13.4.3.2 An Example of Selecting the Appropriate Orthogonal Array

Assume that we have selected three factors, say A, B, and C, to be involved in the experiment and each at two levels, then the degrees of freedom for these three main factors are

$$df_A = 2 - 1 = 1$$

$$df_B = 2 - 1 = 1$$

$$df_C = 2 - 1 = 1$$

And, if we want to evaluate all the degrees of freedom for interaction, then the degrees of freedom for the interactions are

$$df_{A \times B} = df_A \times df_B = 1 \times 1 = 1$$

$$df_{B \times C} = df_B \times df_C = 1 \times 1 = 1$$

$$df_{A \times C} = df_A \times df_C = 1 \times 1 = 1$$

$$df_{A \times B \times C} = df_A \times df_B \times df_C = 1 \times 1 \times 1 = 1$$

The minimum total degrees of freedom required for the experiment is

$$df_{Ln} = df_A + df_B + df_C + df_{A \times B} df_{B \times C} + df_{A \times C} + df_{A \times B \times C}$$

$$= 7$$

Then the minimum trials required in the experiment is

$$n_t = df_{Ln} + 1 = 7 + 1 = 8$$

Since the factors are two-levels, L_8 is the OA suitable for the planned experiments (see Appendix F).

13.4.4   Control and Noise Factors

Taguchi separates factors into two main groups: control factors and noise factors. *Control factors* are those that are set by the manufacturer and cannot be directly changed by the customer. For example, a new design of an electronic device can dictate the material for the packaging or the construction of the lead frame, which cannot be easily modified by the customer. *Noise factors* (see Figure 13.10) include those factors over which the manu-

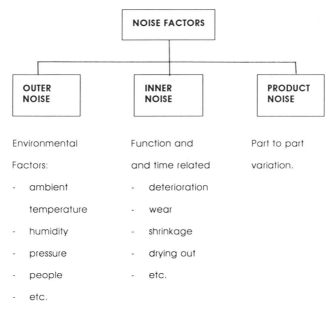

- Outer noises produce variation from outside the product.

- Inner noises produce variation from inside or within the product.

```
                         ┌───────────────┐
                         │ NOISE FACTORS │
                         └───────────────┘
```

OUTER NOISE   INNER NOISE   PRODUCT NOISE

Environmental   Function and   Part to part
Factors: and time related   variation.
- ambient   - deterioration
   temperature   - wear
- humidity   - shrinkage
- pressure   - drying out
- people   - etc.
- etc.

Note: Product may have sensitivity to all three forms of noise simultaneously.

Figure 13.10   Three forms of noise factors.

facturer has no direct control, but which vary with the customer's environment and usage. For example, the temperature or vibration an electronic device receives when being placed in a customer's system (computer, car, airplane, etc.). Customer specifications can also be noted as an example of a noise factor. Noise factors can be divided into three categories:

1. Outer noise
2. Inner noise
3. Product noise

Outer noises introduce variation in the process or product from the outside, inner noises introduce variation in the process or product from the inside, and product noises are related to part-to-part variation. Figure 13.10 describes these three different types of noise.

A product or a process can be influenced simultaneously by all three forms of noise. Improving the process or product design, we can make them robust to all types of noise. This can be accomplished by the introduction of on-line and off-line design of experiments.

13.4.5   Signal-To-Noise Ratios

To optimize the process, we are interested in not only the effect of factors that might impact the average response, but also in the effect on variation. To determine which factors might affect the process variation, Taguchi has created a transformation of the repetition data to another value that is a measure of the variation present. This transformation is reflected through an index called the "Signal-to-noise ratio (S/N ratio)." This ratio consolidates several replicants (at least two) into one value that reflects the amount of variation present in the process. Taguchi developed several S/N ratios that can be used for different types of characteristics:

Lower is better (LB)
Nominal is better (NB)
Higher is better (HB)

Depending on the type of characteristic being evaluated, we select one of these equations to determine the S/N ratio. For example, if the response is yield, we select the higher is better (HB) equation, or if the response is reject rate, we use the lower is better (LB) equation, etc.

Following are the equations to calculate S/N ratios.

1. Lower is better (LB):

$$S/N_{\text{LB}} = -10\log\left(\frac{1}{r}\sum_{i=1}^{r} Y_i^2\right)$$

where r is the number of tests in a trial (number of replicants).

2. Nominal is better (NB):

$$S/N_{\text{NB}} = 10\log\left[\frac{(S_M - V_E)}{rV_E}\right]$$

where

$$S_M = \frac{1}{r}\left(\sum_{i=1}^{r} Y_i\right)^2$$

$$V_E = \frac{1}{r-1}\left(\sum_{i=1}^{r} Y_i^2 - S_M\right)$$

3. Higher is better (HB):

$$S/N_{\text{HB}} = -10\log\left(\frac{1}{r}\sum_{i=1}^{r} \frac{1}{Y_i^2}\right)$$

The S/N ratio is the reciprocal of the variance of the measurement error, so it is maximal for the combination of parameter levels that has the minimum error variance. Its value can be treated as a response of the experiment, which is a measure of the variation within a trial when noise factors are present.

To analyze the S/N ratio values, an ordinary ANOVA technique can be used, which will identify factors that are significant to increasing the average value of S/N and subsequently reducing variation. The methodology of calculating the S/N ratio by using one of the formulas mentioned above will be demonstrated in an example later in this section.

13.4.6   Pure Sum of Squares and Percent Contribution

To further interpret the experimental results, two more columns [pure sum of squares (SS') and percentage contribution (P)] are added to the ANOVA table. Now we will briefly describe these two new elements.

The variance due to a factor or interaction actually contains some amount due to error.[10] Therefore, to more accurately estimate the variation due to a factor, Taguchi introduced the "pure sum of squares" (SS'), which is derived by subtracting the portion due to error from the sum of squares (SS). As an example, for factor A:

$$SS'_A = SS_A - (df_A)(V_E)$$

Since some portion of the sum of squares for a factor and/or an interaction was subtracted out because of error, this amount is added to the error sum of squares in order that the total sum of squares is unchanged.

The percent contribution is a ratio of the pure sum of squares of each significant item to the total sum of squares. It reflects the portion of the total variation observed in an experiment attributed to each significant factor and interaction. The value of the percentage contribution indicates the power of a factor (or interaction) to reduce the variation. Interpreting this value, one can say that if a particular significant factor were controlled precisely, the total variation would be reduced by the amount indicated by the percent contribution. For example, if a factor has a 25% contribution, this means that if we controlled this variable precisely, we would reduce the overall amount of variation by 25%.

It is important to note that when running a design of experiment to improve a manufacturing process, the total variation observed should represent more than 75% of the variation observed in the process.[11] If the percent contribution due to error (unknown and uncontrolled factors) is low (15% or less), then it is assumed that no important factors were omitted from the experiment. If it is a high value (50% or more), then some important factors were definitely omitted, conditions were not precisely controlled, or measurement error was excessive.

We will demonstrate the procedure of performing a Taguchi experiment by using a real case history.

Case History 13.1: Applying the Taguchi Method to Reduce Nonsticking on Pad Problems in the Wire Bonding Process

With the doping of copper in aluminum present on wire bond pads to reduce electromigration of the interconnecting metal used in electronic devices, the wire bonding process has been facing a higher percentage of nonsticking of the wire bonds on the die pads (NSOP). It was decided to conduct a design of experiment to find out the sources of variation and determine the optimum process settings that would minimize the NSOP problems.

Through brainstorming sessions, four factors were identified as the possible major contributors to NSOP problems: bond time, bond force, bond power, and bond temperature. A three-level design was chosen for the experiment (as opposed to a two-level design which affords simplicity) since no previous information on the behavior of the four factors was studied that could have led to an assumption that a linear relationship existed.

The type of experimental design chosen was Taguchi's method. The selection was made considering the time, effort, and cost of performing the design of experiments. Among several orthogonal arrays, the $L_9(3^4)$ would be an appropriate arrangement. In this array, a maximum of four factors could be studied, each at three levels, for a total of nine experimental runs. With the usual factorial experiment, a 3^4 factorial design corresponding to the present case could require 81 runs, which is nine times more than the number of runs needed for the selected Taguchi orthogonal array.

The experimental settings for the four selected factors were as shown in Table 13.16. The response was determined to be relative percent yield. Fifty units were tested per experiment run and repeated twice. The results are shown in Table 13.17. Based on these experimental results, an ANOVA analysis was performed as follows.

Step 1 Make a response table as shown in Table 13.18.

Step 2 Determine the correction factor (CF):

$$CF = \frac{(\Sigma \text{ all data})^2}{n}$$

$$= \frac{(37 + 10 + \cdots + 95)^2}{18}$$

$$= \frac{1465^2}{18}$$

$$= 119{,}235$$

TABLE 13.16   The Experimental Settings

	Level		
Factor	1	2	3
A: Bond force (z pulses)	12	16	20
B: Bond time (msecs)	20	40	60
C: Bond power (mwatts)	80	100	120
D: Temperature (°C)	170	180	190

TABLE 13.17   The Results of the Experiments for NSOP Problems

Run Number	Factor				Response (Yield)		S/N
	A	B	C	D			
1	1	1	1	1	37	10	22.7
2	1	2	2	2	87	85	38.7
3	1	3	3	3	91	95	39.7
4	2	1	2	3	93	59	37.0
5	2	2	3	1	93	91	39.3
6	2	3	1	2	95	95	39.6
7	3	1	3	2	95	97	39.6
8	3	2	1	3	85	67	37.4
9	3	3	2	1	95	95	39.6

Notes: 1. The 1, 2, and 3 in the body of the table are levels of settings. For example, when performing run 1 the process was set with all factors on level 1, which means that the bond force is at 12 z pulses, the bond time is at 20 msecs, the bond power is at 80 mwatts, and the temperature is at 170°C (see Table 13.16).

2. The S/N factor, which is reflected in the right-hand column, is the signal-to-noise (S/N) ratio that will be described later.

Step 3 Determine the sum of squares (SS). The total sum of squares (SS_T) is computed as follows:

$$SS_T = \sum (\text{each datum})^2 - CF$$

$$= (37^2 + 10^2 + \cdots + 95^2) - 119{,}235$$

$$= 9642$$

Based on the results in Table 13.18, the sum of squares for each factor is calculated as follows:

$$SS_A = \frac{(\Sigma A_1)^2}{n} + \frac{(\Sigma A_2)^2}{n} + \frac{(\Sigma A_3)^2}{n} - CF$$

$$= \frac{(405)^2}{6} + \frac{(526)^2}{6} + \frac{(534)^2}{6} - 119{,}235$$

$$= 1741$$

TABLE 13.18   Response Table for the NSOP Experiment

Level	Factor			
	A	B	C	D
1	$\Sigma A_1 = 405$	$\Sigma B_1 = 391$	$\Sigma C_1 = 389$	$\Sigma D_1 = 421$
	$\bar{A}_1 = 67.5$	$\bar{B}_1 = 65.2$	$\bar{C}_1 = 64.8$	$\bar{D}_1 = 70.2$
2	$\Sigma A_2 = 526$	$\Sigma B_2 = 508$	$\Sigma C_2 = 514$	$\Sigma D_2 = 554$
	$\bar{A}_2 = 87.7$	$\bar{B}_2 = 84.7$	$\bar{C}_2 = 85.7$	$\bar{D}_2 = 92.3$
3	$\Sigma A_3 = 534$	$\Sigma B_3 = 566$	$\Sigma C_3 = 562$	$\Sigma D_3 = 490$
	$\bar{A}_3 = 89.0$	$\bar{B}_3 = 94.3$	$\bar{C}_3 = 93.7$	$\bar{D}_3 = 81.7$

Note: ΣA_1 is obtained by summing all the responses for factor A at level 1. From Table 13.17, we find that the responses of runs 1, 2, and 3 were received with factor A set at level 1; therefore,

$$\Sigma A_1 = 37 + 10 + 87 + 85 + 91 + 95 = 405$$

Then

$$\bar{A}_1 = \frac{\Sigma A_1}{\text{number of responses in } A_1}$$

$$= \frac{405}{6}$$

$$= 67.5$$

By using the same method, one can obtain all the values in the response table as above.

In the same way, we can obtain

$$SS_B = 2649$$

$$SS_C = 2659$$

$$SS_D = 1475$$

The error sum of squares (SS_E) is computed as follows:

$$SS_E = SS_T - SS_A - SS_B - SS_C - SS_D$$

$$= 9642 - 1741 - 2649 - 2659 - 1475$$

$$= 1119$$

Step 4 Determine the degrees of freedom (df):

$$\text{df}_T = (\text{total number of responses}) - 1 = 18 - 1 = 17$$

$$\text{df}_A = (\text{number of levels for factor } A) - 1 = 3 - 1 = 2$$

$$\text{df}_B = (\text{number of levels for factor } B) - 1 = 3 - 1 = 2$$

$$\text{df}_C = (\text{number of levels for factor } C) - 1 = 3 - 1 = 2$$

$$\text{df}_D = (\text{number of levels for factor } D) - 1 = 3 - 1 = 2$$

$$\text{df}_E = \text{df}_T - \text{df}_A - \text{df}_B - \text{df}_C - \text{df}_D$$

$$= 17 - 2 - 2 - 2 - 2 = 9$$

Step 5 Determine the variance (V). The variance for each factor is determined by dividing the sum of squares (see Step 3) by the degrees of freedom (see Step 4):

$$V_A = \frac{\text{SS}_A}{\text{df}_A} = \frac{1741}{2} = 870.5$$

$$V_B = \frac{\text{SS}_B}{\text{df}_B} = \frac{2649}{2} = 1324.5$$

$$V_C = \frac{\text{SS}_C}{\text{df}_C} = \frac{2659}{2} = 1329.5$$

$$V_D = \frac{\text{SS}_D}{\text{df}_D} = \frac{1475}{2} = 737.5$$

$$V_E = \frac{\text{SS}_E}{\text{df}_E} = \frac{1119}{9} = 124$$

Step 6 Determine the F ratio (F). The F ratio is determined by dividing the variance of each factor by the error variance:

$$F_A = \frac{V_A}{V_E} = \frac{870.5}{124} = 7.01$$

$$F_B = \frac{V_B}{V_E} = \frac{1324.5}{124} = 10.67$$

$$F_C = \frac{V_C}{V_E} = \frac{1329.5}{124} = 10.70$$

$$F_D = \frac{V_D}{V_E} = \frac{737.5}{124} = 5.93$$

Step 7 Determine pure (or net) sum of squares (SS') for each factor:

$$SS'_A = SS_A - (df_A \times V_E) = 1741 - (2 \times 124.0) = 1493$$

$$SS'_B = SS_B - (df_B \times V_E) = 2649 - (2 \times 124.0) = 2400$$

$$SS'_C = SS_C - (df_C \times V_E) = 2659 - (2 \times 124.0) = 2410$$

$$SS'_D = SS_D - (df_D \times V_E) = 1475 - (2 \times 124.0) = 1226$$

$$SS'_E = SS_T - SS'_A - SS'_B - SS'_C - SS'_D$$

$$= 9642 - 1493 - 2400 - 2410 - 1226 = 2113$$

Step 8 Determine the percent contribution, P (%):

$$P_A = \frac{SS'_A}{SS_T} \times 100 = \left(\frac{1493}{9642} \right) \times 100 = 15.5\%$$

$$P_B = \frac{SS'_B}{SS_T} \times 100 = \left(\frac{2400}{9642} \right) \times 100 = 24.9\%$$

$$P_C = \frac{SS'_C}{SS_T} \times 100 = \left(\frac{2410}{9642} \right) \times 100 = 25.0\%$$

$$P_D = \frac{SS'_D}{SS_T} \times 100 = \left(\frac{1226}{9642} \right) \times 100 = 12.7\%$$

$$P_E = \frac{SS'_E}{SS_T} \times 100 = \left(\frac{2113}{9642} \right) \times 100 = 21.9\%$$

Step 9 Complete the ANOVA summary table. The ANOVA table is completed as shown in Table 13.19. As one can see, all the factors are significant at the $\alpha = 0.05$ level. Furthermore, factor B (bond time) and factor C (bond power) contribute more to the total variation.

Signal-to-Noise (S/N) Ratio
In our example the response is yield which is characterized by the higher the better; therefore, we use the following formula:

$$S/N = -10 \log \left[\left(\frac{1}{r} \right) \sum_{i=1}^{r} \left(\frac{1}{Y_i^2} \right) \right]$$

TABLE 13.19 ANOVA Summary Table for NSOP Experiment

(1) Source of Variation	(2) Sum of Squares (SS)	(3) Degrees of Freedom (df)	(4) Variance (V)	(5) F Ratio (F)	(6) Pure Sum of Squares (SS')	(7) Percent Contribution (P)
A	1741	2	870.5	7.01*	1493	15.5
B	2649	2	1324.5	10.67*	2400	24.9
C	2659	2	1329.5	10.70*	2410	25.0
D	1475	2	737.5	5.93*	1226	12.7
Error	1119	9	124.0		2113	21.9
Total	9642	17				

*Significant at the 95% confidence level.

$F_{0.05,2,9} = 4.26$.

Note: Refer to Appendix E to find the F value for the 95% confidence level with $v_1 = 2$ and $v_2 = 9$ degrees of freedom.

Let's use run 1 to illustrate the calculation of the S/N ratio:

$$S/N_1 = -10\log\left[\left(\frac{1}{2}\right)\left\{\left(\frac{1}{37^2}\right) + \left(\frac{1}{10^2}\right)\right\}\right]$$

$$= 22.7$$

The results of the S/N ratio calculations are shown in the right-hand column of Table 13.17. Based on the S/N ratio values, we perform a regular ANOVA analysis. Note that when we calculate the degrees of freedom, we encounter a special case, that the degrees of freedom for error is zero:

$$df_E = df_T - df_A - df_B - df_C - df_D$$

$$= (9-1) - (3-1) - (3-1) - (3-1)$$

$$= 8 - 2 - 2 - 2 - 2$$

$$= 0$$

In this case, we can use a rule of thumb, which suggests that half of the total degrees of freedom should be pooled. The total degrees of freedom is 8, then half of it is 4. Since the sum of squares of factors A and D are the smallest compared to the other two factors, they are pooled to give an estimate of the error sum of squares. The degree of error now will be 4 from factors A and D. The ANOVA summary table for S/N is shown in Table 13.20.

TABLE 13.20   ANOVA Summary Table for the S/N Ratio

(1) Source of Variation	(2) Sum of Squares (SS)	(3) Degrees of Freedom (df)	(4) Variance (V)	(5) F Ratio (F)	(6) Pure Sum of Squares (SS')	(7) Percent Contribution (P)
A	(51.1)	(2)	—	—	—	—
B	72.8	2	36.42	1.46	23.06	9.6%
C	67.9	2	33.97	1.36	18.15	7.6%
D	(48.5)	(2)	—	—	—	—
Error	99.6	4	24.89		199.15	82.8%
Total	240.4	8				

Parentheses indicate the pooled element.

$F_{0.05, 2, 4} = 6.94$.

Note: Refer to Appendix E to find the F value for the 95% confidence level with $v_1 = 2$ and $v_2 = 4$ degrees of freedom.

The results for the S/N ratio analysis show that all the factors are insignificant, but the error contributes 82.8% to the process variation. This indicates that either the measuring system or other factors that are not included in this experiment have impacted the variation of the process. Further investigation should be conducted to minimize the process variation.

Conclusion

To determine the optimum settings, we can use the response table (see Table 13.18) to evaluate the average response at each level of the factors. Also, response graphs were plotted as shown in Figure 13.11. Since we want the response (percent yield) to be as high as possible, $A_3B_3C_3D_2$ should be selected as the optimum setting. This means the bond force should be set at 20 z pulses, the bond time should be set at 60 msecs, the bond power should be set at 120 mwatts, and the temperature should be set at 180°C.

To reduce the variation, the same analysis should be done based on the S/N ratio results. Although in our case all the factors were shown to be insignificant on the S/N analysis, the response table and graphs are constructed to demonstrate the idea (see Table 13.21 and Figure 13.12).

Note that in selecting the level of factors based on the S/N ratio responses, we should select the one with the highest value. Therefore, $A_3B_3C_3D_2$ should be the optimal setting to reduce the process variation, which is coincidental with the result based on the average response analysis.

Confirmation

A confirmation run was conducted to check the experimental results by setting the process at the optimum conditions for 19 consecutive days,

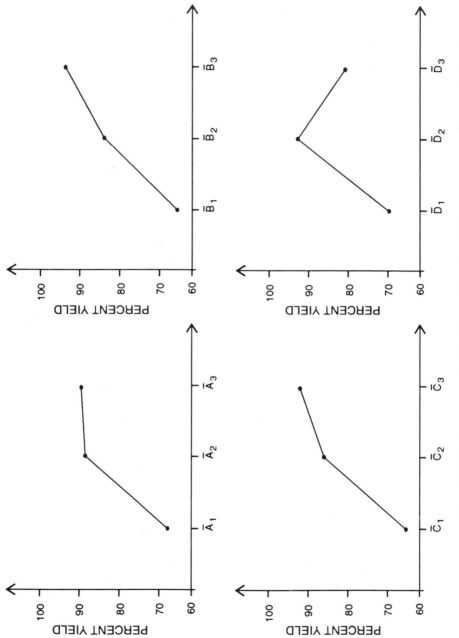

Figure 13.11 Response graphs for different factor levels (based on average response analysis).

TABLE 13.21 Response Table for Selecting the Optimum Settings (Based on the S/N Ratio Analysis)

	Factor			
Level	A	B	C	D
1	$\Sigma A_1 = 101.1$	$\Sigma B_1 = 99.3$	$\Sigma C_1 = 99.7$	$\Sigma D_1 = 101.6$
	$\bar{A}_1 = 33.7$	$\bar{B}_1 = 33.1$	$\bar{C}_1 = 33.2$	$\bar{D}_1 = 33.9$
2	$\Sigma A_2 = 115.9$	$\Sigma B_2 = 115.4$	$\Sigma C_2 = 115.3$	$\Sigma D_2 = 117.9$
	$\bar{A}_2 = 38.6$	$\bar{B}_2 = 38.5$	$\bar{C}_2 = 38.4$	$\bar{D}_2 = 39.3$
3	$\Sigma A_3 = 116.6$	$\Sigma B_3 = 118.9$	$\Sigma C_3 = 118.6$	$\Sigma D_3 = 114.1$
	$\bar{A}_3 = 38.9$	$\bar{B}_3 = 39.6$	$\bar{C}_3 = 39.5$	$\bar{D}_3 = 38.03$

Note: ΣA_1 is computed by summing all the responses for factor A at level 1. From Table 13.17, we find that the S/N ratios of runs 1, 2, and 3 were obtained when factor A was set at level 1; therefore,

$$\sum A_1 = 22.7 + 38.7 + 39.7 = 101.1$$

Then

$$\bar{A}_1 = \frac{\Sigma A_1}{\text{number of } S/N \text{ ratios in } A_1}$$

$$= \frac{101.1}{3}$$

$$= 33.7$$

By using the same method, one can obtain all the values in the response table as above.

accumulating a total of 241,092 processed units. The results show that the quality level of the process has been improved from 4250 ppm ($\bar{p} = 0.425\%$) to 780 ppm ($\bar{p} = 0.078\%$) (see Figure 13.13), which is an approximately sixfold reduction in the average percent defective rate. Through this example, the Taguchi method has been proven to be an effective way of improving the process.

13.4.7 Further Interpretation of the Experimental Results

Once an experiment has been conducted, an ANOVA completed, the significant factor and/or interactions identified, and response graphs plotted, conclusions are made to determine the optimum setting. However, this is not all. There is other useful information that can be obtained from the experiments. We will discuss several techniques to further interpret the experimental results in this section.

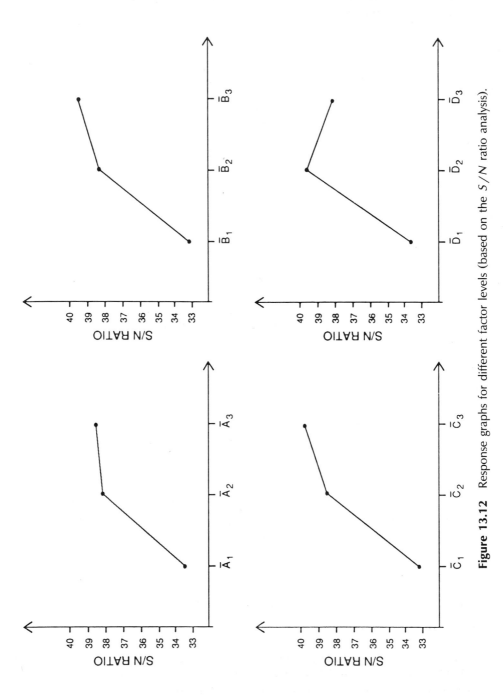

Figure 13.12 Response graphs for different factor levels (based on the S/N ratio analysis).

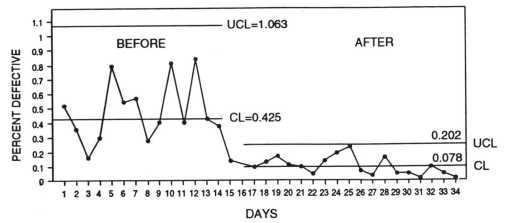

Figure 13.13   The results of applying Taguchi's method for nonstick on pad reduction.

Estimating the Mean
Typically, the experimenter would like to obtain some estimate value of the mean response ($\hat{\mu}$) based on the optimum setting from the experiment results. To estimate the mean, we use the following formula:

$$\hat{\mu} = \left(\sum \text{average response on the significant factor level} \right)$$

$$- \left(\text{number of total significant factors} - 1 \right) \overline{T}$$

where $\overline{T}$ is the average response from the experiment, and can be obtained as follows:

$$\overline{T} = \frac{\sum \text{all the responses}}{\text{number of responses}}$$

To illustrate the calculation, let's use the data from the case history, which is described in the previous section. From Table 13.17,

$$\overline{T} = \frac{(37 + 10 + \cdots + 95)}{18} = 81.4$$

In the case history, four factors are identified as significant. The optimum setting for the process to obtain the highest yield is $A_3 B_3 C_3 D_2$, and $\overline{A}_3 = 89.0$, $\overline{B}_3 = 94.3$, $\overline{C}_3 = 93.7$, and $\overline{D}_2 = 92.3$ (see Table 13.18). Based on these results, we can compute the estimated mean response as follows:

$$\hat{\mu} = \bar{A}_3 + \bar{B}_3 + \bar{C}_3 + \bar{D}_2 - (4-1)\bar{T}$$

$$= 89.0 + 94.3 + 93.7 + 92.3 - (3)81.4$$

$$= 125.1\%$$

However, the percent yield cannot exceed 100%. In this case, we need to apply some kind of transformation to obtain a more reasonable estimate of the mean.

Omega Transformation

Sometimes when the selected response value is in percentage values (such as percent yield, percent loss, or percent defective), and we estimate the process average based on the best setting conditions at the end of the experiment. If the predicted process average calculation results in a yield that is higher than 100% or a reject rate that is less than zero, both cases are meaningless. This suggests that when the values approach 100% or 0%, the additivity may be poor. Experience shows that the additivity is generally good only when the percentages are in a range of 20% to 80%. To resolve this problem, Taguchi describes three methods to avoid meaningless results. These methods are:

1. Inverse sine transformation
2. Assuming arithmetic additivity on the logarithmic scale
3. Omega transformation

Here we will describe only the omega transformation. To describe the principle of the omega transformation, we will utilize the results from the case history.

Applying the Omega Transformation

Taguchi[12] suggests a formula that can be used to transform the original data in decibels (dB), which will allow us to perform arithmetic additivity. The equation of transformation is as follows:

$$-10\log\left(\frac{1}{p}-1\right) = \text{decibel unit}$$

For example, substituting $\bar{T} = 81.4\%$ into the above equation as a value of p, we have

$$-10\log\left(\frac{1}{p}-1\right) = -10\log\left(\frac{1}{81.4\%}-1\right)$$

$$= -10\log\left(\frac{1}{0.814}-1\right)$$

$$= -10\log(0.228)$$

$$= 6.411 \text{ dB}$$

This decibel value can also be found by using Appendix G.[13] In the same way, we use the formula or Appendix G to find:

$$\overline{A}_3 = 89.0\% = 9.08 \text{ dB}$$

$$\overline{B}_3 = 94.3\% = 12.186 \text{ dB}$$

$$\overline{C}_3 = 93.7\% = 11.724 \text{ dB}$$

$$\overline{D}_2 = 92.7\% = 11.038 \text{ dB}$$

From this we get the decibel value for the estimated process mean as follows:

$$\hat{\mu} = \overline{A}_3 + \overline{B}_3 + \overline{C}_3 + \overline{D}_2 - 3\overline{T}$$

$$= 9.08 + 12.186 + 11.724 + 11.038 - 3(6.411)$$

$$= 24.805 \text{ dB}$$

By using Appendix G, we then transfer the decibel values back to a percentage:

$$24.805 \text{ dB} \approx 99.7\%$$

Now we can interpret the experimental results in a meaningful way, which means that if we set the process parameters as suggested in the experiment, we will obtain a yield of 99.7%.

Confidence Interval Around the Estimated Mean

We have estimated the mean response based on the optimum conditions. But how true is this estimation? We would like to have some interval within which the true average would be expected to fall with some stated confidence. The method for calculating a confidence interval (CI) is as follows:

$$\text{CI} = \sqrt{\frac{(F_{\alpha,1,\text{df}_E})(V_E)}{n_{\text{eff}}}}$$

where $F_{\alpha,1,\text{df}_E}$ is the F ratio required, α is the risk, df_E is the degrees of freedom for V_E, V_E is the error variance, and

$$n_{\text{eff}} = \frac{n}{1 + \left(\begin{array}{c}\text{total degrees of freedom associated}\\ \text{with items used in } \mu \text{ estimate}\end{array}\right)}$$

In our case, we selected $\alpha = 0.05$ and $\text{df}_E = 9$, then from Appendix E:

$$F_{0.05,1,9} = 5.12$$

From the ANOVA summary table for responses (see Table 13.19):

$$V_E = 124$$

Since $\mu = \bar{A}_3 + \bar{B}_3 + \bar{C}_3 + \bar{D}_2 - 3\bar{T}$, and $n = 18$:

$$n_{\text{eff}} = \frac{18}{1 + (1+1+1+1)} = 3.6$$

Then

$$\text{CI} = \sqrt{\frac{(F_{0.05,1,9})(V_E)}{n_{\text{eff}}}} = \sqrt{\frac{5.12 \times 124}{3.6}} = 13.3$$

Therefore, with 95% confidence, we can expect that the estimate average response ($\hat{\mu}$) will fall in the range of 99.7% $\pm$ 13.3%, which is 86.4% to 100%.

13.5 NESTED DESIGN

As discussed earlier, improving the process by continually reducing its excess variation is one of the major concerns for companies that are committed to quality. To do this, we need to identify the factors that are most influential on the process variation.

The control chart approach in reducing the process variation works very well to some extent. By analyzing a control chart pattern, we can tentatively determine the cause of disturbing the process. This can be, for example, the difference in operational performance of the incoming materials, a variation in the machine setting procedures or in a shift's performance, etc.

But there is a more formal and accurate way of identifying and estimating the sources of variation. For example, the operating manager of a semiconductor assembly line indicates that there is a high variation in the incoming wafer quality and wants to determine where the major problem is. The sources of variation in this particular case are:

Lot-to-lot variation

Wafer-to-wafer variation

Die-to-die variation

Bump-to-bump variation

Measurement variation

If we had the opportunity to break down the total variation of the incoming wafers by its components, we could concentrate the engineering efforts on the most influential causes of variation. Nested design is one of the most effective methods that can be applied to break down the total variation into mutually exclusive components of variation.

13.5.1   What Is Nested Design?

Nested design is a method of diagnosing sources of variability in a manufacturing process. It provides a breakdown of where the variability occurs within a process. Nested designs are appropriate when each of the factors in an experiment is progressively nested within the preceding factor. This is why nested design is sometimes called hierarchical design. For example, the factors in the wafer quality study are hierarchically nested: wafers within a lot, dies within a wafer, bumps within a die (see Figure 13.14).

A nested design is usually balanced. This is achieved by including an equal number of levels of one factor within each of the levels of the preceding factor. For example, the 16 levels of bumps are nested equally (balanced) in the levels of the die. In other words, every level of die has two levels of bumps. It is important to note that in a nested design we receive more information on factors that are located lower in the hierarchy than on those that are higher. For example, only two lots are selected in the design of Figure 13.14, whereas 16 bumps are included in the experiment, because bumps are in a lower hierarchy. This should be taken into consideration when designing the experiment. There are sometimes situations where it is not desirable to progressively increase the number of levels at the lower stages of the hierarchy. In this case, "staggered" nested design can be used instead of "balanced." Figure 13.15 shows a typical layout of a "staggered" (or unbalanced) nested design.

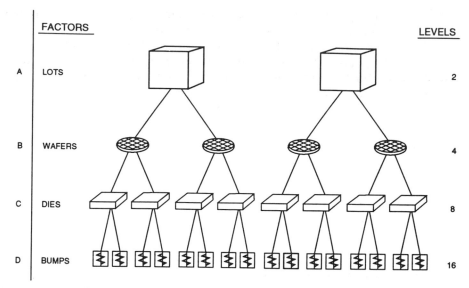

Figure 13.14   Wafer variable study levels.

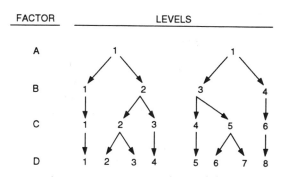

Figure 13.15   Staggered nested design.

It is important to note that in a "staggered nested design" whenever a factor level has only a single level of the next factor nested within it, all subsequent factors also have only a single level nested within them (see Figure 13.15).

13.5.2   Statistical Analysis

The simplest nested design is the two-stage model[14] which is

$$Y_{ijk} = \mu + A_i + B_{j(i)} + E_{(ij)k} \qquad \begin{cases} i = 1, 2, \ldots, a \\ j = 1, 2, \ldots, b \\ k = 1, 2, \ldots, r \end{cases}$$

FACTORS
A

B

REPLICANTS

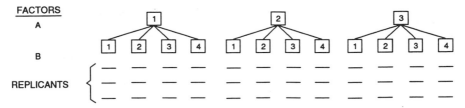

Figure 13.16 A two-stage nested design.
Reprinted with permission from Douglas C. Montgomery, "Design and Analysis of Experiments" (1984), John Wiley & Sons, Inc., p. 358, Figure 12-1.[15]

This model includes a levels of factor A, b levels of factor B nested under each level of A, and r replicates (see Figure 13.16). The subscript $j(i)$ indicates that the jth level of factor B is nested under the ith level of factor A. The subscript $(ij)k$ is used for the error term. This is an example of a balanced design since there are an equal number of levels of B within each level of A, and an equal number of the replicates. It is important to note here that since every level of factor B does not appear with every level of factor A, there can be no interaction between A and B. Our real interest in a nested design of experiment is to break down the total variation in its components.

In the nested design the total sum of squares can be partitioned into a sum of squares due to factor A, a sum of squares due to factor B, under level A, and a sum of squares due to error. Symbolically we can write

$$SS_T = SS_A + SS_{B(A)} + SS_E$$

There are $abr - 1$ degrees of freedom for SS_T, $a - 1$ degrees of freedom for SS_A, $a(b-1)$ degrees of freedom for $SS_{B(A)}$, and $ab(r-1)$ degrees of freedom for error. Note that $abr - 1 = (a-1) + a(b-1) + ab(r-1)$. The sums of squares for each component are determined by the following formulas:

$$SS_A = \sum_{i=1}^{a} \frac{Y_{i..}^2}{br} - \frac{Y_{...}^2}{abr}$$

$$SS_{B(A)} = \sum_{i=1}^{a} \sum_{j=1}^{b} \frac{Y_{ij.}^2}{r} - \sum_{i=1}^{a} \frac{Y_{i..}^2}{br}$$

$$SS_E = \sum_{i=1}^{a} \sum_{j=1}^{b} \sum_{k=1}^{r} Y_{ijk}^2 - \sum_{i=1}^{a} \sum_{j=1}^{b} \frac{Y_{ij.}^2}{r}$$

$$SS_T = \sum_{i=1}^{a} \sum_{j=1}^{b} \sum_{k=1}^{r} Y_{ijk}^2 - \frac{Y_{...}^2}{abr}$$

TABLE 13.22   Analysis of Variance Table for the Two-Stage Nested Design

Source of Variation	Sum of Squares	Degrees of Freedom	Variance
A	$\sum_i \dfrac{Y_{i..}^2}{br} - \dfrac{Y_{...}^2}{abr}$	$a-1$	V_A
B within A	$\sum_i \sum_j \dfrac{Y_{ij.}^2}{r} - \sum_i \dfrac{Y_{i..}^2}{br}$	$a(b-1)$	$V_{B(A)}$
Error	$\sum_i \sum_j \sum_k Y_{ijk}^2 - \sum_i \sum_j \dfrac{Y_{ij.}^2}{r}$	$ab(r-1)$	V_E
Total	$\sum_i \sum_j \sum_k Y_{ijk}^2 - \dfrac{Y_{...}^2}{abr}$	$abr-1$	

Reprinted with permission from Douglas C. Montgomery, "Design and Analysis of Experiments" (1984), John Wiley & Sons, Inc., p. 360, Table 12-2.[16]

The test procedure can be summarized in an analysis of variance table as shown in Table 13.22.

The appropriate statistics for testing the effects of factors A and B depend on whether A and B are fixed or random. If A and B are random, we use the following formulas:

$$E(V_A) = \sigma_E^2 + r\sigma_B^2 + br\sigma_A^2$$

$$E(V_{B(A)}) = \sigma_E^2 + r\sigma_B^2$$

$$E(V_E) = \sigma_E^2$$

13.5.3   An Example of Three-Stage Nested Design

To illustrate the procedure involved in identifying and estimating the sources of variation, let us look at an experiment conducted in the bumping process at the tape automated bonding (TAB) area. The bumping process adds raised metal contacts to chip bond pads and serves to provide a platform for bonding. The engineer wishes to investigate the variability of bump heights (1) due to measurements, (2) among bumps within a die, (3) among dies within a wafer, and (4) among wafers within a lot.

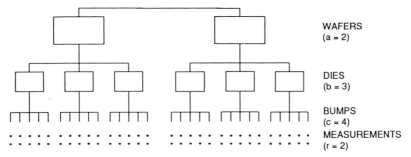

Figure 13.17   A three-stage nested design.

This is a three-stage nested design, with dies nested within wafers, and bumps nested within dies. To estimate the variance components, the engineer conducts the following experiment.

1. $a = 2$ wafers were selected at random from a single lot.
2. $b = 3$ dies were then taken at random from each of the selected wafers.
3. $c = 4$ independent gold bumps were taken at random from the selected die.
4. $r = 2$ measurements were taken for the height of each bump. Figure 13.17 shows the three-level nested design for this process.

13.5.4   Statistical Analysis

The model for the general three-stage nested design is

$$Y_{ijkl} = \mu + A_i + B_{j(i)} + C_{k(ij)} + E_{l(ijk)} \qquad \begin{cases} i = 1, \ldots, a \\ j = 1, 2, \ldots, b \\ k = 1, 2, \ldots, c \\ l = 1, 2, \ldots, r \end{cases}$$

That is, every measurement (Y_{ijkl}) is composed of five elements. In our example, μ is the overall process level (the true value), A_i is the effect of the ith wafer $(i = 1, \ldots, a;\ a = 2)$, $B_{j(i)}$ is the effect of the jth die within the ith wafer $(j = 1, 2, \ldots, b;\ b = 3)$, $C_{k(ij)}$ is the effect of the kth bump within the jth die and ith wafer $(k = 1, 2, \ldots, c;\ c = 4)$, and $E_{l(ijk)}$ is the measurement error $(l = 1, \ldots, r;\ r = 2)$.

In order to analyze which source contributes most of the variability, we need to perform the analysis of variance (see Table 13.23). The data for the experiment are tabulated as in Table 13.24. We will need these figures to complete the ANOVA analysis.

TABLE 13.23   Analysis of Variance for the Three-Stage Nested Design

Source of Variation	Sums of Squares	Degrees of Freedom	Variance
A	$\displaystyle\sum_i \frac{Y_{i...}^2}{bcr} - \frac{Y_{....}^2}{abcr}$	$a-1$	V_A
B (within A)	$\displaystyle\sum_i \sum_j \frac{Y_{ij..}^2}{cr} - \sum_i \frac{Y_{i...}^2}{bcr}$	$a(b-1)$	$V_{B(A)}$
C (within B)	$\displaystyle\sum_i \sum_j \sum_k \frac{Y_{ijk.}^2}{r} - \sum_i \sum_j \frac{Y_{ij..}^2}{cr}$	$ab(c-1)$	$V_{C(B)}$
Error	$\displaystyle\sum_i \sum_j \sum_k \sum_l Y_{ijkl}^2 - \sum_i \sum_j \sum_k \frac{Y_{ijk.}^2}{r}$	$abc(r-1)$	V_E
Total	$\displaystyle\sum_i \sum_j \sum_k \sum_l Y_{ijkl}^2 - \frac{Y_{....}^2}{abcr}$	$abcr-1$	

To do the analysis of variance, take the following steps.
Step 1 Determine the sum of squares:

$$SS_T = \sum_i^a \sum_j^b \sum_k^c \sum_l^r Y_{ijkl}^2 - \frac{Y_{....}^2}{abcr}$$

$$= (25.82^2 + 25.75^2 + \cdots + 23.22^2 + 23.70^2) - \frac{(1187.23)^2}{2 \times 3 \times 4 \times 2}$$

$$= 84.64$$

$$SS_A = \sum_i^a \frac{Y_{i...}^2}{bcr} - \frac{Y_{....}^2}{abcr}$$

$$= \frac{621.24^2 + 565.99^2}{3 \times 4 \times 2} - \frac{(1187.23)^2}{2 \times 3 \times 4 \times 2}$$

$$= 63.60$$

TABLE 13.24 Data for the Example (in microinches)

A	$B(A)$	$C(B)$	Y_{ijkl}		$Y_{ijk.}$	$Y_{ij..}$	$Y_{i...}$	$Y_{....}$
1	A	1	25.82	25.75	51.57	205.93	621.24	1187.23
		2	26.39	24.44	50.83			
		3	26.96	24.60	51.56			
		4	25.46	26.51	51.97			
	B	1	26.75	26.80	53.55	210.74		
		2	26.09	25.85	51.94			
		3	26.31	26.35	52.66			
		4	26.07	26.52	52.59			
	C	1	26.04	26.05	52.09	204.57		
		2	25.63	25.71	51.34			
		3	25.10	25.02	50.12			
		4	25.77	25.25	51.02			
2	A	1	22.31	22.51	44.82	184.48	565.99	
		2	22.17	22.22	44.39			
		3	23.92	23.77	47.69			
		4	23.76	23.82	47.58			
	B	1	23.48	23.66	47.14	186.27		
		2	23.00	22.96	45.96			
		3	23.24	23.25	46.49			
		4	23.13	23.55	46.68			
	C	1	24.94	24.79	49.73	195.24		
		2	24.74	24.82	49.56			
		3	24.52	24.51	49.03			
		4	23.22	23.70	46.92			

A are the wafers; $B(A)$ are the dies within wafers; $C(B)$ are the bumps within dies; Y_{ijkl} is the individual measurement for each bump; $Y_{ijk.}$ is the sum of the measurements within the same bump (for example, $25.82 + 25.75 = 51.57$); $Y_{ij..}$ is the sum of the measurements within the same die (for example, $51.57 + 50.83 + 51.56 + 51.97 = 205.93$); $Y_{i...}$ is the sum of the measurements within the same wafer (for example, $205.93 + 210.74 + 204.57 = 621.24$); and $Y_{....}$ is the sum of all the measurements (for example, $621.24 + 565.99 = 1187.23$).

$$SS_{B(A)} = \sum_i^a \sum_j^b \frac{Y_{ij..}^2}{cr} - \sum_i^a \frac{Y_{i...}^2}{bcr}$$

$$= \frac{(205.93^2 + 210.74^2 + \cdots + 195.24^2)}{4 \times 2} - \frac{(621.24^2 + 565.99^2)}{3 \times 4 \times 2}$$

$$= 10.94$$

$$SS_{C(B)} = \sum_i^a \sum_j^b \sum_k^c \frac{Y_{ijk.}^2}{r} - \sum_i^a \sum_j^b \frac{Y_{ij..}^2}{cr}$$

$$= \frac{(51.57^2 + 50.83^2 + \cdots + 49.03^2 + 46.92^2)}{2}$$

$$- \frac{(205.93^2 + \cdots + 195.24^2)}{4 \times 2}$$

$$= 9.51$$

$$SS_E = \sum_{i=1}^a \sum_{j=1}^b \sum_{k=1}^c \sum_{l=1}^r Y_{ijkl}^2 - \sum_{i=1}^a \sum_{j=1}^b \sum_{k=1}^c \frac{Y_{ijk.}^2}{r}$$

$$= (25.82^2 + 25.75^2 + \cdots + 23.22^2 + 23.70^2)$$

$$- \frac{(51.57^2 + \cdots + 46.92^2)}{2}$$

$$= 0.60$$

Step 2 Determine the degrees of freedom:

$$df_A = a - 1 = 2 - 1 = 1$$

$$df_B = a(b-1) = 2(3-1) = 4$$

$$df_{C(B)} = ab(c-1) = (2)(3)(4-1) = 18$$

$$df_E = abc(r-1) = (2)(3)(4)(2-1) = 24$$

$$df_T = abcr = (2)(3)(4)(2) - 1 = 47$$

Step 3 Determine the variance:

$$V_A = \frac{SS_A}{df_A} = \frac{63.60}{1} = 63.60$$

$$V_{B(A)} = \frac{SS_{B(A)}}{df_{B(A)}} = \frac{10.94}{4} = 2.73$$

$$V_{C(B)} = \frac{SS_{C(B)}}{df_{C(B)}} = \frac{9.51}{18} = 0.53$$

$$V_E = \frac{SS_E}{df_E} = \frac{0.60}{24} = 0.025$$

TABLE 13.25   ANOVA Table for the Results from the Example

Source	SS	df	V	E
A: Wafers	63.60	1	63.60	23.30
B: Dies	10.94	4	2.73	5.15
C: Bumps	9.51	18	0.53	21.20
D: Error	0.60	24	0.025	
Total		47		

Step 4 Determine the F ratio:

$$F_A = \frac{V_A}{V_{B(A)}} = \frac{63.60}{2.73} = 23.30$$

$$F_{B(A)} = \frac{V_{B(A)}}{V_{C(B)}} = \frac{2.73}{0.53} = 5.15$$

$$F_{C(B)} = \frac{V_{C(B)}}{V_E} = \frac{0.53}{0.025} = 21.20$$

We can summarize the above results in an ANOVA table (see Table 13.25).

13.5.5   Conclusion from the ANOVA Table

Comparing the F ratios from the ANOVA table with the F table values (see Appendix E), we can conclude that:

1. On a wafer basis, since $F_{(0.01, 1, 4)} = 21.20$, and $F_{\text{wafer}} = 23.30 > 21.20$, there is significant variability between wafers even at the 0.01 risk level.
2. On a die basis, since $F_{(0.01, 4, 18)} = 4.58$, and $5.15 > 4.58$, there is also significant variability between dies at the 0.01 risk level.
3. On a bump basis, since $F_{(0.01, 18, 24)} = 2.74$, and $21.20 > 2.74$, there is very high significant variability between bumps at the 0.01 risk level.

13.5.6   Estimating the Variance Components

In order to estimate the variance components the expected variance for the nested random effects model used were

$$V_A = bcr\sigma_A^2 + cr\sigma_B^2 + r\sigma_C^2 + \sigma_e^2$$

$$V_{B(A)} = cr\sigma_B^2 + r\sigma_C^2 + \sigma_e^2$$

$$V_{C(B)} = r\sigma_C^2 + \sigma_e^2$$

$$V_E = \sigma_e^2$$

The estimates of the variance components can be found after replacing the expected values of the variances by the observed variances, and then solving the equations for σ_e^2, σ_C^2, σ_B^2, and σ_A^2. This gives

$$\sigma_e^2 = V_E = 0.025$$

$$\sigma_A^2 = \frac{V_A - V_{B(A)}}{bcr} = \frac{63.60 - 2.73}{3 \times 4 \times 2} = 2.54$$

$$\sigma_B^2 = \frac{V_{B(A)} - V_{C(B)}}{cr} = \frac{2.73 - 0.53}{2 \times 4} = 0.28$$

$$\sigma_C^2 = \frac{V_{C(B)} - V_E}{r} = \frac{0.53 - 0.025}{2} = 0.25$$

and

$$\sigma_e = \sqrt{0.025} = 0.16$$

$$\sigma_A = \sqrt{2.54} = 1.59$$

$$\sigma_B = \sqrt{0.28} = 0.53$$

$$\sigma_C = \sqrt{0.25} = 0.50$$

13.5.7   What Is the Ratio of Measurement Error to the Total Tolerance (P/T Ratio)?

Now that we know the measurement error (σ_e), we need to determine if this error is negligible. It is assumed that if 6 standard deviations of the precision is less than 10% to 15% of the total range of the product being measured by the instrument, the measurement error is considered negligible. In our case, since the total tolerance is 8 microinches, then

$$P/T \text{ ratio} = \frac{6\sigma_e}{\text{total tolerance}} = \frac{6 \times 0.16}{8} = 0.12 = 12\%$$

The result shows that the measurement error is negligible.

13.5.8 Comparing the Error of Measurement with the Variability of the Product

Overall variability includes both product variability and measurement variability, and can be expressed as

$$\sigma_o = \sqrt{\sigma_{\text{wafer}}^2 + \sigma_{\text{die}}^2 + \sigma_{\text{bump}}^2 + \sigma_e^2}$$

$$= \sqrt{2.54 + 0.28 + 0.25 + 0.025}$$

$$= \sqrt{3.095}$$

$$= 1.76$$

and the true product variability is

$$\sigma_p = \sqrt{\sigma_{\text{wafer}}^2 + \sigma_{\text{die}}^2 + \sigma_{\text{bump}}^2}$$

$$= \sqrt{2.54 + 0.28 + 0.25}$$

$$= \sqrt{3.07}$$

$$= 1.75$$

According to J. M. Juran, if the sigma of the measuring method is less than one-tenth of the sigma observed, then the effect of error upon the sigma of the product will be less than 1%. In our example,

$$\frac{\sigma_e}{\sigma_o} \times 100 = \frac{0.16}{1.76} = 0.091 = 9.1\%$$

Therefore, the measuring method has very little effect of impacting the observed variability of the product.

REFERENCES

1. J. M. Juran and Frank M. Gryna, "Juran's Quality Control Handbook," 4th ed. (1988), McGraw-Hill Book Company, pp. 26.7–26.10, Table 26.3.
2. Phillip J. Ross, "Taguchi Techniques for Quality Engineering" (1988), McGraw-Hill Book Company, p. 44.
3. Douglas C. Montgomery, "Design and Analysis of Experiments," 2nd ed. (1984), John Wiley & Sons, Inc., pp. 189–192.
4. Douglas C. Montgomery, "Design and Analysis of Experiments," 2nd ed. (1984), John Wiley & Sons, Inc., pp. 261–262.

5. Genichi Taguchi, "Introduction to Quality Engineering" (1986), Asian Productivity Organization, p. 181.

6. Genichi Taguchi, "On Line Quality Control During Production" (1981), Japanese Standards Association, p. 1.

7. Genichi Taguchi, Elsayed A. Elsayed, and Thomas Hsiang, "Quality Engineering in Production Systems" (1989), McGraw-Hill Book Company, pp. 12–15.

8. George E. P. Box, William G. Hunter, and J. Stuart Hunter, "Statistics for Experimenters" (1978), John Wiley and Sons, Inc., p. 374.

9. G. Taguchi and S. Konishi, "Orthogonal Arrays and Linear Graphs" (1987), American Supplier Institute, Inc., pp. 1–3 and 36–37.

10. C. R. Hicks, "Fundamental Concepts in the Design of Experiments" (1982), Holt, Rinehart & Winston.

11. Phillip J. Ross, "Taguchi Techniques for Quality Engineering" (1988), McGraw-Hill Book Company, p. 116.

12. Genichi Taguchi, "System of Experimental Design," Vol. 1 (1987), UNIPUB/ Kraus International Publications, pp. 112–115.

13. Genichi Taguchi, "System of Experimental Design," Vol. 2 (1987), American Supplier Institute, Inc., pp. 1109–1112.

14. Douglas C. Montgomery, "Design and Analysis of Experiments" (1984), John Wiley & Sons, Inc., pp. 358–359.

15. Douglas C. Montgomery, "Design and Analysis of Experiments" (1984), John Wiley & Sons, Inc., p. 358, Figure 12-1.

16. Douglas C. Montgomery, "Design and Analysis of Experiments" (1984), John Wiley & Sons, Inc., p. 360, Table 12-2.

Chapter 14

On-Line Design of Experiments

Today, as never before, to stay in business and make a profit a company needs to have the capability to introduce new ideas, new products, and new processes. This should be done systematically on a continuous basis, and with the shortest cycle time possible. These conditions are crucial if we want to receive maximum benefits from research and development activities.

There is a Russian expression, "The spoon has value when eating lunch." A new product will provide a high profit if a company brings this product to the market at the right time.

Let's take a look at an example of how a new idea can be introduced in an operation (see Figure 14.1). Introducing a new idea can be a long, time-consuming effort and, because of this, sometimes a new idea becomes old even before a return on investment is realized. Therefore, it is important to focus efforts on reducing to a minimum the cycle time needed to introduce new products and processes.

The reduction of R&D cycle time is a separate subject. At this point, we are only concerned with the last stage of the flowchart (when the product is already formally released to the line). Let's assume that a new process was released to the manufacturing floor and this was achieved within the shortest possible cycle time. What is sometimes observed at this stage are yields considerably lower than expected, and costs per unit much higher than the marketing department can accept, and quality levels lower than when the process was running on the pilot line.

Why does this happen? Is it because the idea was bad? Or, is it because the R&D department did a poor job? Most of the time this is not the cause. This kind of result happens mainly because the process was developed in a totally different environment and with different conditions than exist in full-scale production. The process parameters which seemed optimal in the

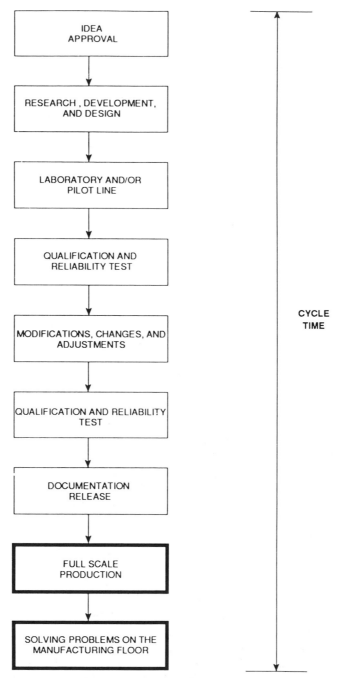

Figure 14.1 A flow chart for introducing a new product or process.

laboratory or in small-scale production (on a pilot line) may not be suitable for full-scale production, and then more time, money, and engineering staff are needed to take care of the ensuing problems.

One way to adjust the process to the new manufacturing environment is to apply off-line design of experiments to determine the best process settings. But in real life, the R&D department is already busy with the development of a new idea and there is no time left to spend on improvement of the "just finished" process. The manufacturing engineering department may be trying to fix the simple, obvious problems by using trial-and-error methods. The manufacturing staff is normally not responsible for process optimization. Its main responsibility is to follow the specification requirements and other documentation developed by the R&D departments. There are often some contributing efforts from the manufacturing engineering department, but its resources are limited and it doesn't have the required conditions and knowledge to perform the necessary off-line design of experiments because it needs to support the product floor.

So sometimes it takes years before a new process is optimized, and by the time this happens, that is, when the yield reaches asymptotic limits, when the cost is reduced substantially, and the quality is excellent, the product could already be obsolete, and therefore final optimization may have occurred too late.

How about "old processes?" Can we say that they run under the best possible conditions? We put the term "old processes" in quotes because usually there is no such process. This is because from the time a process is released to the floor, a great many things are happening continuously. The process flow is changed, some equipment has been modernized or replaced, new material has been approved, the workforce is changed, etc.; and this is happening on a continuous basis. But the specification requirements which were developed years ago are sometimes not revised, and the process keeps running under the "old" condition. Can we say that the "old process" runs at the best possible operating conditions?

At AMD we were seeking a program that would allow us to improve the process directly on the manufacturing floor without jeopardizing the quality and productivity. We were looking for methods that would allow us to involve the operating staff in process improvement. Who knows the existing process better than the people who keep their hands on the pulse of the process? We were interested in delegating the ownership and credit for process improvement to the operating personnel, and then using our engineering staff as advisors and focusing them on take caring of preventing problems and the successful introduction of new processes and products. How can this be done?

While working on the development of new educational programs for the AMD staff, we became familiar with the concept called evolutionary operation (EVOP). This concept was developed by Dr. G. E. P. Box 34 years ago mainly for the chemical industry. The philosophy behind this approach is to run the process in such a way that while manufacturing the product, the process will also generate information about itself, which can be used for continuous process improvement.

The EVOP program is tailormade to be run by operating personnel. After other investigations the Manufacturing Services Division of AMD decided to develop a special program that would allow us to introduce EVOP principles in our manufacturing plant. The purpose of this chapter is to describe some EVOP principles and share our experiences of applying EVOP in the electronics industry.

14.1   EVOLUTIONARY OPERATION (EVOP)

14.1.1   What is it?

Evolutionary operation (EVOP) was developed by George E. P. Box (1957), who said,

> " · · · processes should be run so as to generate not only product, but also information on how the product can be improved."[1]

The basic philosophy behind this method is that for a process to be effective it must produce two things: (1) a product and (2) information on how to improve the product.

EVOP is a simple but powerful tool which has been developed in such a way that it can be applied directly on the manufacturing floor and conducted by the operating personnel themselves.

In most cases EVOP becomes a continuing investigative routine for a plant and replaces the normal *static* operation. However, it is important to emphasize that evolutionary operation is not a substitute for fundamental investigation. The use of EVOP should be as a tool for on-line process improvement and design of experiment (DOE) should be used as a method for off-line investigation. These two approaches complement each other very well.

For example, DOE is sometimes used to select the most important controllable variables for setting the EVOP program; and EVOP often indicates areas where more fundamental techniques should be applied.

14.1.2   General Description

To conduct the EVOP experiments, aside from the best known (static) conditions, slight deviations are introduced into the process by means of a carefully prescribed pattern (Figure 14.2). The pattern requires that small changes be made in the selected controllable variables within the engineering specification requirements. The center point (position 0 in Figure 14.2) represents the best-known conditions at a given time. Positions 1, 2, 3, and 4

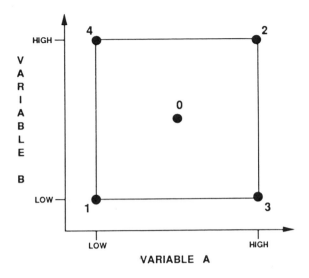

NOTE: O INDICATES THE BEST KNOWN CONDITIONS

Figure 14.2 Evolutionary operation pattern for two controllable variables.

represent changes in the two variables. These changes (deviations) are considered to be small enough not to jeopardize the production quality but large enough to accumulate information about the product. Continuing to run production under the conditions described by the pattern will accumulate information while the product is being manufactured. This information will be used by the EVOP team to evaluate changes in operating conditions which, when optimized and implemented, will result in process improvements.

Each completed set of five experimental runs operated on conditions 0, 1, 2, 3, and 4 is called a *cycle* and a combined number of n cycles is called a *phase*. We usually move to a new phase when significant results are achieved. The new phase brings with it a new set of operating conditions based on the results from the previous phase.

14.1.3 Feedback System

One of the most important elements of the EVOP program is the constant information feedback system (Figure 14.3).

Traditional process input elements such as people, equipment, materials, methods, and environment form a static process. Adding one more element to the input (information about the product) changes the process condition from static to evolutionary. This additional input is obtained from the EVOP program.

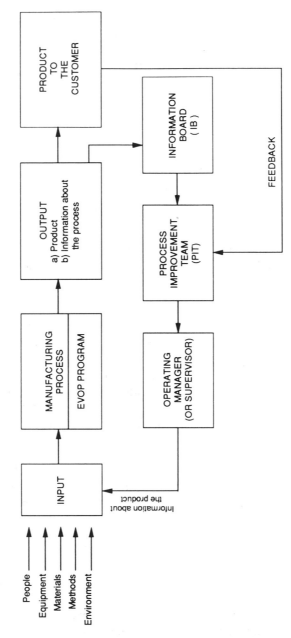

Figure 14.3 The continuous EVOP feedback system.

Because the evolutionary operation program is applied directly on the manufacturing floor it forces the process to produce information about itself without upsetting the production output.

Now the process has two kinds of output: (1) product output that will be shipped to the customer and (2) information output about the process in the form of process responses that are used for process improvement. This information is obtained from the product output by means of process information.

The information generated by the EVOP program is continually placed on a summary sheet called the information board (IB) and is used by the process improvement team (PIT) to make decisions about running the process. Based on these decisions, recommendations are given to the operating manager to make process changes.

14.1.4 Interpreting the Information Board

The information board (IB) contains the major source of information needed to run the EVOP program used by operating personnel. Knowing how to read and interpret this information is very important for the decision-making process.

The information boards used at AMD are basically designed as described by Box and Hunter.[2] Slight modifications are made in every plant to tailor the board to its own needs.

Let's use an information board from a project conducted on a chemical process optimization, and describe all the elements on the board, and then try to interpret the combined information (see Figure 14.4). Note that the top of this information board represents the results of an EVOP program which is in phase 2 and has already made three cycles. The replica of the EVOP layout makes it easy to associate experimental conditions with the response it produced. The information board represents the responses, the main and interaction effects (with their 95% confidence limits), and the change-in-mean effect (with its 95% confidence limits).

The confidence limits are expressed as a plus or minus value associated with each effect (see Table 14.1). If the absolute value of an effect exceeds the confidence limit, the effect is considered significant. Usually a 95% confidence limit is used, which is equivalent to two sigma error limits. The calculation of the effect is made in such a way that the sign before the effect indicates the direction to move the level of the factor to maximize the response.

If a maximum response is desired (as in this case), the level of the factor is moved in the same direction as the sign. In our example, the negative sign on the acid temperature effect indicates that a lower temperature will increase the efficiency of the process. If a minimum response is desired, the level of the factor is moved in the direction opposite the sign.

Phase II, Cycle 3

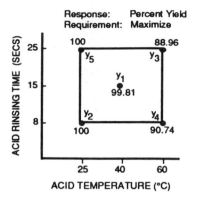

Response: Percent Yield
Requirement: Maximize

	Effects	Two Sigma Error Limits
Temperature Effect	-10.15	±5.58
Rinsing Time Effect	-0.89	±5.58
T X R Interaction Effect	-0.89	±5.58
Change in Mean Effect	-3.91	±4.96

Figure 14.4   EVOP information board.

The interaction effect indicates whether the effect of one factor is different at different levels of the other factor. In our example the interaction effect is not significant because its value is smaller than that of the two sigma error limits. This means that the factors do not interact.

The change-in-mean (CIM) effect gives a measure of the difference between the average response obtained by running the EVOP experiment

TABLE 14.1   Analysis of Effects

Variable	Effect	Two Sigma Limits	Conclusion
Acid temperature effect	− 10.15	± 5.58	Significant
Rinsing time effect	− 0.89	± 5.58	Not significant
$T \times R$ interaction effect	− 0.89	± 5.58	Not significant
Change-in-mean effect	− 3.91	± 4.96	Not significant

and the response obtained by running the process at the center point conditions. From this information we can know whether the conditions being run during the current phase give an average result better or worse than the reference conditions. In other words, the CIM measures the direct cost incurred by obtaining information in any particular phase. In our example the CIM effect is not significant.

Given the information shown in the information board (see Figure 14.4), what decision should be made by the process improvement team or by the operating manager? In this example the decision is not difficult to make since we are dealing with only one response and strong significance for the effect of at least one variable has been demonstrated. But what if we are dealing with two or three responses and the significance of the responses is not obvious? Should the operating manager always wait until the estimated effect is found to be significant?

The answer to this question can be found in Box and Hunter's paper[3] where it says, "To wait for something to fall outside the interval estimate is a very inefficient use of the information provided by evolutionary operation. The most important objection to such a model of operation is that it cuts out the interpretive role of the production supervisor."

The information board should only be used as an information channel and the decision of what direction to change the process should be made by the people who know the process best.

14.1.5 Current Best-Known Conditions

There will exist a set of best-known conditions when we introduce the EVOP program and move from one phase to each new one at any given stage. At phase 1 this set of conditions is usually derived from the engineering specifications. But when a new phase is begun the current "best-known conditions" will change. In other words, each new phase will have a new set of current best-known conditions within that particular phase.

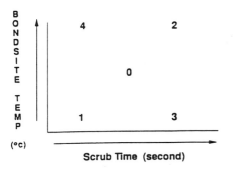

Figure 14.5 An EVOP plan.

Moving from phase to phase, we need to be able to compare results with some type of reference. This can be done by returning to the current best-known conditions once in every cycle. Because of this, most of the time we will need to add an additional (reference) point to the factorial design.

In some cases the current best-known conditions correspond to one of the standard design points. To illustrate the idea of locating reference points, we

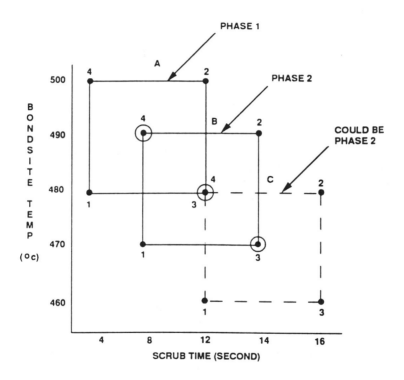

A: Phase 1 -   Current best known conditions as
                 a center of the design.

B: Phase 2 -   Current best known conditions as
                 a center of the design.

C: Could be   The current best-know
    Phase 2 -   conditions are included as a part
                 of the new design (no center point).

Figure 14.6 Setting the current best-known conditions.

will use an example as in Figure 14.5. The current best-known conditions were set at the center of the design since there was no information available at the beginning of the experiment indicating the direction in which improvement could occur. The information obtained from the results of phase 1 (see Figure 14.6) suggests that the current best-known condition for phase 2 be established with parameters of 480°C for bond site temperature and 12 seconds for scrub time which represents the right lower point of design of phase 1. These conditions could be included as a part of the new design but the EVOP committee decides to use those conditions at the center of the design. This is a more practical decision. The inclusion of reference conditions will help to make sure that the results obtained at any new phase are related to the controlled small changes and not affected by external influences such as changes in the incoming material or any other unexpected and unexplained impacts.

Including the best-known conditions in each cycle as a reference makes it possible to compare the average performance results achieved by introducing small changes with the results achieved by running the process without changes. This comparison is provided by subtracting the average response of reference conditions from the average response over all runs in the EVOP cycle. The resulting contrast is called the change-in-mean (CIM) effect.

14.1.6   Change-in-Mean (CIM) Effect

Only after the best-known conditions are included in each cycle as a reference point is the CIM effect calculated. And it doesn't matter how the reference point is introduced; it can be introduced as an additional point or it can be included as one of the points in the factorial design. There is only a slight difference in the formulas of calculation (see Figure 14.7). The purpose of the CIM effect is to provide a measure of the temporary *direct* cost for obtaining information during the present phase.

14.1.7   EVOP in the Electronics Industry

The evolutionary operation method was developed mainly in reference to "process industries" such as chemical, paper, paint, plastic, glass, steel, and other industries where the quality of their product depends on a "process."

Through the years EVOP has spread to other industries where the best operating conditions cannot be established immediately. However, the application of EVOP in the electronics industry remains rare especially in assembly plants. Why? What are the requirements that make a process more favorable to the application of EVOP? According to Gerald J. Hahn and Arthur F. Dershowitz, the authors of an EVOP survey,[4] a process should

The Location of the Reference Point	Form of Design	Formulas for Computation of the CIM Effect	Remarks
(1) The reference point is included in one of the corners of the design.	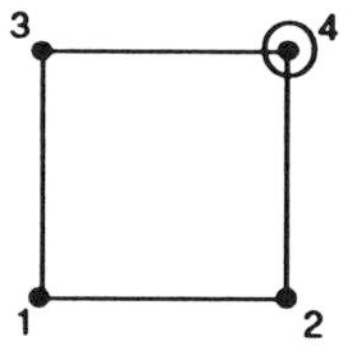	**CIM** $= 1/4\,(\bar{y}_1 + \bar{y}_2 + \bar{y}_3 + \bar{y}_4) - \bar{y}_4$ $= 1/4\,(\bar{y}_1 + \bar{y}_2 + \bar{y}_3 - 3\bar{y}_4)$	Information is available as to the direction in which improvement should be sought.
(2) The reference point is located at the center point of the design.	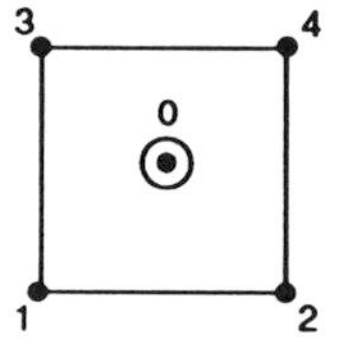	**CIM** $= 1/5\,(\bar{y}_0 + \bar{y}_1 + \bar{y}_2 + \bar{y}_3 + \bar{y}_4) - \bar{y}_0$ $= 1/5\,(\bar{y}_1 + \bar{y}_2 + \bar{y}_3 + \bar{y}_4 - 4\bar{y}_0)$	No information is available as to the direction in which improvement should be sought.
(3) The reference point is outside the design.	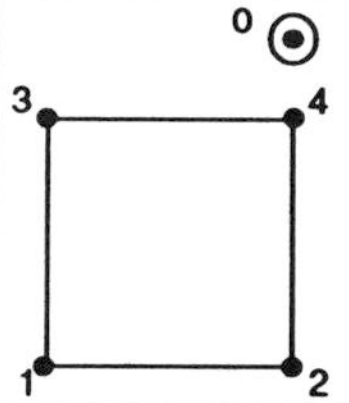	The same as (2).	The engineering specifications are broad and there is a low risk of jeopardizing the quality.

Note: $\bar{y}$ is the average response after completing n cycles.

Figure 14.7 Change-in-mean effect computation.

have the following characteristics:

1. The process should involve high-volume production over a reasonably extensive period of time.
2. The potential benefits of the product improvement should be great.
3. The process variables should have the ability to fluctuate readily and stabilize rapidly after a process change.
4. Rapid measurements of the response variables should be possible.

All these characteristics definitely belong to a major group of electronic processes.

Also, experience in the electronics industry shows that even after applying different methods of design of experiment (DOE) in the laboratory and on the pilot line there is still a need for additional experimental work directly on the full-scale process to adjust the controlled parameters to the new manufacturing environment and conditions. EVOP can be used successfully for this.

There is one more argument that favors applying EVOP methods. In the electronics industry there are some processes where output quality is measured in parts per million (ppm).

Let's use the hermetic package seal process as an example. The capability of this process is in the range of 5 to 20 ppm. To measure the quality of this kind of process, we need to have a large quantity of samples. Now let's assume that we want to introduce an off-line design of experiment to improve this process or to try a new method or material. By using a classical design of experiment, we would need to jeopardize a lot of finished goods to run the experiment. We would get the results a lot faster than by running an on-line EVOP, but at what price? Sometimes we just can't afford such an experiment.

In contrast, using EVOP to improve or investigate the hermetic seal process will allow us to run the process, produce quality parts, and at the same time obtain information on how to improve the hermetic sealing process. We are not limited by the quantity of samples needed for every run; we run with the production schedule anyway.

There are no methodological or technical reasons not to apply EVOP in the electronics industry. There is only one organizational reason which explains why we don't use this powerful and simple method. It is that we don't normally assign ownership of process improvements to the operations people who are closely involved with the manufacturing process on a day-to-day basis. So if the operations personnel don't feel that they should run production and at the same time improve the process, there is no demand for a tool like EVOP which is especially tailored for this purpose.

Since AMD manufacturing management has empowered factory operations personnel with process improvement authority and responsibilities and given them the full recognition for the results, a demand for special knowledge became apparent. This change in responsibility has made it possible to introduce a specially developed educational course on how to achieve on-line process improvements.

You can still find some pessimists who prefer to work in a static production mode of operation, but as more success stories of applying EVOP become known, the pessimism about EVOP decreases dramatically.

14.1.8   The Application of EVOP

Now that the reader has become familiar with the basic concepts and terminology of the EVOP method, we can provide further illustration of the application of this method. The best way to do this is by using real case histories. The following examples are based on project activity at the AMD plants as a result of introducing the EVOP method. And since all case histories used the same method of calculating the main effects, interaction effects, and change-in-mean effects, we decided to describe the EVOP method by using only one example each for two and three variables which are illustrated in the appendix at the end of the chapter (see Appendix 14.1). In the case histories the reader will only find the results of the experiment which are reflected on the information board. To understand the methodology of calculation, we would recommend the reader refer to Appendix 14.1 before reading the first case history. After this all other case histories will be self-explanatory.

Case History 14.1: Voids Reduction

The Problem

Die attach is an important operation in semiconductor assembly. A major quality concern with eutectic die attach is the amount of voids underneath the die. Voids are empty spaces in the form of trapped air and are an undesirable consequence of scrubbing the die onto the package substrate. They appear as white spots seen on an X-ray image of the attached die and are quantified as a percentage of the total die area.

A limited percentage of voids does not directly affect yield, but is a concern because the voids affect die performance and reliability, especially in terms of thermal conductivity.

The Starting Point

The problem of die attach voids was brought up for discussion by the Process Improvement Committee (PIC) and a decision was made to run an EVOP program on the die attach process. This means that not only has permission been received to do so, but also that a commitment was made by the operating manager to support this program and assign responsible people to the effort.

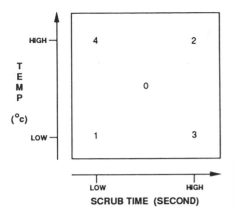

Figure 14.8 An EVOP plan. (*Note:* The numbers represent the run order. 0 is the reference point.)

In one of the brainstorming sessions where special techniques (such as fishbone diagrams) had been used, a list of control variables believed to be associated with voids formation were developed. At the same time by using the cause-and-effect matrix two variables believed to be the most influential were identified. These two variables have been used to start running the EVOP program.

Preparing the Program

The two most important control variables to start with were (1) *bond site temperature* and (2) *die scrub time*. As we now know, the EVOP program can be run by using more than one response variable. But in this program the *percentage of voids* was chosen as the single response variable.

As already mentioned, the most important aspect of the evolutionary operation program is that small changes in the control variables are introduced according to a prescribed pattern. This pattern (see Figure 14.8) represents an EVOP procedure which is based on a simple factorial design with two levels and two factors (2^2) plus a center point.

The ranges for die scrub time and bond site temperature were selected in such a way as to ensure that additional nonconforming parts would not be produced when running the program. The center point (0), which is also called the reference point, is based on the best-known conditions of time and temperature. The four surrounding points (1, 2, 3, and 4) explore the advantages to be gained by moving the operating points in any direction on the plan. These are the points representing the limits of the deliberate changes which were introduced in the process while running the EVOP program. The center point provides a means of comparing the results of all five operating conditions versus the constant conditions at the center point.

In setting the levels of the factors, it was considered that if we set them too far apart it was possible that quality problems would occur and if the

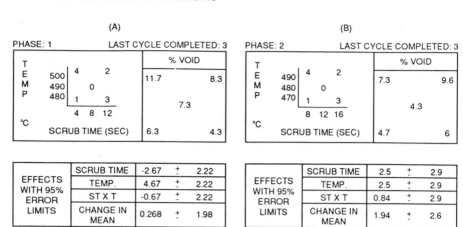

Figure 14.9 EVOP information board.

levels of the factors were set too close no measurable effects would be obtained. So the levels of the factors were selected not to downgrade the product, but to obtain the information needed with the optimal quantity of cycles.

To determine the sequence of operating conditions, the technique employed in EVOP was used. Five sets of conditions were carried out in numerical sequence, starting with run 0 (see Figure 14.8). A complete set of five conditions is considered to be a cycle.

At the end of each cycle, using specially designed EVOP worksheets (see Appendix 14.1), the main effects, interaction effects, and change-in-mean effects of the factors are calculated, and the 95% confidence interval is determined. The results are placed on the information board for decision making (see Figure 14.9).

To make it easier for the reader to understand and interpret the results reflected on the information board, here is some additional theory.

The effects with 95% ($2s$) confidence limits can be seen on the information board. The main effect of a factor is the change in response produced by a change in the factor from the low level to the high level. The calculation of the effect is made in such a way that the sign before the effect indicates the direction to move the level of the factor to maximize the response. If a minimum response is desired, the level of the factor is moved in the direction opposite the sign.

The 95% confidence limits are expressed as a plus or minus value with each effect; e.g., temperature: 4.67 ± 2.22. This means that the temperature effect is 4.67 and its 95% confidence limits are ± 2.22. If the effect is to be considered significant, its absolute value should be larger than the $2s$ limits value. In our example the temperature effect is significant.

The interaction effect indicates whether the effect of one factor is different at different levels of the other factor. We should compare this value with its confidence limits to determine whether or not the interaction effect is significant. If there is a significant effect, the factors are said to interact.

The information board in Figure 14.9 shows that the interactions of the variables are not significant, so the interpretations of the main effects can be made separately.

The change in mean gives a measure of the operating effect by the EVOP method compared to conducting all runs at the center point conditions.

Figure 14.9(A) shows the conditions of phase 1 and the final results after three cycles. The center point in the first phase was the best-known operating conditions: bond site temperature = 490°C and die scrub time = 8 seconds. The significance of the scrub time effect and the temperature effect was already evident in cycle 2 of this phase, but a third cycle was conducted to confirm the results and similarly a strong significance was noted.

It was apparent that the best condition having the least amount of voids occurred at a lower bond site temperature with longer die scrub time (condition 3). Different responses for the three cycles of phase 1 are shown at the bottom of Figure 14.9(A).

To further determine other possible optimum conditions, it was decided to move to phase 2 where the lower right corner of phase 1 becomes the center point for phase 2 (Figure 14.10). The same procedure was used in phase 2, but no significance is noted from the summary results [see Figure 14.9(B)]. This means that at this stage of investigation the best-known conditions for

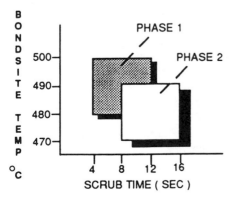

Figure 14.10 Phase relations.

scrub time and bond site temperature should be considered as: (1) scrub time = 12 seconds and (2) bond site temperature = 480°C, which represents the setting of the points on the lower right corner of phase 1 (see Figure 14.10).

Conclusions and Results

The example described above demonstrates how a simple technique can be used to determine the best settings of control variables. Figure 14.11 shows that even before applying the EVOP program the die attach process was capable of producing 100% of the product within the specification requirements. The specification requirements allow the devices to have a maximum of 57% voids in relation to the die area. In fact, before process improvement, the percentage of voids to the die area in the worst case was determined to

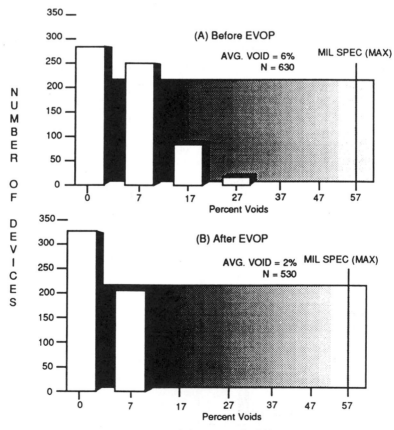

Figure 14.11 Frequency distribution of devices with different percentage of voids to the die area (A) before EVOP and (B) after EVOP.

be 27% and the average voids was 6% [see Figure 14.11(A)]. After optimizing the control parameters, the worst-case condition of voids was reduced to 7% of the die area and the average percentage of voids was reduced to 2%. In comparison with the previous 6%, it is a threefold improvement [see Figure 14.11(B)]. But the EVOP program does not stop here.

New variables are waiting to be optimized. The Process Improvement Committee decided to have a followup EVOP meeting to discuss the results from the study and determine the new direction of continuous die attach process improvement.

Case History 14.2: Solder Dip Process Improvement

The Problem
Analyzing the solder dip quality, the process improvement team found that the major cause of reduced yield was excess solder buildup on the package shorting bar rail ends. It was decided to conduct an EVOP experiment to determine the best process settings for the parameters that influence the excess solder buildup.

Experience showed that the most critical parameters that impact the excess solder problem were *spin blower speed* and the *distance of the air knives above the carrier* of the units. So the experiment was started with these two controllable variables to determine the optimal conditions. As a response variable, it was decided to use *yield*.

The Experiment
A two-factor two-level evolutionary operations technique with a center point was designed for this experiment (see Figure 14.12). The factor on the *Y* axis is the spin blower speed and the factor on the *X* axis is the height of the air

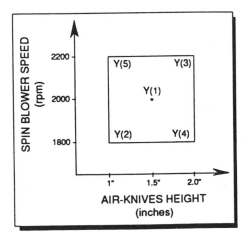

Figure 14.12 The 2^2 factorial design for phase 1 experiment.

knives. The four corners are all the possible combinations of the two factors at the two levels.

The center point represents the best-known operating condition at the time the experiment was started and the four combinations of the factors were set in the allowable specification range. It was decided to use a sample size of 3000 units per run.

Results for Phase 1

After running the first cycle, no estimate of the standard deviation could be made and not much information was derived. At the end of the second cycle, it was possible to estimate the experimental error, but since none of the effects nor their interaction exceeded the confidence limits, no changes in operating conditions were planned at this stage. Cycles 3 and 4 also did not show significant changes except that the speed effect became closer to the confidence limits.

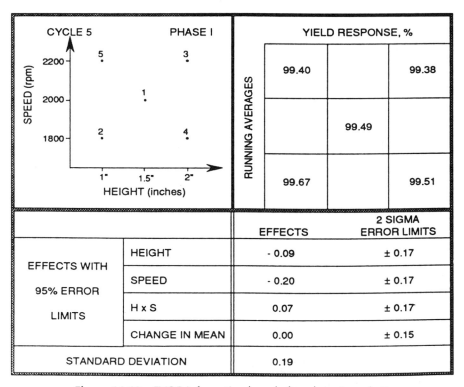

Figure 14.13   EVOP information board after phase 1, cycle 5.

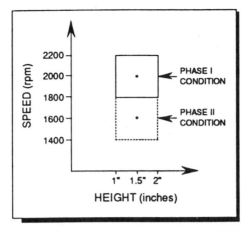

Figure 14.14 New design for phase 2 EVOP.

At cycle 5 (see Figure 14.13) the effect of speed (-0.20) exceeded its confidence limits (± 0.17). This means that now the effect can be considered significant and *the negative sign indicates that the process should be moved to a lower speed to produce better results*. At this stage we also observed that the second variable (height effect) is not significant and no interaction between these two variables exists.

The conclusions at phase 1 are:

1. The speed effect has a significant impact on the excess solder buildup.
2. The height effect is not significant.
3. The interaction effect (between speed and height) is not significant.
4. A second phase with a new center point at speed 1600 rpm and height 1.5 inches (see Figure 14.14) should be performed.

Results for Phase 2
At the end of the second cycle of phase 2, the following results were observed (see Figure 14.15):

1. The speed effect (0.125) exceeded the confidence limits (± 0.055) which indicates that it significantly impacts the process.
2. The height effect (-0.045) is within its confidence limits (± 0.055); therefore, it is not significant.
3. The change-in-mean effect and the interaction effect are not significant.

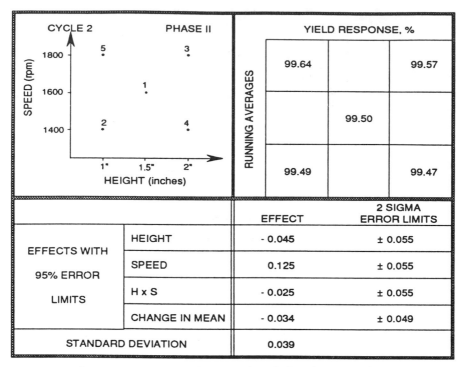

Figure 14.15   EVOP information board after phase 2, cycle 2.

Conclusions

The yield response is higher when the blower speed is 1800 rpm and the air knives' height is 1 inch away from the pallet which represents the upper left corner at phase 2. Running the process at this condition, we have achieved an improvement in the yield (see Figure 14.16) which confirms the statement above. However, the EVOP results from phase 2 (see Figure 14.15) indicate that a further improvement can be achieved if we set the process with a speed somewhere between the center points of phase 1 and phase 2. This conclusion is based on the fact that at phase 2, cycle 2 the speed effect shows a significant positive value (0.125) which indicates that now the speed can be increased.

In order to achieve further improvements, a phase 3 has been proposed by the PIC to the operating supervisor, by setting the process at new operating conditions (see Figure 14.17).

Comments

This experiment demonstrates the use of the EVOP as an ongoing process improvement activity which allows us to optimize the operating conditions on a continuous basis.

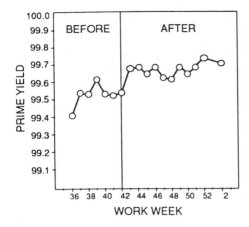

Figure 14.16 "Before and after" chart for the solder dip process.

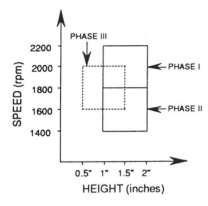

Figure 14.17 Phase 3 EVOP design.

This case study is an example of how the negative sign of the effect should be interpreted. If an effect is negative, and the response is to be maximized, the variable should be changed in a negative direction.

14.1.9 EVOP as a Confirmation Method

Evolutionary Operations brings information out from the production process, clearly and regularly. Its proper function is that of an information producer, and not of a decision maker.[5]

Using design of experiments (DOE), an engineer determined optimal levels of the parameters for a process. This was done off line in laboratory conditions. Once the operating conditions were determined, he calculated the allowable specification ranges for changes in the operating conditions. Then, taking into consideration the high quality requirements of this operation, the engineer considered a very narrow specification range. By narrowing the specification range we usually increase the cost of the production. The objective of the engineer was to find the optimum. Some kind of on-line feedback was needed to ensure that the engineer did not overreact. In this case, the EVOP is the right technique to use (see Case History 14.3).

Case History 14.3: Confirming the Specifications of the Die Attach Process

Data analysis of the die attach process for molded packages shows that the two most important elements of this process are *epoxy thickness* and *die tilting* (see Figure 14.18). If these two elements are not set and controlled properly, numerous problems will be created in the wire bond process, such as nonsticking pad, irregular ball size, wire sweeping, cratering on bond pads, wires shorting to each other, and so on.

Because of this, the engineering department developed tight specification requirements for these parameters and established a process that met these requirements. However, some questions came up. Do other, better operating conditions exist? Can we say that the process is working at the best operating conditions? To answer these questions, it was decided to perform an EVOP program.

Setting up the Program
Control variables:

1. Die attach force, F (in grams)
2. Tool gap setting, G (in mils)

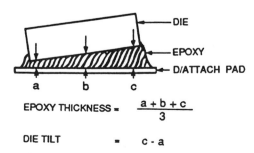

Figure 14.18   The meaning of "epoxy thickness" and "die tilt."

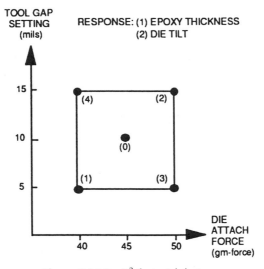

Figure 14.19 2^2 factorial design.

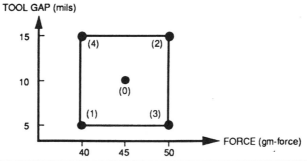

Response	Epoxy Thickness		Die Tilt	
Spec.	Target = 1 mil range = 0.5 - 1.5 mil		Maximum 0.5 mil	
Mean	0.793 0.977 0.825 0.818 0.738		0.512 0.756 0.527 0.487 0.430	
2 S.E. Limit	+/- 0.127		+/- 0.136	
Phase mean Force effect Tool gap effect Force x tool gap effect Change in mean effect	0.830 0.048 +/- 0.127 0.103 +/- 0.127 0.135 +/- 0.127 0.012 +/- 0.114		0.542 0.073 +/- 0.136 0.156 +/- 0.136 0.171 +/- 0.136 0.056 +/- 0.122	

Figure 14.20 Information board after phase 1, cycle 3.

Response variables:

1. Epoxy thickness: (a) target $= 1$ mil; (b) range $= 0.5$ to 1.5 mils
2. Die tilt: (a) maximum allowable $= 0.5$ mil

The layout of the program is shown in Figure 14.19.

Results

After cycle 3 the following effects show significance (see the information board in Figure 14.20):

1. Tool gap setting—at the die tilt response.
2. Interaction of force and tool gap setting—at the die tilt response.
3. Interaction of force and tool gap setting—at the epoxy thickness response.

Interpretation

The results show that the tool gap setting is significant. But in our case, where there is a powerful interaction between two controllable variables, this has little meaning. We need to interpret the effect in conjunction with the interactions [see Figures 14.21(A) and (B)].

Conclusion

Considering the interaction effects, we have two options to optimize the process:

1. *Option 1:* Set the process at condition 3 (i.e., A_2B_1), which is force (50 grams) and tool gap setting (5 mils). The information board shows that in cycle 3 at condition 3 the die tilt response is 0.430 mil, which is lower than the maximum specification requirement (0.5 mil). The second response, epoxy thickness (0.738 mil) is still within the specification limits (0.5 to 1.5 mils).
2. *Option 2:* Set the process at the center point. In this case, we achieve a better response for epoxy thickness (0.818 mil), but the response for die tilt (0.487 mil) comes closer to the maximum specification (0.5 mil).

When we have a strong interaction, as in this case, it is impossible to achieve the highest epoxy thickness and the lowest die tilt. Some tradeoff has

A×B Interaction for Epoxy Thickness
(A: Force, B: Tool Gap Settings)

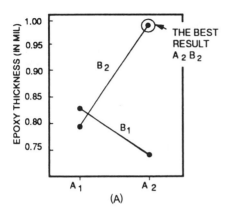

(A)

A X B Interaction for Epoxy Die Tilt
(A: Force, B: Tool Gap Settings)

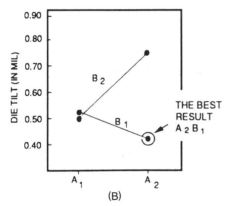

(B)

Figure 14.21 (A) $A \times B$ interaction for epoxy thickness (*A*, force; *B*, tool gap settings). (B) $A \times B$ interaction for epoxy die tilt (*A*, force; *B*, tool gap settings).

to be made. The process improvement team (PIT) decided to leave the process setting on the existing condition (center point).

Remarks

We always used to expect new information from an experiment that would allow us to make changes in the process. This time the information obtained from the EVOP program suggests "don't touch the process" because our engineering judgment was right. Can we say that this is a successful study? Sure we can. Now we feel more confident about the process and we also know that our specification requirements were designed correctly. This experiment proved again that EVOP is a powerful technique, not only to improve the process, but also to verify whether we made the right decision.

14.1.10   Engineering Specifications and EVOP

One of the major concerns that arises during the introduction of the EVOP program is how to be assured that the deliberate changes in process variables will not generate additional product outside the engineering specifications. Because of this, sometimes there is resistance to these changes. We might hear from some operations managers, "We have very tight specifications which make it difficult to maintain the process yield, so there is no room to introduce deliberate deviations from the existing process settings." This type of concern is understandable and should not be ignored. But, by investigating the particular situations, we can find that most of the specifications are our manufacturing requirements and not the customers' specifications. These specifications are usually tighter than those required by the customer. So to resolve the problem, it is reasonable to request that the manufacturing engineer revise an artificially tight existing specification requirement, once it has been determined that the optimum process conditions lie outside the historical internally imposed specifications. This is the usual practice in the AMD organization.

But what about a situation where we are dealing with the customer specifications? In this case, taking into consideration that "the customer is always right," we will not start negotiations with the customers even if there is a lot of room for negotiation. Instead, we need to be sure that we are dealing with a process that is in a state of statistical control. If not, an effort to reduce the variability by introduction of a process capability study is essential. Activity control will reduce the process variation and will make more room for EVOP (see Figure 14.22).

Figure 14.22(A) represents the distribution of a process that is not in a state of statistical control and, because of the existence of unnatural variation in the system, the process spread consumes all of the allowable specification range. So there is no room for introducing additional variation by EVOP without jeopardizing the quality.

Bringing the process into a state of statistical control will reduce the process spread [see Figure 14.22(B)] and make it possible to run an EVOP program, which means introducing deliberate variations in the process parameters to determine the best settings.

Taking it a step further, the next question to arise could be, "What if the process is already in a state of statistical control and the customer's specification spread equals the exact process spread ($Cp = Cpk = 1$)?" [see Figure 14.22(C)]. In this case there are still opportunities to continue to reduce the process spread, not by SPC anymore, but by application of on-line design of experiments.

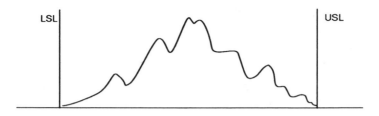

(A) The process is not in a state of statistical control and the spread consumes all the allowable specification randge.

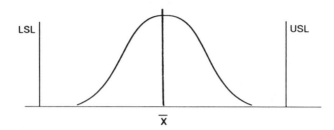

(B) The process is brought in a state of statistical control and the spread is reduced.

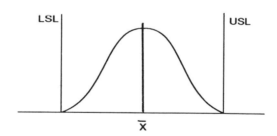

(C) The process is in a state of statistical control, but the natural spread consumes all of the customer's specification range.

Figure 14.22 Different situations of process variability.

Now we come to a situation where EVOP is needed but cannot be introduced because of the narrow specifications. We had a similar situation in one of the AMD plants.

The Process Improvement Committee made a decision to run an EVOP program introducing very conservative deviations from the best-known conditions. A temporary measure of additional testing and inspection was made

TABLE 14.2   Existing Parameter Windows

Power	55 to 66 pulses
Time	15 to 25 pulses
Force	20 to 30 grams

Note: A pulse is a fixed increment of change controlled within the bonding machine.

for extra protection. After a short time of running the EVOP program, further reduction of process variation was achieved, extra testing and inspection were stopped, and all this made it possible to introduce EVOP on a continuous basis. The yield has been significantly increased.

In conclusion, we should say that the specification requirements usually are not a problem in introducing EVOP. What is necessary is to involve people from different departments who have different responsibilities and skills in EVOP activities. This will allow utilization of power of EVOP in increasing productivity.

Case History 14.4: Wire Bond Process Optimization by Using Three-Variable EVOP

Introduction

Wire bonding is a process of connecting the terminals of the circuitry on a microchip (die) to a package lead frame. In this ceramic package process, fine aluminum wires are attached to the die terminals (called bonding pads) by a precise bonding tool. The wire is then drawn into a loop by the same tool and the other end of the wire is attached to the frame, thus forming the electrical connection between the pad and the lead.

A good quality bond requires that the wire stick to the surface while maintaining a good quality shape (without cracks, etc.). This can be controlled by three variables: bond force, power, and time. These parameters can be programmed into the wire bonder for each wire of the integrated circuit to be bonded, both on the pad and the lead.

The bonding process operates within preset windows for each parameter (see Table 14.2). However, even with these existing limits, the optimum settings for the parameters are not known. It was decided to conduct a design of experiment to optimize the process parameters and improve the process capability.

Setting up the experiments

An evolutionary operations (EVOP) program was introduced to investigate three variables simultaneously which means to conduct a 2^3 design of experiment directly on the line. Since bond force, power, and time are recognized as the most important process parameters that impact the process

Overall Layout

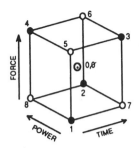

Figure 14.23 Overall layout.

quality, they have been selected as the first three controllable variables in the experiment. As a dependent (or response) variable, it has been decided to use the "number of defective units per sample."

Figure 14.23 shows the layout of the 2^3 factorial design represented as a cube. The center point of this cube is the current best-known condition (the reference) which is: 60 pulses for power, 20 pulses for time, and 25 grams for

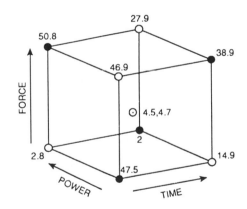

			2 STANDARD ERROR LIMITS
MAIN EFFECT	A: TIME	-16.1	±5.83
	B: POWER	-16.2	±5.83
	C: FORCE	24.3	±5.83
INTERACTION EFFECT	A X B	4.2	±5.83
	A X C	1.9	±5.83
	B X C	12.4	±5.83
	A X B X C	-11.7	±5.83
CHANGE - IN - MEAN EFFECT		19.5	±5.21

Figure 14.24 EVOP information board after phase 1, cycle 3 (relative number of defects observed).

force. With this condition the experiment was started. The parameter settings were then varied with steps of two units as follows:

FACTORS	LEVEL 1	LEVEL 2
A = TIME	18 pulses	22 pulses
B = POWER	58 pulses	62 pulses
C = FORCE	23 grams	27 grams

The changes in the variables are assumed to be small enough so that serious disturbances in product quality will not occur, yet large enough so that potential improvement in the process performance will eventually be discovered.

The numbers at the vertexes of the cube refer to the sequence in which the runs were made. To eliminate removable extraneous variation while running the experiments, the cube comprising a full cycle of the experiment was broken into two blocks. Runs 0, 1, 2, 3, and 4 shown by filled dots comprise block I; and runs Ø, 5, 6, 7, and 8 shown by open dots comprise

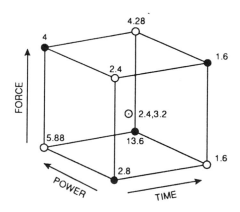

			2 STANDARD ERROR LIMITS
MAIN EFFECT	A: TIME	1.52	± 1.32
	B: POWER	5.08	± 1.32
	C: FORCE	-2.92	± 1.32
INTERACTION EFFECT	A X B	2.52	± 1.32
	A X C	-1.76	± 1.32
	B X C	-2.72	± 1.32
	A X B X C	-1.96	± 1.32
CHANGE - IN - MEAN EFFECT		1.36	± 1.2

Figure 14.25 EVOP information board after phase 2, cycle 3 (relative number of defects observed).

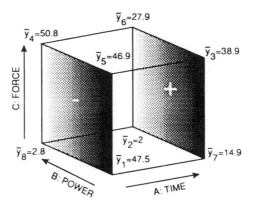

Figure 14.26 "High" and "low" level faces for the variable time (relative number of defects observed).

block II. The reference condition was run once in each block (i.e., twice per cycle) and is numbered 0 and $\emptyset$, respectively.

Production runs were performed for each setting with a constant sample size. Only one machine and 28 lead devices were used for all the runs.

Results and Interpretation
The results of the EVOP experiment are presented on an information board (see Figures 14.24 and 14.25). This information is used by the EVOP committee to make decisions about the process. Before interpreting the results reflected on the board, we'll review some terminology.

Main Effects
The main effect of a factor is defined as the average change in response produced by a change from the low level to the high level of the factor. To illustrate how the main effects can be calculated, we will use variable A (time). In this case, the main effect is a comparison of the average response on the "high time" face of the cube with the average response on the "low time" face (see Figure 14.26). This can be calculated by using

$$\text{Main effect A} = \frac{1}{4}\left[(\bar{y}_3 - \bar{y}_5) + (\bar{y}_7 - \bar{y}_1) + (\bar{y}_6 - \bar{y}_4) + (\bar{y}_2 - \bar{y}_8)\right]$$

$$= \frac{1}{4}\left[(38.9 - 46.9) + (14.9 - 47.5) + (27.9 - 50.8) + (2.0 - 2.8)\right]$$

$$= -16.1$$

The sign before the effect indicates the direction to move the levels of the factor to maximize the response. If a minimum response is desired (as in our case), the level of the factor should be moved in the direction opposite to the sign. The main effects for the other two variables can be determined in the same way.

TABLE 14.3   2 × 2 Contingency Table of Responses for Factors A and B

	B_1	B_2
A_1	$(47.5 + 46.9)/2 = 47.2$	$(2.8 + 50.8)/2 = 26.8$
A_2	$(14.9 + 38.9)/2 = 26.9$	$(2.0 + 27.9)/2 = 15.0$

Standard Errors

We summarized the responses from the EVOP in terms of main effects, interaction effects, and change-in-mean effects. These values are, of course, subject to *experimental errors*.

To measure the reliability of the responses, we will use the two standard error limits (2 S. E. limits). These limits indicate a range that almost certainly includes the true value. For example (see the bottom portion of Figure 14.24), the time effect after cycle 3 (phase 1) is -16.1 and its 2 S. E. limits are ± 5.83. This effect can be interpreted as significant because its absolute value exceeds the error limits ($\pm 2s$). But considering that this effect interacts significantly with power and force effects, we should interpret these variables jointly.

Interaction Effect

The interaction effect shows whether the effect of one factor is different at different levels of the other factor. Before interpreting the main effect of a given factor, we must look to see whether there are any appreciable interactions that involve this factor. In our example we can observe a significant interaction between all factors because the $\pm 2s$ errors are smaller than the interaction effects (see the bottom portions of Figures 14.24 and 14.25).

Because of the interactions, the effects of time, power, and force were assessed jointly rather than individually. This has been done by setting out 2×2 tables of averages for the factors concerned (as an example, see Table 14.3). (We only studied the two-factor interactions. The three-factor interaction is not usually of much practical importance by itself.)

Yate's Algorithm

The EVOP experiment is mainly done by supervisors and operators directly on the line. Because of this, the calculation of the main effects, their interactions, and the confidence limits should be very simple. Dr. Box has developed special worksheets where he applied Yates' algorithm to calculate the main effects, and used the range method to estimate the standard errors. With minimal modifications we have used his worksheets to simplify all of the calculations in this experiment (see Appendix 14.1).

Interpretation of the Effects

Now we can bring together all of the statistics and interpret the results from the EVOP experiment. The result after three cycles of operation in phase 1 are shown at the vertexes of the cube (see Figure 14.24). The estimated main

(A: TIME, B: POWER, 1 = LOW, 2 = HIGH)

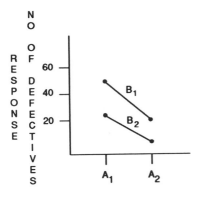

Figure 14.27 Response graphs for $A \times B$ interaction.

effects and interactions with 2.S.E. limits are given below the cube. Our attention should be focused on those effects that are reasonably large in magnitude compared with their standard errors. The results of Figure 14.24 show that all of the three variables and their interactions are significant.

As mentioned earlier, considering the significance of the interaction effects, we must investigate the variables jointly by setting out 2×2 tables of average response for the variables concerned. For example, for variable A (time) and variable B (power) we can set up a table as in Table 14.3.

From Table 14.3 we have constructed a graph which reflects the interaction between variable A (time) variable B (power) as in Figure 14.27. It shows that the best result (the lowest number of defectives) is at condition A_2B_2.

Similarly, we proceeded with the other variables and at phase 1 we came to the conclusion that the operating condition that gave the best result is $A_2B_2C_1$, which means time = 22 pulses, power = 62 pulses, and force = 23 grams.

However, the EVOP committee decided to proceed with a new phase to see whether the process could be further optimized. The best condition from phase 1 was used as the center point. The settings for the phase 2 experiment were as follows:

FACTORS	LEVEL 1	LEVEL 2
A = TIME	20 pulses	24 pulses
B = POWER	60 pulses	64 pulses
C = FORCE	21 grams	25 grams

Figure 14.25 shows the information board after phase 2, cycle 3. Now the results show the best condition is $A_2B_1C_1$, which means time = 24 pulses, power = 60 pulses, and force = 21 grams. A confirmation run was made by setting up the parameters at the best condition after phase 2.

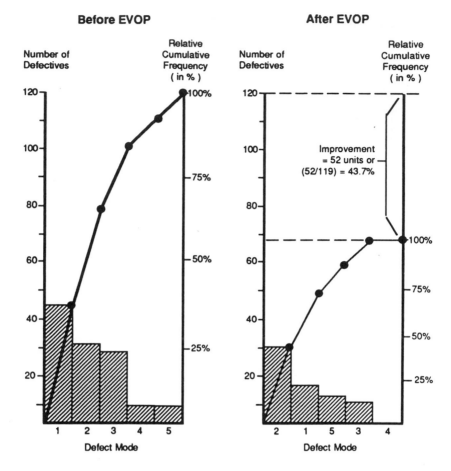

	BEFORE EVOP		AFTER EVOP	
Defective Mode	QTY.	%	QTY.	%
1. Lifted Metal	45	37.8 %	15	22.4 %
2. Lifted Bond on Pad	32	26.9 %	31	46.3 %
3. Tearing Bond	28	23.5 %	10	14.9 %
4. Wire Reduction	7	5.9 %	0	0.0 %
5. Others	7	5.9 %	11	16.4 %
TOTAL	119		67	

Figure 14.28   Pareto analysis before and after EVOP.

Conclusion

Based on the results of the confirmation run, a Pareto analysis was made to compare the number of defective units before and after the experiment. The results showed an improvement of 43.7% (see Figure 14.28). It can be seen that the defect mode 1 problem was significantly reduced. The slight reduction in defect mode 2 could be an indication that other parameters should be worked on, such as wire material, etc. The EVOP committee has determined to do more experiments and make the process more robust.

14.2 SIMPLEX EVOP

Before describing simplex EVOP, let's briefly summarize Dr. Box's philosophy of plant operation. This is a method of deliberately running a process at different specific conditions to produce products of slightly different qualities. Then the most advantageous set of conditions to run the process is selected to produce the highest-quality product in the most economical manner. Box's EVOP is based on a factorial approach philosophy. The range of the factors is small and the design is repeated until there are indications that the design could be moved with advantage to a slightly different area of the response surface being studied. The new design will usually contain at least one point from the original and will be run for a number of cycles until a further decision about moving to a new process setting can be made.

Simplex EVOP which was proposed by Spendley[6] et al. is one of the most popular alternative methods for on-line process improvement. This is a method whereby k process variables are perturbed by means of a simplex involving $k + 1$ points. After each run the point in the simplex that yields the poorest result is replaced by its reflection, forming a new simplex. This approach differs basically from the factorial approach philosophy in that a move is made after each run, rather than waiting for some evidence of significant process improvement. The basic idea of simplex EVOP can be understood by considering the case for which there are just two variables X_1 and X_2 and it is desired to improve the response (let's say yield). The points labeled 1, 2, and 3 in Figure 14.29 show an initial design arranged in the form of an equilateral triangle. The yield results obtained from these three runs were 82, 85, and 86, respectively. Since 82 is the lowest result a point labeled 4 is added opposite to vertex 1 and forms a new equilateral triangle with vertexes 2 and 3. Then a run is made of the new conditions giving a yield of 88. The lowest of the runs 2, 3, and 4 is now number 2, namely a yield of 85. The next run is therefore performed at point 5, again opposite the vertex giving the lowest yield value. Runs 3, 4, and 5 now form a third equilateral triangle. Since the yield at point 5 is 90, the next run (6) would be made opposite vertex 3 and positioned so that runs 4, 5, and 6 form an equilateral triangle, and so on. It is obvious that in real application the vertexes will not always go up and, of course, we cannot expect that the progress will always be continuous (see vertex 11). Sometimes we have situations when we observe

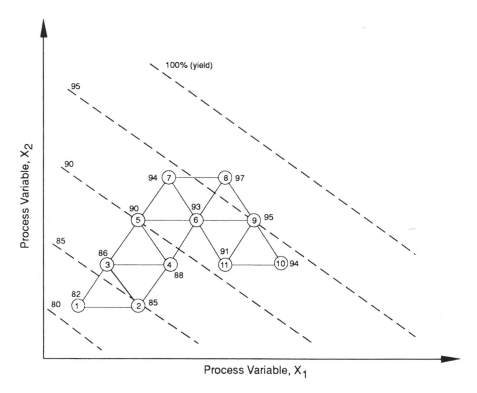

Figure 14.29 A two-dimensional simplex superimposed on a contour map of response lines (percentage yields shown).

a return to the previous positions. The major advantage of simplex EVOP is that a large number of quantitative factors can be investigated at the same time with a minimum number of experimental runs.

However, in practical application the most popular design in simplex EVOP is with two or three variables. This is because of the simplicity of graphical representation. The following examples of applying simplex EVOP will help make this approach of continuous process improvement more understandable.

14.2.1 Description of the Simplex EVOP Procedure

This description of the simplex procedure is in the form of major action steps followed by two illustrative examples which will clarify the simplex EVOP method.

Step 1 Define the dependent variable (the response) to be improved, such as yield, cost, profit, quality, etc. Any experiment can give a number of process performance characteristics which may make economical or technical sense to optimize. However, from the first experience of applying simplex

EVOP we came to the conclusion that for simplicity, it is more practical to determine one, the most influential (highest leverage) response variable and use it to make decisions in the direction of improvement, even though the program allows us to use more than one response. At the same time a special form is developed to keep track of all other auxiliary responses, which should be used for later consideration when determining the optimal conditions.

Step 2 Select the most influential process variables (factors) to be varied in the study, such as pressure, temperature, time, etc. As in the EVOP program using special techniques, the process improvement team (PIT) makes up a list of variables, which have an impact on the selected main responses. From this list the team chooses only the most important factors. This selection is based mainly on the prior knowledge of the engineering and operations staff. Sometimes when there is no such prior knowledge a preliminary off-line experiment is made. A two-level fractional factorial for this purpose is adequate.

Step 3 Determine the constraints (boundaries) such as specifications, safety limits, quality levels, etc. The constraints may be dictated by the equipment limitations (see Case History 14.5), customers, or internal specification requirements, etc. Any undesirable results (e.g., instability of the output) can also be treated as constraints.

Step 4 Locate the initial simplex. To run the simplex experiment, it is necessary to locate the initial values for each of the factors. These will often be the factor levels in current use unless prior experimentation has indicated a better region for starting the simplex experiments. Then with the desired step size, the vertex which comprises the initial simplex is located. For a two-factor simplex the figure is a triangle and therefore requires three points. Since two factors can be followed graphically, all that is necessary is to lay out the two factors as the x and y coordinates and locate the vertexes of the simplex on the graph. A three-factor simplex is a tetrahedron, so factors A, B, and C require vertexes 1, 2, 3, and 4 (see Figure 14.30).

Step 5 Start the experiments to search for the optimum response. After the first simplex has been run, the worst experimental value is eliminated and a new vertex is located by reflecting the simplex in the direction opposite from the undesirable result. For the two-factor case this may be done graphically (see Case History 14.5). For three or more factors, the coordinates for the new point are found by simple calculation. A detailed procedure of calculation is given in Appendix 14.2.

If a given vertex is retained after $k + 1$ successive simplexes (where k is the number of factors), it is possible that it was a spurious value caused by error so the value should be replaced by a new observation at that point.

If the response at a new point is also the worst value in the simplex, causing it to reflect back onto the previous position, then the second worst reading should be eliminated instead of the worst. If a peak has been reached this will cause the simplex to circle the maximum so that the cessation of progress will be verified and the region around the maximum defined.

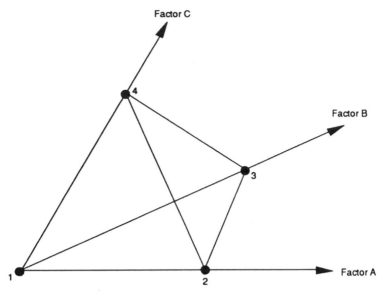

Figure 14.30 A graphical representation of three-factor simplex.

Step 6 Reduce the step size and start a new simplex experiment to locate the most optimum response. (A large step should be used first to decrease the effort required to reach the general region of optimum.) To accomplish this, the coordinates of the simplex giving the best value are taken as the new experimental origin, and the remaining vertexes are located by applying the smaller step.

Case History 14.5: Tin-Plating Process Optimization

The Problem
Tin plating is the process of coating lead frames with at least 300 microinches of tin to protect them from corrosion and to provide good board solderability. There are five major elements that represent the tin-plating: cosmetic requirements, other visual inspection criteria, solderability, adhesion tests, and tin thickness. For this example the first four components are practically satisfactory. The only element that needs attention is tin thickness optimization.

The Starting Point
When the investigation started, the process average defective for tin thickness was 5500 ppm (which means that on the average 0.55% of the parts had a tin-plating thickness lower than the specification requirement). At this time the erratic pattern on the control chart indicates that the process is not in a state of statistical control. A team consisting of members from the produc-

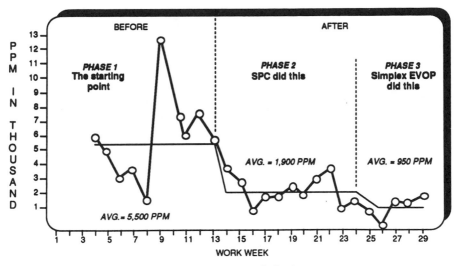

Figure 14.31   Process optimization steps.

tion, maintenance, and process engineering departments is formed to bring the process into control.

Equipment improvement, operator retraining, and other engineering actions help to identify and eliminate the special causes and improve the process average to a level lower than 2000 ppm (see phases 1 and 2 in Figure 14.31). This is a significant improvement, but 2000 ppm is still high.

Where Do We Go From Here?

When the process is brought into control, the 2000-ppm process average represents its inherent capability. So if we want to continue to improve, we need to change the system. The simplex evolutionary operation (simplex EVOP) is the right approach to use because the process variables can be perturbed only one time per shift.

The Simplex EVOP Experiment

To conduct the experiment, two controllable variables are selected: anode-to-cathode distance and anode-to-cathode surface area ratio. Tin thickness failure rate (in %) is selected to be the response variable.

Starting from the best-known conditions, that is 10 cm distance as per equipment design and a ratio greater than 1:1, we perform the runs at the close of every shift. Only one plating bath and the most stable package type (28 leads) are used.

Increments of 0.5 surface area ratio and 2 cm distances are taken for the experiment. In each run, the response is noted. The conditions used along

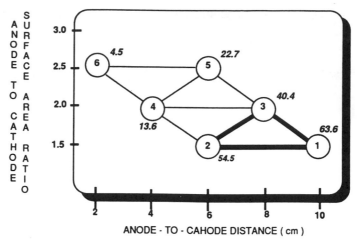

Figure 14.32   Simplex EVOP runs with resulting relative percentage tin thickness failure rate.

with their corresponding responses, are then laid out on the corners of the simplex triangle. The triangle with heavier lines in Figure 14.32 represents the initial simplex. The fourth run is performed at conditions which are mirror images of the least favorable run (run 1), the one with the highest percentage of defective tin thicknesses. Other runs are conducted in a similar fashion until run 6. The experiment ends at run 6 since zero anode-to-cathode distance is impossible. The parameters eventually shift from the highest to the lowest percentage of tin thickness failure.

Results
The simplex EVOP runs and results can be seen in Figure 14.32 and Table 14.4.

TABLE 14.4   Simplex EVOP Data

Run	Anode-to-Cathode Distance (cm)	Anode-to-Cathode Surface Area Ratio	Relative Tin Thickness Failure Rate (in %)
1	10	1.5:1	63.6
2	6	1.5:1	54.5
3	8	2.0:1	40.4
4	4	2.0:1	13.6
5	6	2.5:1	22.7
6	2	2.5:1	4.5
7	0	2.0:1	N/A

Conclusion

Theoretically, it turns out that a 2 cm anode-to-cathode distance and a 2.5:1 anode-to-cathode surface area ratio are the most optimum combination of the two variables.

Effect on Tin Thickness

The EVOP team is very eager to implement the optimum conditions which were determined in the experiment. Full conversion of the tanks is then scheduled for the following week.

However, upon installation of the fixed anode bar support, maintenance technicians verify that the minimum practicable distance is 4 cm for accurate alignment of the equipment transporter (pickup arm and carrier). So the final combination is set at 4 cm anode-to-cathode distance and 2.0:1 surface area ratio.

After implementation of the above combination, 50% improvement is observed (see the trend chart, Figure 14.31). Now these two optimized variables can be excluded from the tin thickness failure cause list so that we can focus our attention on the more complicated problem of plating solution efficiency.

Case History 14.6: A Three-Variable Simplex EVOP Approach to Reduce Tin Flakes on Cerdip Package Lids

The Problem

For the purpose of protecting lead frames of cerdip packages from corrosion and also providing for good board solderability, a coating of tin is applied by an electrolytic process known as tin plating. However, with more stringent requirements on device top mark illegibility, there is growing concern about eliminating loose tin flakes that may be deposited on the package causing top mark defects.

The process improvement team in the tin plate area had identified the following process parameters as affecting the tin flake/tin deposit through a consensus-based rank order of possible causes:

1. The *temperature* of the sulfuric acid (microprocessor controlled)
2. The immersion *time* in the sulfuric acid (also microprocessor controlled)
3. The *specific gravity* of the acid (monitored with hydrometer)

With the objective of minimizing the response variable, i.e., the *percentage of units with tin flakes*, simplex EVOP was employed to quickly optimize the process parameters that have been identified. A secondary response would be the visual appearance of the tin-plated surface, and a tertiary response would be the cost of manufacture.

For evaluating a function of three variables, a simplex of four vertexes was first run to determine the response at each vertex. The simplex is then moved away from the *largest* response value in the direction of the steepest descent by substituting the vertex that has the highest occurrence of tin flakes with one located by reflection through the centroid of the other vertexes.

The following rules are applied as the procedure and are repeated indefinitely.

1. The scales for the separate factors were chosen so that the unit change in each is of equal interest.
2. In any simplex the point that gives the least acceptable response is rejected and replaced by its mirror image with coordinates given by *twice the average of the coordinates of the common points minus the coordinate of the rejected point.*
3. If (in the simplex on three factors) an observation has occurred in four successive simplexes and is not eliminated under rule 2 above, the observation will be discarded and experimentation is repeated at that point. This is to reduce the risk of circling around a spuriously good result and accepting this result as if it were the genuine optimum.

TABLE 14.5   Simplex 1

		Variable			Response
		Temperature	Time	Specific Gravity	Percentage Tin Flake
Coordinates	1	70	5	1.5	100%
of simplex	2	80	5	1.5	0%
EVOP	3	75	8	1.5	83.3%
	4	75	6.5	1.4	100%
(A) Sum of retained coordinates		225	18	4.5	
(B) Twice average (A)		150	12	3	
(C) Coordinates of discarded point		75	6.5	1.4	
(D) Coordinates of next point (B)−(C)		75	5.5	1.6	
Sum of values for whole simplex					283.3%
Average value for whole simplex					71%

TABLE 14.6 Simplex 2

		Variable			Response
		Temperature	Time	Specific Gravity	Percentage Tin Flake
Coordinates of simplex EVOP	1	70	5	1.5	100%
	2	80	5	1.5	0%
	3	75	8	1.5	87%
	5	75	5.5	1.6	0%
(A) Sum of retained coordinates		230	18.5	4.6	
(B) Twice average (A)		153.3	12.3	3.1	
(C) Coordinates of discarded point		70	5	1.5	
(D) Coordinates of next point (B)−(C)		83.3	7.3	1.6	
Sum of values for whole simplex					187%
Average value for whole simplex					47%

4. Where the new point in a simplex has the least acceptable response, the simplex is not returned to its mirror image, but moved from the next lowest point. This is to reduce the risk of circling around a spuriously bad result.

Tables 14.5 through 14.8 show the results of experiments. The detailed calculations are demonstrated in Appendix 14.2.

Looking to simplex 1 (see Table 14.5), while there were two points that had the highest percentage of tin flakes, the process improvement team (PIT) had chosen to discard point 4 because the tin flaking was observed to be worse visually than that of point 1. The average response for simplex 1 is 71%.

In Simplex 2 (see Table 14.6), point 5 is the mirror image of point 4. Point 1 was discarded with a response of 100% tin flaking. The average response for simplex 2 is down to 47%.

In simplex 3 (see Table 14.7), point 6 is the mirror image of point 1. Point 3 was discarded with a response of 87% tin flaking. The average response for simplex 3 is down to 22%.

TABLE 14.7 Simplex 3

		Variable			Response
		Temperature	Time	Specific Gravity	Percentage Tin Flake
Coordinates of simplex EVOP	2	80	5	1.5	0%
	3	75	8	1.5	86.7%
	5	75	5.5	1.6	0%
	6	83.3	7.3	1.6	0%
(A) Sum of retained coordinates		238.3	17.8	4.7	
(B) Twice average (A)		158.9	11.9	3.13	
(C) Coordinates of discarded point		75	8	1.5	
(D) Coordinates of next point (B)−(C)		83.9	3.9	1.63	
Sum of values for whole simplex					86.7%
Average value for whole simplex					22%

TABLE 14.8 Simplex 4

		Variable			Response
		Temperature	Time	Specific Gravity	Percentage Tin Flake
Coordinates of simplex EVOP	2	80	5	1.5	0%
	5	75	5.5	1.6	0%
	6	83.3	7.3	1.6	0%
	7	83.9	3.9	1.63	0%
(A) Sum of retained coordinates					
(B) Twice average (A)					
(C) Coordinates of discarded point					
(D) Coordinates of next point (B)−(C)					
Sum of values for whole simplex					0%
Average value for whole simplex					0%

In simplex 4 (see Table 14.8), point 7 is the mirror image of point 3. All points are now showing the desired response of 0% tin flaking, and the average response for the whole simplex is 0%.

The process improvement team at the tin plate area had succeeded in moving as quickly as possible across the response surface to reach the optimum of 0% tin flakes in four simplexes. However, the simplex experiments did not stop there. The team had decided to perform more simplexes that would maintain the 0% tin flaking response, and at the same time move in the direction that would be more economical to run the process; i.e., find the lowest temperature, the shortest immersion time, and the lowest specific gravity of the acid that would still result in 0% tin flaking. Even after that, simplex EVOP was still maintained as a routine procedure for continuous feedback, and simplexes were moved away from danger points that occurred occasionally as the response surface drifted slightly in time.

14.3 INTRODUCING AN EVOP PROGRAM

The procedure for introducing an EVOP program consists of four major stages: planning, preparation, execution, and decisional (see Figure 14.33). The step-by-step procedure is described below.

1. Planning Stage
A. Develop and introduce an EVOP educational and training program for all levels of participants (see Appendix 14.3). *Note:* This should be done after a brief introduction of the EVOP concept to the management to obtain its support and commitment.

B. Develop EVOP organizational structure.

1. Establish process improvement committees (PIC) in every plant or product line. *Note:* The committees' function is not only for EVOP activities. These committees are involved in all stages related to process improvement, process control, process capability study, and on-line and off-line design of experiments.
2. Form process improvement teams (PIT) for every process.

C. Develop a plan that demonstrates where, when, and who will introduce the EVOP program. *Note:* This is normally done by the PIC.

2. Preparation Stage
A. Use brainstorming sessions to identify the major controllable variables to be investigated on a particular process (or process step).

B. Select two or three of the most important controllable variables to be used in the initial study.

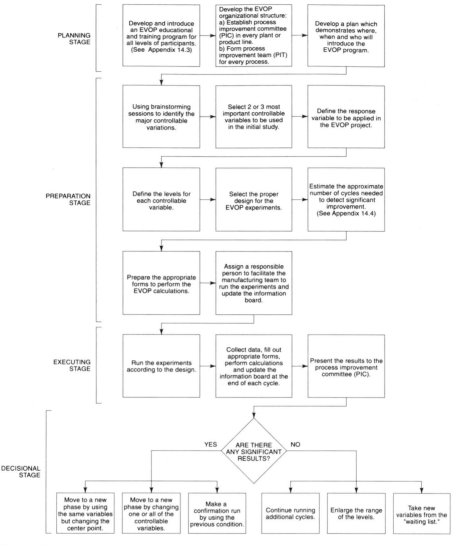

Figure 14.33 A Flowchart of introducing the EVOP Program.

C. Define the response variables to be applied in the EVOP program. *Note:* Practice the use of more than one response variable which will include technical and economic aspects.

D. Determine the levels for each controllable variable. *Note:* Make sure that the levels (the deviations from the center point) are optimal (large enough to generate significant effects and small enough not to jeopardize the quality of the product).

E. Select the proper design for the EVOP (2^2 or 2^3, with or without a center point).

F. Estimate the approximate number of cycles needed to detect significant improvement (see Appendix 14.4).

G. Prepare appropriate forms to perform the EVOP calculations.

H. Assign a responsible person who is knowledgeable in EVOP to facilitate the manufacturing team, which will run the project and update the information board.

3. Execution Stage

A. Run the experiment according to the design.

B. Collect data, fill out the appropriate forms, perform calculations, and update the information board at the end of each cycle.

C. Present the results to the Process Improvement Committee (PIC). *Note:* The results are usually submitted to the PIC only when (a) a significant result is achieved or (b) the number of cycles exceeds the planned number and no results are achieved.

4. Decision Stage

A. Determine whether there are any significant results. When one or more of the effects are significant, usually the following decisions can be made.

1. Move to a new phase by using the same variables but changing the center point.
2. Move to a new phase by changing one or all of the controllable variables.
3. Make a confirmation run using the optimum conditions to ensure that the results are correct.

B. When no significant results are achieved after completion of the planned number of cycles usually the following decisions can be made.

1. Continue running additional cycles to see if significant results can be achieved.
2. Enlarge the range of levels if the conditions allow you to do so (i.e., so as not to jeopardize the quality).
3. Take new variables from the "waiting list."

14.4 THE EVOP EDUCATIONAL PROGRAM

For any new strategy to be accepted and implemented, an educational program related to this strategy should be developed and introduced. And this educational program should be developed in such a way that it will clearly deliver the concepts and benefits of the new strategy and also how and why the strategy should be adopted.

Experience shows that people will never seriously accept any educational program if this program is not related to their own interests.

For the AMD plants, EVOP became an important element of the process improvement system. In this system EVOP was adopted as a strategy of on-line process improvement. This strategy became especially important after achieving the positive results of bringing the manufacturing process into a state of statistical control.

Then we wondered, "If the process is in control, what is next?" The application of statistical process control allowed us to achieve significant process improvements which were reflected in the increased yield and reduction of the ppm (parts per million) defective levels. But how do we continue the improvement? As soon as the process is brought into a state of statistical control, SPC by itself does not help. Other methods are needed to continue the process of improvement. So SPC develops a demand for EVOP, and when people are waiting for some new tools, that is the right time to organize training and give them the knowledge. But how should the EVOP concept be introduced?

Doctor Box developed the EVOP design of experiment in such a way that it can be introduced to manufacturing people in a short course of 20 hours without going into the theory behind this strategy. In this course, the manufacturing managers and supervisors together with the engineering support groups become familiar with the fundamentals of variation, confidence intervals, 2^2 or 2^3 factorial design, and other elements which help them better understand the EVOP formulas and calculations (see Appendix 14.3 for the full program outline). The most important part of this course was the EVOP game which was conducted in the last four hours of the course. This game allowed the students to simulate a real situation of a manufacturing process and perform an EVOP program. The class was divided into three groups where every group was responsible for a particular response.

At the end of the course the participants received an EVOP handout which included all forms and instructions for performing a 2^2 and 2^3 on-line design of experiment with one or more responses.

Based on this material, an assignment was given to the students to develop an EVOP program in their own working areas. This was agreed to and supported by their management. After the program was completed an EVOP workshop was conducted to discuss the results. There were a lot of interesting mistakes that allowed the instructor to emphasize some specifics of EVOP which were not understood in the beginning (theoretical section) of the education.

Having a workshop as a part of the educational program is a very efficient approach. When managers and supervisors became familiar with the method and enthusiastic about the initial results, they tailored the program to the needs of the operators. The supervisors became EVOP instructors for their people.

One year later a two-day international conference was held where the manufacturing people presented 35 EVOP programs with impressive economic and technical results. A special issue of the AMD-MSD periodical

Target was published where the EVOP approach was introduced and the results of the first steps of application were presented.

From this time EVOP became a continuous mode of on-line process improvement. As it often happens, when people have accepted a strategy and they can see positive results, they want to know more about this philosophy. So a second level of EVOP education has been prepared due to the employees' requests. This program includes new details and modifications based on new experiences from application and other specific needs.

Finally, it is important to note that the EVOP educational program is continuous. The process of education continued directly on the manufacturing floor by resolving and explaining problems related to EVOP and by discussing the results from the information boards at the process improvement sessions.

14.5 THE STRATEGY DEPENDS ON THE OBJECTIVES

Evolutionary operation (EVOP) and design of experiments (DOE) can be considered as two different experimental programs which should be used

TABLE 14.9   The Difference Between EVOP and DOE

EVOP	DOE
1. Design to be run mainly by operating personnel who are running the product line. *Note:* Some assistance is needed periodically from the engineering department.	1. Performed by engineers and skilled technicians. *Note:* Often there is a shortage of this category of people so the demand is higher than the supply.
2. Is conducted on the manufacturing floor during actual production.	2. Is usually performed in laboratory or pilot line conditions during product development.
3. Usually for simplicity it is practical to use only 2^2 or 2^3 factorial designs. Only small perturbations of the process variables are introduced so that the quality of the product will not be jeopardized. Because of this it takes more time to determine the effects.	3. Can use many factors simultaneously and introduce relatively large perturbations of the process variables to determine their effects rapidly.
4. The experiment does not result in the production of unusable products.	4. Most of the time the experiment results in the production of unusable products.
5. The experimental results usually conform to the real product results.	5. The experimental results need to be tuned-up for full scale production.
6. Practically no extra cost for running the program.	6. The experiments cost money.

according to what the objectives of study are. To help the experimenter determine whether to use EVOP or DOE, the difference between these two programs is listed in Table 14.9.

14.6 OTHER EVOP TECHNIQUES

Since Box first proposed evolutionary operations as a philosophy for continuous process improvement, various EVOP modifications and other techniques have been suggested by different authors. In this chapter we described the classical Box EVOP and also simplex EVOP as proposed by Spendley et al. These two methods have been adapted by the AMD plants as a mode for continuous process improvement. Other on-line process improvement techniques such as rotating square evolutionary operation (ROVOP) and plant experiment (PLEX) have not been broadly applied, and are only occasionally used in the AMD plants. Because of the space limitations in this book, we will briefly describe these techniques to give the reader some impression about them.

14.6.1 Rotating Square Evolutionary Operation[7]

Rotating square evolutionary operation (ROVOP) is restricted by two or three factors so that the experimental design remains as simple as with EVOP. Using more than three factors would make the ROVOP very complicated.

The basic procedure is to start with very small variants (see Figure 14.34). This permits effects to converge more quickly to some predetermined levels of significance and provides information on the quadratic effects. The experimentation continues until a response surface, shown to be valid at the predetermined level of significance, is developed. At that point the pattern is moved, consecutively, in the direction of optimization and ROVOP continues until no further movement is indicated.

When the process improvement team decides to move to new cycle conditions, at least two points of the new cycle should lie on or within the bounds of the experimental space already explored. Also the range between the high and low levels of the variable to be changed should be reduced by at least a factor of $\sqrt{2}$; the levels of variables shown to be insignificant need not be reduced (see Figure 14.35).

The advantage of this method is that the starting levels are not so influential and can be very close to the existing operating conditions. This is very important when we are dealing with a process where the specification requirements are very tight. ROVOP is very useful if the process is already operating close to its optimum. This is because the rotating pattern (at 45°)

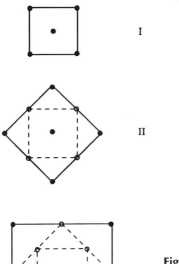

I

II

Figure 14.34 The first three ROVOP cycles.
III Reprinted with permission from C. W. Lowe, "Some Techniques of Evolutionary Operation," Transactions of Institution of Chemical Engineers, Vol. 42, 1964, p. 7337.

immediately provides quadratic terms. There are also some disadvantages that don't make ROVOP as popular as EVOP. For example, the data analysis is more complicated than for EVOP, requires a computer, and only quantitative factors can be included in such an experiment. However, now that there is a tendency to generate more variable than attribute data and since computer application on the line has became routine, ROVOP may be resurrected.

14.6.2 Plant Experimentation (PLEX)

Plant experimentation (PLEX) has the same purpose as EVOP. Both of these methods are developed for process improvement. However, their executions differ significantly. EVOP is an ongoing, on-line design of experiment for continuous process improvement by introduction of small changes in the variable levels. This experiment is usually performed by the operations people and with little or no additional engineering or technical support. In contrast, PLEX may be one or a series of short duration "one-shot" experimental designs under continuous engineering and technical supervision. If EVOP is recognized as a method that is slow in returning information, PLEX is considered a method that speeds up the return of information by concentrating on a sequence of short-duration studies with the variable levels

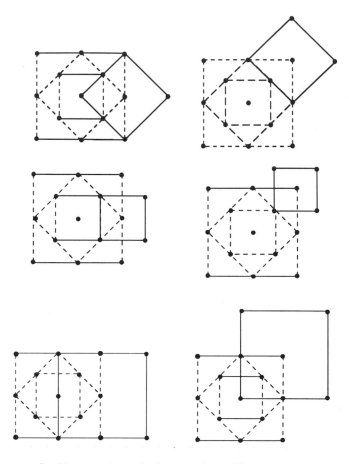

Possible movements toward optimum operating conditions on
cycle of ROVOP, providing overlap with earlier operating
conditions.

Figure 14.35 Possible movements toward optimum operating conditions on the cycle
of ROVOP, providing overlap with earlier operating conditions.
Reprinted with permission from C. W. Lowe, "Some Techniques of Evolutionary
Operation," Transactions of Institution of Chemical Engineers, Vol. 42, 1964, p. 7337.

changing over wider, more informative ranges than those felt safe for EVOP.
Because of this, PLEX has a higher risk of upsetting the process and this is
why continuous engineering participation is required.

Taking all this into consideration, PLEX should be applied only when the
need for fast information is defined and economically justified. PLEX may be
justified, for example, when we have a strong need of increasing manufacturing capacity, or when there is a serious quality problem to be immediately
resolved. The reason for applying the more risky method of PLEX should be

made obvious to all the people who participate in the experiment. This will reduce the natural human resistance to tampering with the process.

Procedure

When a particular problem is identified, an improvement team is formed (usually from different departments with different expertise capable of solving the problem). In the first step a list of all process variables is made and the critical parameters determined. Based on the situation a suitable type of design of experiment is selected in a limited amount of time so as not to jeopardize the product. As soon as the "one-shot" experiment is done, the process is usually returned to its previous operational conditions until the results of the experiment can be fully digested and incorporated into the operating procedure, if beneficial (see Figure 14.36). After the first experiment is done, usually a second "shot" becomes required. This may be a repetition of the same type of experiment, but with different conditions, or this may be an absolutely new type of experiment, depending on the circum-

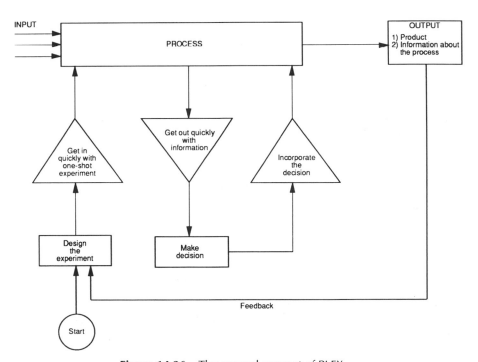

Figure 14.36   The general concept of PLEX.

stances. The "in and out" experiments are continued until the problem is resolved (see Figure 14.36). The PLEX procedure is based on factorial design which can be considered as an off-line experiment. But we refer to PLEX as an on-line activity mainly because it involves experiments performed directly in the plant.

APPENDIX 14.1
Calculations for Two- and Three-Variable EVOP Programs

Considering that EVOP is a design of experiment that is performed directly on the line by manufacturing personnel, a special worksheet has been developed by G. E. P. Box and N. R. Draper which reduces all of the calculations to a routine that can easily be handled, even by people who have had no statistical training.

The EVOP worksheet basically contains four parts requiring different calculations.

Part I: Calculation of Average

This part of the worksheet allows us to record the results we obtain from running the process at different operating conditions in each cycle. It also permits us to accumulate the data from previous cycles in order to derive an average response for each operating condition.

Part II: Calculation of Effects

In this section the main effect, interaction effect, and change-in-mean effect are calculated based on the new average responses which we derive in part I.

Part III: Calculation of Standard Deviation

Here the standard deviation of the experimental error is estimated based on the results of part I.

Part IV: Calculation of 2 Sigma Error (S.E.) Limits

In order to have a 95% confidence in the significance of the effects, we compute the 2 sigma error limits in this part for main, interaction, and change-in-mean effects. If the absolute value of an effect is larger than its 2 sigma error limits, then it is recognized as having a significant impact on the process.

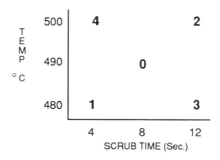

Conditions	0	1	2	3	4
CYCLE 1	5	4	6	3	14
CYCLE 2	8	7	10	5	11
CYCLE 3	9	8	9	5	10

Figure 14.A1   Lay-out and responses for the three cycles of phase 1.

WORKSHEET FOR TWO-VARIABLE EVOP

To illustrate the calculations for two-variable EVOP worksheets, we use the data in Case History 14.1 (see Figure 14.A1). In this example the effects of factor A (eutectic die attach scrub time, in seconds) and factor B (die attach temperature, in °C) are studied. The positions 0, 1, 2, 3, and 4 represent five different combinations of the two variables. For example, condition 1 represents when the scrub time is set at 4 seconds and the temperature is set at 480°C; condition 2 represents when the scrub time is set at 12 seconds and the temperature is set at 500°C; etc. For this example it was decided to make the response the percentage of voids observed by x-ray techniques under the silicon die. The results of three cycles are listed in the table of Figure 14.A1. At the end of each cycle we should perform the calculation in the EVOP worksheet to see if there is any significant effect.

The completed worksheet for the first cycle is shown in Table 14.A1. The results for cycle 1 are entered in line (iii), "New Observations." Since there is no previous cycle, line (i), (ii), and (iv) are left blank in the cycle 1 worksheet. The entries of line (v), "New Sums," and line (vi), "New Averages," are the same as line (iii), "New Observations," because there is only a single set of observations available at this point.

The entries under "Calculation of Standard Deviation" are left blank since the standard deviation cannot be estimated from the data of a single cycle. A prior estimate of the standard deviation calculated from plant records can be used in cycle 1, but such an estimate is not reliable. To make a final decision, more cycles should be run to obtain an estimate of the

TABLE 14.A1

Cycle: n = 1   Phase: 1   Response: Percent Voids

Calculation of Averages						Calculation of Standard Deviation
Operating Conditions	(0)	(1)	(2)	(3)	(4)	
						Prior Estimate S' =
(i) Previous cycle sum						Previous sum S =
(ii) Previous cycle average						Previous average S =
(iii) New observations	5	4	6	3	14	New S = range x $f_{5,n}$ =
(iv) Differences [(ii) - (iii)]						Range of (iv) =
(v) New Sums [(i) + (iii)]	5	4	6	3	14	New Sum S =
(vi) New averages [$\bar{y}_i$=(v)/n]	5	4	6	3	14	New Average S = New sum S/(n-1)
Calculation of Effects						Calculation of 2 S.E. Limits
Factor A effect = 1/2 ($\bar{y}_2+\bar{y}_3-\bar{y}_1-\bar{y}_4$) = -4.5						For main and interaction effects
Factor B effect = 1/2 ($\bar{y}_2+\bar{y}_4-\bar{y}_1-\bar{y}_3$) = 6.5						(2/$\sqrt{n}$)S =
AxB interaction effect = 1/2 ($\bar{y}_1+\bar{y}_2-\bar{y}_3-\bar{y}_4$) = -3.5						For change-in-mean effect
Change-in-mean effect = 1/5 ($\bar{y}_1+\bar{y}_2+\bar{y}_3+\bar{y}_4-4\bar{Y}_0$) = 1.4						(1.78/$\sqrt{n}$)S =

standard deviation. It's found that the standard deviation usually does not change greatly from one phase of EVOP to another. Therefore, the standard deviation from the previous phase (if any) can be used as a prior estimate at the first cycle of the new phase. And as the results from cycle 2 are available, we can follow the steps listed below to fill in the worksheet. Calculations for all subsequent cycles follow the same pattern. Tables 14.A2 and 14.A3 show the worksheets for cycles 2 and 3.

Step 1: Perform the Calculation of Averages

A. Copy the entries from lines (v) and (vi) of the worksheet of the previous cycle onto line (i), "Previous Cycle Sum," and line (ii), "Previous Cycle Average."

B. The results of the present cycle are entered in line (iii) as "New Observations."

C. Derive the entries in line (iv) by subtracting the "New Observations" in line (iii) from the "Previous Cycle Average" in line (ii). Note that the appropriate algebraic sign should be attached with the result.

D. Add the entries in line (i) to that in line (iii) to obtain the "New Sums" in line (v).

E. Obtain the "New Averages" in line (vi) by dividing the entry in line (v) by the number of cycles, n.

TABLE 14.A2

Cycle: n = 2   Phase: 1   Response: Percent Voids

Calculation of Averages						Calculation of Standard Deviation
Operating Conditions	(0)	(1)	(2)	(3)	(4)	Prior Estimate $S^* =$
(i) Previous cycle sum	5	4	6	3	14	Previous sum S =
(ii) Previous cycle average	5	4	6	3	14	Previous average S =
(iii) New observations	8	7	10	5	11	New S = range x $f_{5,n}$ = 7.0 x 0.30 = 2.1
(iv) Differences [(ii) - (iii)]	-3	-3	-4	-2	+3	Range of (iv) = 7.0
(v) New Sums [(i) + (iii)]	13	11	16	8	25	New Sum S = 2.1
(vi) New averages $[\bar{y}_i=(v)/n]$	6.5	5.5	8	4	12.5	New Average S = New sum S/(n-1) = 2.1

Calculation of Effects	Calculation of 2 S.E. Limits
Factor A effect = $1/2\,(\bar{y}_2+\bar{y}_3-\bar{y}_1-\bar{y}_4)$ = -3 Factor B effect = $1/2\,(\bar{y}_2+\bar{y}_4-\bar{y}_1-\bar{y}_3)$ = 5.5 AxB interaction effect = $1/2\,(\bar{y}_1+\bar{y}_2-\bar{y}_3-\bar{y}_4)$ = -1.5 Change-in-mean effect = $1/5\,(\bar{y}_1+\bar{y}_2+\bar{y}_3+\bar{y}_4-4\bar{y}_0)$ = 0.8	For main and interaction effects $(2/\sqrt{n})S = (2/\sqrt{2})2.1 = 2.97$ For change-in-mean effect $(1.78/\sqrt{n})S = (1.78/\sqrt{2})2.1 = 2.64$

TABLE 14.A3

Cycle: n = 3   Phase: 1   Response: Percent Voids

Calculation of Averages						Calculation of Standard Deviation
Operating Conditions	(0)	(1)	(2)	(3)	(4)	Prior Estimate $S^* =$
(i) Previous cycle sum	13	11	16	8	25	Previous sum S =2.1
(ii) Previous cycle average	6.5	5.5	8	4	12.5	Previous average S =2.1
(iii) New observations	9	8	9	5	10	New S = range x $f_{5,n}$ = 5.0 x 0.35 = 1.75
(iv) Differences [(ii) - (iii)]	-2.5	-2.5	-1	-1	2.5	Range of (iv) = 5.0
(v) New Sums [(i) + (iii)]	22	19	25	13	35	New Sum S = 3.85
(vi) New averages $[\bar{y}_i=(v)/n]$	7.33	6.33	8.33	4.33	11.67	New Average S = New sum s/(n-1) = 1.925

Calculation of Effects	Calculation of 2 S.E. Limits
Factor A effect = $1/2\,(\bar{y}_2+\bar{y}_3-\bar{y}_1-\bar{y}_4)$ = -2.67 Factor B effect = $1/2\,(\bar{y}_2+\bar{y}_4-\bar{y}_1-\bar{y}_3)$ = 4.67 AxB interaction effect = $1/2\,(\bar{y}_1+\bar{y}_2-\bar{y}_3-\bar{y}_4)$ = -0.67 Change-in-mean effect = $1/5\,(\bar{y}_1+\bar{y}_2+\bar{y}_3+\bar{y}_4-4\bar{y}_0)$ = 0.268	For main and interaction effects $(2/\sqrt{n})S = (2/\sqrt{3})1.925 = 2.22$ For change-in-mean effect $(1.78/\sqrt{n})S = (1.78/\sqrt{3})1.925 = 1.98$

TABLE 14.A4   Value of $f_{k,n}$

Value of f$_{k,n}$

n=	2	3	4	5	6	7	8	9	10
K = 5	0.30	0.35	0.37	0.38	0.39	0.40	0.40	0.40	0.41
9	0.24	0.27	0.29	0.30	0.31	0.31	0.31	0.32	0.32
10	0.23	0.26	0.28	0.29	0.30	0.30	0.30	0.31	0.31

Step 2: Perform the Calculation of Effects
For factors A and B compute the main effect, the interaction effect, and the change-in-mean effect based on the formulas. *Note:* $\overline{Y}_1$ represents the average response (from row vi) for operation condition 1.

Step 3: Perform the Calculation of Standard Deviation
   A. Obtain the "Previous Sum S" and "Previous Average S" from the "New Sum S" and "New Average S" of the previous cycle, respectively. (Since cycle 1 does not have values for "New Sum S" and "New Average S," therefore "Previous Sum S" and "Previous Average S" of cycle 2 are left blank.)
   B. Derive the difference between the largest and smallest values recorded in line (iv) and enter the result in the "Range of (iv)" [e.g., in cycle 2, range $= +3 - (-4) = 7.0$].
   C. Multiply the value of "Range of (iv)" by the factor $f_{k,n}$ to obtain the value for "New S". This is an estimated standard deviation (S) contributed by this cycle. The values of $f_{k,n}$ for n as large as 10 cycles are given in Table 14.A4. (k is the total number of operating conditions and n is the number of cycles. In our example, $k = 5$ and in cycle 2, $f_{5,2} = 0.30$.)
   D. Compute the value of "New Sum S" by adding the value of "New S" to "Previous Sum S." (For example, in cycle 3, "Previous Sum S" $= 2.1$, "New S" $= 1.75$; therefore, "New Sum S" $= 2.1 + 1.75 = 3.85$.)
   E. Divide the value of "New Sum S" by $n - 1$ to obtain "New Average S." [For example, in cycle 3, "New Average S" $= 3.85/(3-1) = 1.925$.]

Step 4: Perform the Calculation of 2 S.E. Limits
Substitute the value of "New Average S" from the result of Step 3 for S in the equations to derive the 2 sigma error limits for the main and interaction effects and the error limits for the change-in-mean effect.

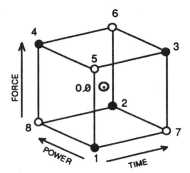

Figure 14.A2   Layout of the 2^3 factorial design.

Response (Number of Defectives) for Phase 1

	Block I					Block II			
Condition	0	1	2	3	4	Ø	5	6	7	8
Cycle 1	3.6	59.2	1.6	45.6	53.6	4.8	47.6	22.8	14.0	2.8
Cycle 2	4.4	35.6	3.6	41.6	59.2	5.6	35.6	34.4	9.6	3.6
Cycle 3	5.6	47.6	0.8	29.6	39.6	3.6	57.6	26.4	21.2	2.0

WORKSHEETS FOR THREE-VARIABLE EVOP

In a three-variable evolutionary operation program, a cycle is broken into two blocks of five runs as shown in Figure 14.A2. Runs 0, 1, 2, 3, and 4 (represented by the filled dots) comprise block 1, and Ø, 5, 6, 7, and 8 (represented by open dots) comprise block II. The reference condition was run once in each block (i.e., twice per cycle), which is numbered by 0 and Ø, respectively.

The calculations in the three-variable EVOP worksheets are similar to those used in the two-variable scheme, but two worksheets will be used for each cycle, one for block I and one for block II.

The data in Case History 14.4 are used as an illustration of how to construct and use the worksheets for a three-variable EVOP program. Figure 14.A2 shows the layout of the 2^3 design and responses (number of defectives) for the three cycles in phase 1.

Tables 14.A5 through 14.A.7 show the worksheets for phase 1. The entries on these sheets are largely self-explanatory. Note that in this example we used a previously estimated standard deviation which is marked with an asterisk (*). This value will be used in the first cycle. For the second cycle the standard deviation will be computed based on the data of the current experiments.

TABLE 14.A5 Three-Variable EVOP Calculation Sheet

Block I

Cycle: n = 1   Phase: 1   Response: Number of Defectives

Calculation of Averages						Calculation of Standard Deviation
Operating Conditions	(0)	(1)	(2)	(3)	(4)	
(i) Previous sum for block I						Previous sum S (all blocks) =
(ii) Previous average for block I						
(iii) New observations for block I	3.6	59.2	1.6	45.6	53.6	New S = range x $f_{5,n}$=
(iv) Differences [(ii) - (iii)]						Range of (iv) =
(v) New Sums [(i) + (iii)]	3.6	59.2	1.6	45.6	53.6	New Sum S (all blocks) =
(vi) New averages [$\bar{y}_i$=(v)/n]	3.6	59.2	1.6	45.6	53.6	

Block II

Cycle: n = 1   Phase: 1   Response: Number of Defectives

Calculation of Averages						Calculation of Standard Deviation
Operating Conditions	(Ø)	(5)	(6)	(7)	(8)	Prior estimate of S = 14.8*
(i) Previous sum for block II						Previous sum S =
(ii) Previous average for block II						
(iii) New observations for block II	4.8	47.6	22.8	14.0	2.8	New S = range x $f_{5,n}$=
(iv) Differences [(ii) - (iii)]						Range of (iv) =
(v) New Sums [(i) + (iii)]	4.8	47.6	22.8	14.0	2.8	New Sum S (all blocks) =
(vi) New averages [$\bar{y}_i$=(v)/n]	4.8	47.6	22.8	14.0	2.8	New Average S = New Sum S/2(n-1) =

Calculation of Effects Using Yates' Algorithm							Calculation of 2 S.E. Limits
	$\bar{y}_i$	(i)	(ii)	(iii)	Divider	Effect	Nature of Effect	
(1)	59.2	73.2	77.6	247.2	8	30.9	Mean	
(7)	14.0	4.4	169.6	-79.2	4	-19.8	A	For main and interaction Effects:
(8)	2.8	93.2	-46.4	-85.6	4	-21.4	B	$(1.41/\sqrt{n})S = 20.9^*$
(2)	1.6	76.4	-32.8	15.2	4	3.8	AB	
(5)	47.6	-45.2	-68.8	92.0	4	23.0	C	For change-in-mean: $(1.26/\sqrt{n})S = 18.6^*$
(3)	45.6	-1.2	-16.8	13.6	4	3.4	AC	
(4)	53.6	-2.0	44.0	52.0	4	13.0	BC	
(6)	22.8	-30.8	-28.8	-72.8	4	-18.2	ABC	

Calculation of Change-in-Mean	Value of $f_{5,n}$

Reference mean = $(\bar{y}_0 + \bar{y}_\varnothing)/2 = (3.6 + 4.8)/2 = 4.2$

Phase Mean = $(\bar{y}_0 + \bar{y}_\varnothing + \bar{y}_1 + ... + \bar{y}_8)/10 = (3.6+4.8+59.2....+2.8)/10 = 25.56$

Change-in-Mean = Phase Mean - Reference Mean = 25.56 - 4.2 = 21.36

n =	2	3	4	5	6
	0.30	0.35	0.37	0.38	0.39

* Asterisks indicate use of the prior estimate.

TABLE 14.A6   Three-Variable EVOP Calculation Sheet

Block I

Cycle: n = 2   Phase: 1   Response: Number of Defectives

Calculation of Averages						Calculation of Standard Deviation
Operating Conditions	(0)	(1)	(2)	(3)	(4)	
(i) Previous sum for block I	3.6	59.2	1.6	45.6	53.6	Previous sum S (all blocks) =
(ii) Previous average for block I	3.6	59.2	1.6	45.6	53.6	
(iii) New observations for block I	4.4	35.6	3.6	41.6	59.2	New S = range x $f_{5,n}$ = 29.2X0.30=8.76
(iv) Differences [(ii) - (iii)]	-0.8	23.6	-2.0	4.0	-5.6	Range of (iv) = 29.2
(v) New Sums [(i) + (iii)]	8.0	94.8	5.2	87.2	112.8	New sum S (all blocks) = 8.76
(vi) New averages [$\bar{y}_i$=(v)/n]	4.0	47.4	2.6	43.6	56.4	

Block II

Cycle: n = 2   Phase: 1   Response: Number of Defectives

Calculation of Averages						Calculation of Standard Deviation
Operating Conditions	(∅)	(5)	(6)	(7)	(8)	Prior estimate of S = 14.8*
(i) Previous sum for block II	4.8	47.6	22.8	14.0	2.8	Previous sum S = 8.76
(ii) Previous average for block II	4.8	47.6	22.8	14.0	2.8	
(iii) New observations for block II	5.6	35.6	34.4	9.6	3.6	New S = range x $f_{5,n}$= 23.6X0.30=7.08
(iv) Differences [(ii) - (iii)]	-0.8	12.0	-11.6	4.4	-0.8	Range of (iv) = 23.6
(v) New Sums [(i) + (iii)]	10.4	83.2	57.2	23.6	6.4	New Sum S (all blocks) = 15.84
(vi) New averages [$\bar{y}_i$=(v)/n]	5.2	41.6	28.6	11.8	3.2	New Average S = New sum S/2(n-1) = 7.92

Calculation of Effects Using Yates' Algorithm							Calculation of 2 S.E. Limits
	$\bar{y}_i$	(i)	(ii)	(iii)	Divider	Effect	Nature of Effect	
(1)	47.4	59.2	65.0	235.2	8	29.4	Mean	
(7)	11.8	5.8	170.2	-52.0	4	-13.0	A	For main and interaction Effects:
(8)	3.2	85.2	-26.2	-53.6	4	-13.4	B	(1.41/ $\sqrt{n}$)S = 7.9
(2)	2.6	85.0	-25.8	-4.8	4	-1.2	AB	
(5)	41.6	-25.6	-53.4	105.2	4	26.3	C	For change-in-mean: (1.26/ $\sqrt{n}$)S = 7.06
(3)	43.6	-0.6	-0.2	0.4	4	0.1	AC	
(4)	56.4	2.0	25.0	53.2	4	13.3	BC	
(6)	28.6	-27.8	-29.8	-54.8	4	-13.7	ABC	

Calculation of Change-in-Mean	Value of $f_{5,n}$					
Reference mean = $(\bar{y}_0 + \bar{y}_\emptyset)$ /2 = (4.0 + 5.2) /2 = 4.6	n =	2	3	4	5	6
Phase Mean = $(\bar{y}_0 + \bar{y}_\emptyset + \bar{y}_1 + ... + \bar{y}_8)$ /10 = (4.0+5.2+47.4....+3.2) /10 = 24.44		0.30	0.35	0.37	0.38	0.39
Change-in-Mean = Phase Mean - Reference Mean = 24.44 - 4.6 = 19.84						

* Asterisks indicate use of the prior estimate.

TABLE 14.A7 Three-Variable EVOP Calculation Sheet

Block I

Cycle: n = 3   Phase: 1   Response: Number of Defectives

Calculation of Averages						Calculation of Standard Deviation
Operating Conditions	(0)	(1)	(2)	(3)	(4)	
(i) Previous sum for block I	8.0	94.8	5.2	87.2	112.8	Previous sum S (all blocks) = 15.84
(ii) Previous average for block I	4.0	47.4	2.6	43.6	56.4	
(iii) New observations for block I	5.6	47.6	0.8	29.6	39.6	New S = range x $f_{5,n}$ = 18.4x0.35=6.44
(iv) Differences [(ii) - (iii)]	-1.6	-0.2	1.8	14.0	16.8	Range of (iv) = 18.4
(v) New Sums [(i) + (iii)]	13.6	142.4	6.0	116.8	152.4	New sum S (all blocks) = 22.28
(vi) New averages [$\bar{y}_i$=(v)/n]	4.5	47.5	2.0	38.9	50.8	

Block II

Cycle: n = 3   Phase: 1   Response: Number of Defectives

Calculation of Averages						Calculation of Standard Deviation
Operating Conditions	(∅)	(5)	(6)	(7)	(8)	
(i) Previous sum for block II	10.4	83.2	57.2	23.6	6.4	Prior estimate of S = 14.8*
(ii) Previous average for block II	5.2	41.6	28.6	11.8	3.2	Previous sum S = 22.28
(iii) New observations for block II	3.6	57.6	26.4	21.2	2.0	New S = range x $f_{5,n}$= 18.2x0.35=6.37
(iv) Differences [(ii) - (iii)]	-1.6	-16.0	2.2	-9.4	1.2	Range of (iv) = 18.2
(v) New Sums [(i) + (iii)]	14.0	140.8	83.6	44.8	8.4	New Sum S (all blocks) = 28.65
(vi) New averages [$\bar{y}_i$=(v)/n]	4.7	46.9	27.9	14.9	2.8	New Average S = New sum S/2(n-1) = 7.16

Calculation of Effects Using Yates' Algorithm							Calculation of 2 S.E. Limits
	$\bar{y}_i$	(i)	(ii)	(iii)	Divider	Effect	Nature of Effect	
(1)	47.5	62.4	67.2	231.7	8	29.0	Mean	
(7)	14.9	4.8	164.5	-64.3	4	-16.1	A	
(8)	2.8	65.8	-33.4	-64.7	4	-16.2	B	For main and interaction Effects:
(2)	2.0	78.7	-30.9	16.9	4	4.2	AB	$(1.4\sqrt{/n})S = 5.83$
(5)	46.9	-32.6	-57.6	97.3	4	24.3	C	
(3)	38.9	-0.8	-7.1	-7.5	4	-1.9	AC	For change-in-mean: $(1.2\sqrt{/n})S = 5.21$
(4)	50.8	-8.0	31.8	49.5	4	12.4	BC	
(6)	27.9	-22.9	-14.9	-46.7	4	-11.7	ABC	

Calculation of Change-in-Mean	Value of $f_{5,n}$
Reference mean = $(\bar{y}_0 + \bar{y}_\varnothing)/2 = (4.5 + 4.7)/2 = 4.6$	n = 2 3 4 5 6
Phase Mean = $(\bar{y}_0 + \bar{y}_\varnothing + \bar{y}_1 + ... + \bar{y}_8)/10 = (4.5+4.7+47.5....+2.8)/10 = 24.09$	0.30 0.35 0.37 0.38 0.39
Change-in-Mean = Phase Mean - Reference Mean = 24.09 - 4.6 = 19.49	

* Asterisks indicate use of the prior estimate.

The main effects, interaction effects, and change-in-mean effects are only estimated at the end of the complete cycle by combining the information of blocks I and II. Here we will introduce the Yates algorithm to derive the value of effects.

STEPS FOR CALCULATING EFFECTS USING THE YATES ALGORITHM (See Corresponding Area in Table 14.A5)

1. Enter the "New Averages" for each operating condition in the order as shown in the first column under the heading "$\bar{Y}_i$."
2. Perform the Yates algorithm to obtain column (i). In our example on the block II worksheet for cycle 1 (see Table 14.A5), the first four figures in column (i) are obtained as follows: 73.2 is the sum of 59.2 and 14; 4.4 is the sum of 2.8 and 1.6; 93.2 is the sum of 47.6 and 45.6; and 76.4 is the sum of 53.6 and 22.8. The next four numbers are obtained as follows: -45.2 is equal to $14 - 59.2$; -1.2 equals $1.6 - 2.8$; -2.0 equals $45.6 - 47.6$; and -30.8 equals $22.8 - 53.6$. Repeat these operations with the elements of column (i) to obtain column (ii), and with the elements of column (ii) to obtain column (iii).
3. Divide the entries in column (iii) by "Divider" to obtain "Effects." Note that the Yates algorithm is a simplified way to calculate the effects. The results should be the same as those calculated the classic way. For example,

Main effect A

$$= \frac{[(\text{Sum of the average responses when } A \text{ is set at high level}) - (\text{Sum of the average responses when } A \text{ is set at low level})]}{4}$$

at cycle 1,

Main effect A

$$= \frac{1}{4}\left[\left(\bar{Y}_3 + \bar{Y}_7 + \bar{Y}_6 + \bar{Y}_2\right) - \left(\bar{Y}_5 + \bar{Y}_1 + \bar{Y}_4 + \bar{Y}_8\right)\right]$$

$$= \frac{1}{4}[(45.6 + 14 + 22.8 + 1.6) - (9.8 + 59.2 + 53.6 + 2.8)] = -19.8$$

FACTORS FOR ESTIMATING THE 2 S.E. LIMITS

For this we use a factor $1.41/\sqrt{n}$ and multiply by the standard deviation S to derive the 2 S.E. limits for the main effects and interaction effects, and a factor of $1.26/\sqrt{n}$ is used to derive the 2 S.E. limits for the change-in-mean

TABLE 14.A8 Rules for Locating the Coordinates for the Initial Simplex

Factors	Coordinates of Points
1. A	$A_1; A_2$
2. A, B	$A_1B_1; A_2B_1; \dfrac{(A_1+A_2)}{2}B_2$
3. A, B, C	$A_1B_1C_1; A_2B_1C_1; \dfrac{(A_1+A_2)}{2}B_2C_1;$ $\dfrac{(A_1+A_2)}{2}\dfrac{(B_1+B_2)}{2}C_2$
4. A, B, C, D	$A_1B_1C_1D_1; A_2B_1C_1D_1; \dfrac{(A_1+A_2)}{2}B_2C_1D_1;$ $\dfrac{(A_1+A_2)}{2}\dfrac{(B_1+B_2)}{2}C_2D_1; \dfrac{(A_1+A_2)}{2}\dfrac{(B_1+B_2)}{2}\dfrac{(C_1+C_2)}{2}D_2$

effect. These figures are just the appropriate values of the student's t statistic divided by the square root of n (the number of cycles), and they depend on the sample size (i.e., the number of operating conditions) used to estimate the error limits of the effects.

For example, to estimate the main and interaction effects, we use only eight operating conditions (sample size = 8), but to estimate the change-in-mean effect, we need to use all 10 operating conditions (sample size = 10). Therefore, the factor to estimate the 2 S.E. limits for the change-in-mean effect is different from the factor for the main and interaction effects. As for how to derive the factors, please refer to Box and Hunter (1959),[8] Bingham (1963),[9] or Box 1957.[10]

APPENDIX 14.2
Procedure of Locating the Coordinates of a Simplex

A simplified system of locating the coordinates for simplex experimentation is described in this procedure. To locate the coordinates for the initial simplex, we should apply the rules demonstrated in Table 14.A8.

Using the data in Case History 14.5 as an example, we have the variables and levels for the experiments as shown in Table 14.A9.

Follow the rules in Table 14.A8. The coordinates of the first simplex in this example are located in Table 14.A10.

After the first simplex is located, experiments can be run according to the settings determined above. As soon as we receive the responses from the experiments, we can move to the next simplex by rejecting the least desirable point and replacing it by its mirror image.

TABLE 14.A9   Variables and Levels for Simplex Experiments

Factor	Level 1	Level 2
A	70	80
B	5	8
C	1.5	1.4

TABLE 14.A10   The Coordinates of the First Simplex in the Example

Point Number	Coordinates in Symbols	Coordinates In Figure		
		A	B	C
Point 1	$A_1B_1C_1$	70	5	1.5
Point 2	$A_2B_1C_1$	80	5	1.5
Point 3	$\frac{(A_1+A_2)}{2}B_2C_1$	75	8	1.5
Point 4	$\frac{(A_1+A_2)}{2}\frac{(B_1+B_2)}{2}C_2$	75	6.5	1.4

To calculate the coordinate of the new point, we can use a worksheet as shown in Table 14.A11, and proceed as following:

Step 1 Enter the coordinates of the simplex points and responses into the appropriate positions.

Step 2 Compute the "Sum of Retained Coordinates" in line (a). For example, for factor A, we want to discard point 4 and retain the rest. Therefore,

$$\text{Sum of retained coordinates} = 70+80+75 = 225$$

Step 3 Calculate the "Twice Average of (a)" in line (b). For factor A,

$$\frac{(70+80+75)}{3} \times 2 = 150$$

Step 4 Enter the "Coordinates of Discarded Point" in line (c). In this case, point 4 will be discarded, and the coordinate for factor A is 75.

Step 5 Derive the "Coordinates of Next Point" by subtracting the entries of line (c) from the entries of line (b). For factor A,

$$150-75 = 75$$

TABLE 14.A11   Worksheet for Simplex EVOP

Factor		A	B	C	D	Response 1	Response 2
Coordinates of	1	70	5	1.5		100	
simplex points	2	80	5	1.5		0	
	3	75	8	1.5		83.3	
	4	75	6.5	1.4		100	
(a) Sum of retained coordinates		225	18	4.5			
(b) Twice average of (a)		150	12	3.0			
(c) Coordinates of discarded point		75	6.5	1.4			
(d) Coordinates of next point [b − c]		75	5.5	1.6			
Sum of values for whole simplex						283.3	
Average value for whole simplex						71	

Step 6 Compute the "Sum of Values for Whole Simplex." In our case,

$$100 + 0 + 83.3 + 100 = 283.3$$

Step 7 Compute the "Average Value for Whole Simplex."

$$\frac{283.3}{4} = 71$$

This gives some indication of how the experiment progresses.

APPENDIX 14.3
EVOP Training Program

Below is a 20-hour educational course for managers and supervisors who will participate in the introduction of the EVOP program. This course is an extension to the overall SPC educational program, so it is assumed that the students already have the necessary basic SPC knowledge to understand the EVOP concept.

Box EVOP

1. EVOP as a method of continuous process improvement
2. The EVOP vocabulary
3. EVOP as a factorial design
4. Basics about a response surface
5. Confidence interval and test of significance
6. How to read the information board
7. How to use the Box EVOP worksheet
8. Where to start the EVOP program
9. Organizational aspects
10. Advantages and disadvantages

Simplex EVOP

1. The concept of simplex EVOP
2. How to design a simplex EVOP
3. Advantages and disadvantages
4. Where is the best place to use Simplex EVOP

Note: In the period of education, a four-hour EVOP game is introduced to simulate a "real" situation when introducing an EVOP program.

APPENDIX 14.4
Number of Cycles Needed to Detect a Significant Effect

During the EVOP experiment you can make a number of cycles but still not have a significant effect. Did you pick the wrong variables? Or was the deviations from the center point too small? Should you change the operating conditions or continue to make more cycles? These questions arise when performing an EVOP experiment, especially when the specification requirements do not allow you to make significant deviations from the best-known conditions. Box and Draper proposed a table[11] from which we can obtain some idea of the number of cycles that might be needed for a given phase of an EVOP program. In this table (see Table 14.A12) the quantity k is the proportional increase in standard deviation produced by changing the variables. In accordance with the factorial design of the EVOP program, the symbol α represents the chance of mistakenly detecting the effects of the variables when no effects exist, and similarly β represents the risk of failing to detect effects which do exist. The entries in the body of Table 14.A12 indicate the number of cycles necessary to achieve the stated values of α and β. It allows us to determine the number of cycles for 2^2 and 2^3 factorials. It is

TABLE 14.A12 Number of Cycles Required to Detect (with Probability $1 - \beta$, Using an α-Level Test) Main Effects that Increase the Standard Deviation of a Process from σ to $k\sigma$ in Two-Factor and Three-Factor EVOP Schemes without Center Points

Design	α	β	k						
			1.2	1.3	1.4	1.5	1.6	1.7	1.8
2^2 factorial	0.10	0.10	5.6	3.6	2.6	2.0	1.6	1.3	1.1
	0.10	0.05	6.8	4.4	3.1	2.4	1.9	1.6	1.3
	0.05	0.10	6.9	4.4	3.2	2.4	1.9	1.6	1.4
	0.05	0.05	8.2	5.2	3.8	2.9	2.3	1.9	1.6
2^3 factorial	0.10	0.10	3.2	2.1	1.4	1.1	0.9	0.8	0.6
	0.10	0.05	3.9	2.5	1.8	1.4	1.1	0.9	0.8
	0.05	0.10	3.9	2.5	1.8	1.4	1.1	0.9	0.8
	0.05	0.05	4.6	3.0	2.1	1.6	1.3	1.1	0.9

Reprinted with permission from George E. P. Box and Norman R. Draper, "Evolutionary Operation" (1969), John Wiley & Sons, Inc., p. 212, Table A4.1.[11]

understandable that the number of cycles must be an integer, so suitable rounding should be performed when necessary. To apply the table, select the α and β values depending on how precise an estimate is required. The k value's selection depends on the allowable specification spread. If the specifications are narrow we cannot afford a broad deviation from the best-known conditions. In this case, for example, the k factor may be 1.2 which means we allowed the standard deviation to be increased no more than 20%. If the specification spread allowed a relatively large change in the standard deviation, then the k value selected can be as large as 1.8. For example, if α and β are determined to be 0.10 and 0.05 respectively and an increase in the standard deviation of 30% (i.e., $k = 1.3$) is acceptable, then two or three cycles of the three-variable EVOP should be enough to obtain a significant effect. In general, the practical application of EVOP shows that when the ranges of the variables are well chosen, three or four cycles can provide a good chance of revealing at least the sign of important effects.

REFERENCES

1. G. E. P. Box, "Evolutionary Operation: A Method for Increasing Industrial Productivity," Applied Statistics, Vol. 6, No. 2 (1987), p. 81.

2. G. E. P. Box and J. S. Hunter, "Condensed Calculations for Evolutionary Operation Programs," Technometrics, Vol. 1, No. 1 (February 1959), p. 80.

3. J. S. Hunter, "Optimize Your Chemical Process with Evolutionary Operation," Chemical Engineering (September 1960), p. 197.

4. G. J. Hahn and A. F. Dershowitz, "Evolutionary Operation Today—Some Survey Results and Observations," Applied Statistics, Vol. 23, No. 2 (1974), p. 216.

5. J. S. Hunter, "Optimize Your Chemical Process with Evolutionary Operation," Chemical Engineering (1960), p. 197.

6. W. Spendley, G. R. Hext, and F. R. Himsworth, "Sequential Application of Simplex Designs in Optimization and Evolutionary Operation," Technometrics, Vol. 4, No. 4 (November 1962), pp. 441–461.

7. C. W. Lowe, "Some Techniques of Evolutionary Operation," Transaction of Institution of Chemical Engineers, Vol. 42 (1964), p. 334.

8. G. E. P. Box and J. S. Hunter, "Condensed Calculations for Evolutionary Operation Programs," Technometrics, Vol. 1, No.1 (1959), 77–95.

9. R. S. Bingham, "EVOP for Systematic Process Improvement," Industrial Quality Control, Vol. 20, No. 3 (1963), pp. 17–23.

10. G. E. P. Box, "A Method of Increasing Industrial Productivity," Applied Statistics, Vol. 6, No. 2 (1957), pp. 81–101.

11. G. E. P. Box and N. R. Draper, "Evolutionary Operation" (1969), John Wiley & Sons, Inc., p. 212, Table A4.1.

Sample TCPI Audit Checklist

SPC System Definition and Implementation

1. Is the SPC system defined and documented? ___
2. Are the critical nodes defined and documented? ___
3. Are the criteria for the determination of node criticality defined and documented? ___
4. Does a process flowchart exist that identifies the critical nodes? ___
5. Does the flowchart indicate the critical parameters monitored? ___
6. Does the flowchart indicate the type of control charts used? ___
7. Are on-line and/or off-line control techniques documented and implemented? ___
8. Have the individual and functional responsibilities of the SPC system been clearly defined? ___
9. Have the SPC goals been defined and documented? ___
10. Has an SPC implementation plan been documented? ___
11. Have critical process-related materials been determined and characterized? ___
12. Are the critical characteristics of process-related materials being monitored? ___
13. Is top management involved in the TCPI activities? ___
14. Is there a procedure for the systematic transmission of a TCPI report to the appropriate levels of management? ___
15. Do TCPI reports lead to corrective actions when needed in a prompt manner? ___
16. Is there management commitment to continuous improvement through the use of statistical methods? ___

TCPI Education

1. Has the TCPI education program been documented? ___
2. Has the TCPI education program been implemented? ___
3. Does the education program involve personnel from various levels (from managers to operators)? ___
4. Are TCPI education records properly documented and updated? ___
5. Are there any periodic reports on the education status? ___
6. Is TCPI education curriculum revised to reflect current practiced TCPI techniques? ___

SPC Requirements at Each Critical Node

Note: _____

1. Are the control charts implemented as indicated on the flowchart? ___
2. Are there documented procedures for using the control charts? ___
3. Are control charts maintained by and readily accessible to the operators? ___
4. Are the right type of control charts being used? ___
5. Are data collected at the point of manufacture on a real-time basis? ___
6. Are the control charts kept up to date? ___
7. Are control limits shown on every control chart? ___
8. Are process averages and control limits correctly calculated and do they reflect the process? ___
9. Are the control charts interpreted in accordance with the eight tests of special causes? ___
10. Are causes and corrective actions noted for out-of-control and significant pattern situations? ___
11. Have reasonable and effective time limits been established for responding to deficiencies? ___
12. Has corrective action been taken to address deficiencies? ___
13. Has a procedure been implemented for establishing and adjusting control limits statistically? ___

Appendix B

Factors for Constructing Control Charts

	$\bar{X}$ and R Charts				$\bar{X}$ and s charts			
	Chart for Averages ($\bar{X}$)	Chart for Ranges (R)			Chart for Averages ($\bar{X}$)	Chart for Standard Deviations (s)		
Subgroup Size n	Factors for Control Limits A_2	Divisors for Estimate of Standard Deviation d_2	Factors for Control Limits D_3	Factors for Control Limits D_4	Factors for Control Limits A_3	Divisors for Estimate of Standard Deviation c_4	Factors for Control Limits B_3	Factors for Control Limits B_4
2	1.880	1.128	—	3.267	2.659	0.7979	—	3.267
3	1.023	1.693	—	2.574	1.954	0.8862	—	2.568
4	0.729	2.059	—	2.282	1.628	0.9213	—	2.266
5	0.577	2.326	—	2.114	1.427	0.9400	—	2.089
6	0.483	2.534	—	2.004	1.287	0.9515	0.030	1.970
7	0.419	2.704	0.076	1.924	1.182	0.9594	0.118	1.882
8	0.373	2.847	0.136	1.864	1.099	0.9650	0.185	1.815
9	0.337	2.970	0.184	1.816	1.032	0.9693	0.239	1.761
10	0.308	3.078	0.223	1.777	0.975	0.9727	0.284	1.716
11	0.285	3.173	0.256	1.744	0.927	0.9754	0.321	1.679
12	0.266	3.258	0.283	1.717	0.886	0.9776	0.354	1.646
13	0.249	3.336	0.307	1.693	0.850	0.9794	0.382	1.618
14	0.235	3.407	0.328	1.672	0.817	0.9810	0.406	1.594
15	0.223	3.472	0.347	1.653	0.789	0.9823	0.428	1.572
16	0.212	3.532	0.363	1.637	0.763	0.9835	0.448	1.552
17	0.203	3.588	0.378	1.622	0.739	0.9845	0.466	1.534
18	0.194	3.640	0.391	1.608	0.718	0.9854	0.482	1.518
19	0.187	3.689	0.403	1.597	0.698	0.9862	0.497	1.503
20	0.180	3.735	0.415	1.585	0.680	0.9869	0.510	1.490
21	0.173	3.778	0.425	1.575	0.663	0.9876	0.523	1.477
22	0.167	3.819	0.434	1.566	0.647	0.9882	0.534	1.466
23	0.162	3.858	0.443	1.557	0.633	0.9887	0.545	1.455
24	0.157	3.895	0.451	1.548	0.619	0.9892	0.555	1.445
25	0.153	3.931	0.459	1.541	0.606	0.9896	0.565	1.435

Reprinted with permission from ASTM STP-15D "Manual on the Presentation of Data and Control Chart Analysis," American Society for Testing and Materials (1976), p. 83.

Appendix **C**

Area under Normal Distribution (*Z* Table)

Appendix C.   Area Under Normal Distribution (Z Table)

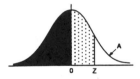

Z	0.00	0.01	0.02	0.03	0.04	0.05	0.06	0.07	0.08	0.09
0.00	5.000E-01	4.960E-01	4.920E-01	4.880E-01	4.840E-01	4.801E-01	4.761E-01	4.721E-01	4.681E-01	4.641E-01
0.10	4.602E-01	4.562E-01	4.522E-01	4.483E-01	4.443E-01	4.404E-01	4.364E-01	4.325E-01	4.286E-01	4.247E-01
0.20	4.207E-01	4.168E-01	4.129E-01	4.090E-01	4.052E-01	4.013E-01	4.974E-01	4.936E-01	4.897E-01	4.859E-01
0.30	3.821E-01	3.783E-01	3.745E-01	3.707E-01	3.669E-01	3.632E-01	3.594E-01	3.557E-01	3.520E-01	3.483E-01
0.40	3.446E-01	3.409E-01	3.372E-01	3.336E-01	3.300E-01	3.264E-01	3.228E-01	3.192E-01	3.156E-01	3.121E-01
0.50	3.085E-01	3.050E-01	3.015E-01	2.981E-01	2.946E-01	2.912E-01	2.877E-01	2.843E-01	2.810E-01	2.776E-01
0.60	2.743E-01	2.709E-01	2.676E-01	2.643E-01	2.611E-01	2.578E-01	2.546E-01	2.514E-01	2.483E-01	2.451E-01
0.70	2.420E-01	2.389E-01	2.358E-01	2.327E-01	2.297E-01	2.266E-01	2.236E-01	2.207E-01	2.177E-01	2.148E-01
0.80	2.119E-01	2.090E-01	2.061E-01	2.033E-01	2.005E-01	1.977E-01	1.949E-01	1.922E-01	1.894E-01	1.867E-01
0.90	1.841E-01	1.814E-01	1.788E-01	1.762E-01	1.736E-01	1.711E-01	1.685E-01	1.660E-01	1.635E-01	1.611E-01
1.00	1.587E-01	1.562E-01	1.539E-01	1.515E-01	1.492E-01	1.469E-01	1.446E-01	1.423E-01	1.401E-01	1.379E-01
1.10	1.357E-01	1.335E-01	1.314E-01	1.292E-01	1.271E-01	1.251E-01	1.230E-01	1.210E-01	1.190E-01	1.170E-01
1.20	1.151E-01	1.131E-01	1.112E-01	1.093E-01	1.075E-01	1.056E-01	1.038E-01	1.020E-01	1.003E-01	9.853E-02
1.30	9.680E-02	9.510E-02	9.342E-02	9.176E-02	9.012E-02	8.851E-02	8.691E-02	8.534E-02	8.379E-02	8.226E-02
1.40	8.076E-02	7.927E-02	7.780E-02	7.636E-02	7.493E-02	7.353E-02	7.214E-02	7.078E-02	6.944E-02	6.811E-02
1.50	6.681E-02	6.552E-02	6.426E-02	6.301E-02	6.178E-02	6.057E-02	5.938E-02	5.821E-02	5.705E-02	5.592E-02
1.60	5.480E-02	5.370E-02	5.262E-02	5.155E-02	5.060E-02	4.947E-02	4.846E-02	4.746E-02	4.648E-02	4.551E-02
1.70	4.457E-02	4.363E-02	4.272E-02	4.182E-02	4.093E-02	4.006E-02	3.920E-02	3.836E-02	3.754E-02	3.673E-02
1.80	3.593E-02	3.515E-02	3.438E-02	3.363E-02	3.288E-02	3.216E-02	3.144E-02	3.074E-02	3.005E-02	2.938E-02
1.90	2.872E-02	2.807E-02	2.743E-02	2.680E-02	2.619E-02	2.559E-02	2.500E-02	2.442E-02	2.385E-02	2.330E-02
2.00	2.275E-02	2.222E-02	2.169E-02	2.118E-02	2.088E-02	2.018E-02	1.970E-02	1.923E-02	1.876E-02	1.831E-02
2.10	1.786E-02	1.743E-02	1.700E-02	1.659E-02	1.618E-02	1.578E-02	1.539E-02	1.500E-02	1.463E-02	1.426E-02
2.20	1.390E-02	1.355E-02	1.321E-02	1.287E-02	1.255E-02	1.222E-02	1.191E-02	1.160E-02	1.130E-02	1.101E-02
2.30	1.072E-02	1.044E-02	1.017E-02	9.903E-03	9.642E-03	9.387E-03	9.137E-03	8.894E-03	8.658E-03	8.424E-03
2.40	8.198E-03	7.976E-03	7.760E-03	7.549E-03	7.344E-03	7.143E-03	6.947E-03	6.756E-03	6.589E-03	6.387E-03
2.50	6.210E-03	6.036E-03	5.868E-03	5.703E-03	5.543E-03	5.386E-03	5.234E-03	5.085E-03	4.940E-03	4.799E-03
2.60	4.661E-03	4.527E-03	4.396E-03	4.269E-03	4.145E-03	4.024E-03	3.907E-03	3.792E-03	3.681E-03	3.572E-03
2.70	3.467E-03	3.364E-03	3.264E-03	3.167E-03	3.072E-03	2.980E-03	2.890E-03	2.803E-03	2.718E-03	2.635E-03
2.80	2.555E-03	2.477E-03	2.401E-03	2.327E-03	2.256E-03	2.186E-03	2.118E-03	2.052E-03	1.988E-03	1.926E-03
2.90	1.866E-03	1.807E-03	1.750E-03	1.695E-03	1.641E-03	1.589E-03	1.538E-03	1.489E-03	1.441E-03	1.395E-03

Appendix C.   Area Under Normal Distribution (continued)

Z	0.00	0.01	0.02	0.03	0.04	0.05	0.06	0.07	0.08	0.09
3.00	1.350E-03	1.306E-03	1.264E-02	1.223E-02	1.183E-02	1.114E-03	1.107E-03	1.070E-03	1.035E-03	1.001E-03
3.10	9.676E-04	9.354E-04	9.042E-04	8.740E-04	8.447E-04	8.163E-04	7.888E-04	7.622E-04	7.364E-04	7.114E-04
3.20	6.871E-04	6.637E-04	6.410E-04	6.190E-04	5.977E-04	5.770E-04	5.571E-04	5.378E-04	5.191E-04	5.010E-04
3.30	4.835E-04	4.665E-04	4.501E-04	4.343E-04	4.189E-04	4.041E-04	3.898E-04	3.759E-04	3.625E-04	3.495E-04
3.40	3.370E-04	3.249E-04	3.132E-04	3.019E-04	2.909E-04	2.804E-04	2.702E-04	2.603E-04	2.508E-04	2.416E-04
3.50	2.327E-04	2.242E-04	2.159E-04	2.079E-04	2.002E-04	1.927E-04	1.855E-04	1.786E-04	1.719E-04	1.655E-04
3.60	1.592E-04	1.532E-04	1.474E-04	1.418E-04	1.364E-04	1.312E-04	1.282E-04	1.214E-04	1.167E-04	1.123E-04
3.70	1.079E-04	1.038E-04	9.974E-06	9.587E-05	9.214E-05	8.855E-06	8.509E-06	8.175E-06	7.854E-05	7.543E-06
3.80	7.248E-05	6.961E-05	6.685E-05	6.420E-06	6.165E-05	5.919E-05	5.682E-05	5.455E-06	5.238E-06	5.025E-06
3.90	4.822E-05	4.627E-05	4.440E-05	4.260EE-05	4.086E-05	3.920E-05	3.780E-06	3.608E-06	3.458E-06	3.316E-06
4.00	3.179E-05	3.048E-05	2.921E-05	2.800E-05	2.684E-05	2.572E-05	2.465E-05	2.362E-05	2.263E-05	2.168E-05
4.10	2.076E-05	1.989E-05	1.905E-05	1.824E-05	1.747E-05	1.672E-05	1.801E-05	1.533E-06	1.467E-05	1.404E-06
4.20	1.344E-05	1.286E-05	1.231E-05	1.177E-05	1.126E-05	1.077E-05	1.031E-06	9.857E-06	9.426E-06	9.014E-06
4.30	8.619E-06	8.240E-06	7.878E-06	7.530E-06	7.196E-06	6.879E-06	6.574E-06	6.282E-06	6.002E-05	5.734E-06
4.40	5.478E-06	5.233E-06	4.998E-06	4.773E-06	4.558E-06	4.353E-06	4.156E-08	3.968E-08	3.787E-06	3.615E-08
4.50	3.451E-06	3.293E-06	3.143E-06	2.999E-06	2.861E-08	2.730E-08	2.604E-06	2.484E-06	2.369E-06	2.259E-06
4.60	2.154E-06	2.054E-06	1.959E-06	1.867E-06	1.780E-06	1.697E-06	1.617E-06	1.541E-06	1.469E-06	1.399E-06
4.70	1.333E-06	1.270E-06	1.210E-06	1.153E-06	1.098E-06	1.046E-06	9.956E-07	9.480E-07	9.026E-07	8.593E-07
4.80	8.181E-07	7.787E-07	7.411E-07	7.054E-07	6.712E-07	6.367E-07	6.077E-07	5.782E-07	5.500E-07	5.235E-07
4.90	4.976E-07	4.733E-07	4.501E-07	4.280E-07	4.070E-07	3.869E-07	3.678E-07	3.496E-07	3.323E-07	3.159E-07
5.00	3.002E-07	2.853E-07	2.711E-07	2.575E-07	2.447E-07	2.324E-07	2.206E-07	2.097E-07	1.991E-07	1.891E-07
5.10	1.796E-07	1.705E-07	1.619E-07	1.537E-07	1.459E-07	1.385E-07	1.314E-07	1.247E-07	1.184E-07	1.123E-07
5.20	1.066E-07	1.011E-07	9.591E-08	9.098E-08	8.629E-08	8.184E-08	7.762E-08	7.360E-08	6.979E-08	6.617E-08
5.30	6.273E-08	5.947E-08	5.637E-08	5.343E-08	5.064E-08	4.799E-08	4.548E-08	4.309E-08	4.083E-08	3.868E-08
5.40	3.664E-08	3.471E-08	3.288E-08	3.114E-08	2.949E-08	2.792E-08	2.644E-08	2.503E-08	2.370E-08	2.244E-08
5.50	2.124E-08	2.010E-08	1.903E-08	1.801E-08	1.704E-08	1.613E-08	1.526E-08	1.444E-08	1.366E-08	1.292E-08
5.60	1.222E-08	1.156E-08	1.093E-08	1.034E-08	9.776E-09	9.244E-09	8.741E-09	8.264E-09	7.812E-09	7.385E-09
5.70	6.980E-09	6.598E-09	6.235E-09	6.893E-09	5.568E-09	5.262E-09	4.971E-09	4.697E-09	4.437E-09	4.191E-09
5.80	3.959E-09	3.739E-09	3.532E-09	3.335E-09	3.150E-09	2.974E-09	2.808E-09	2.651E-09	2.503E-09	2.363E-09
5.90	2.230E-09	2.105E-09	1.987E-09	1.875E-09	1.769E-09	1.670E-09	1.576E-09	1.487E-09	1.402E-09	1.323E-09
6.00	1.248E-09	1.177E-09	1.110E-09	1.047E-09	9.876E-10	9.314E-10	8.783E-10	8.281E-10	7.808E-10	7.361E-10
6.10	6.940E-10	6.542E-10	6.166E-10	5.812E-10	5.478E-10	5.163E-10	4.865E-10	4.585E-10	4.320E-10	4.070E-10
6.20	3.835E-10	3.613E-10	3.403E-10	3.206E-10	3.020E-10	2.844E-10	2.679E-10	2.523E-10	2.376E-10	2.237E-10
6.30	2.107E-10	1.983E-10	1.867E-10	1.785E-10	1.655E-10	1.558E-10	1.466E-10	1.380E-10	1.299E-10	1.223E-10
6.40	1.151E-10	1.083E-10	1.019E-10	9.586E-11	9.020E-11	8.486E-11	7.983E-11	7.510E-11	7.064E-11	6.645E-11
6.50	6.250E-11	5.878E-11	5.529E-11	5.199E-11	4.889E-11	4.597E-11	4.323E-11	4.065E-11	3.821E-11	3.593E-11
6.60	3.377E-11	3.175E-11	2.984E-11	2.805E-11	2.637E-11	2.478E-11	2.329E-11	2.189E-11	2.057E-11	1.933E-11
6.70	1.816E-11	1.706E-11	1.603E-11	1.506E-11	1.415E-11	1.329E-11	1.249E-11	1.173E-11	1.102E-11	1.035E-11
6.80	9.719E-12	9.127E-12	8.572E-12	8.049E-12	7.559E-12	7.097E-12	6.664E-12	6.257E-12	5.874E-12	5.515E-12
6.90	5.178E-12	4.860E-12	4.562E-12	4.283E-12	4.020E-12	3.773E-12	3.541E-12	3.323E-12	3.119E-12	2.927E-12
7.00	2.747E-12	2.577E-12	2.418E-12	2.269E-12	2.129E-12	1.997E-12	1.874E-12	1.758E-12	1.649E-12	1.547E-12
7.10	1.451E-12	1.361E-12	1.277E-12	1.198E-12	1.223E-12	1.063E-12	9.879E-13	9.264E-13	8.688E-13	8.147E-13
7.20	7.639E-13	7.163E-13	6.716E-13	6.297E-13	5.904E-13	5.535E-13	5.189E-13	4.864E-13	4.560E-13	4.275E-13
7.30	4.007E-13	3.756E-13	3.520E-13	3.300E-13	3.092E-13	2.898E-13	2.716E-13	2.546E-13	2.386E-13	2.235E-13
7.40	2.095E-13	1.963E-13	1.839E-13	1.723E-13	1.615E-13	1.513E-13	1.417E-13	1.328E-13	1.244E-13	1.166E-13
7.50	1.092E-13	1.023E-13	9.581E-14	8.975E-14	8.407E-14	7.874E-14	7.375E-14	6.908E-14	6.470E-14	6.060E-14
7.60	5.675E-14	5.315E-14	4.977E-14	4.661E-14	4.365E-14	4.087E-14	3.827E-14	3.584E-14	3.356E-14	3.142E-14
7.70	2.942E-14	2.755E-14	2.579E-14	2.415E-14	2.261E-14	2.116E-14	1.981E-14	1.855E-14	1.736E-14	1.625E-14
7.80	1.522E-14	1.424E-14	1.333E-14	1.248E-14	1.168E-14	1.093E-14	1.023E-14	9.579E-15	8.965E-15	8.391E-15
7.90	7.853E-15	7.349E-15	6.878E-15	6.437E-15	6.024E-15	5.637E-15	5.275E-15	4.937E-15	4.620E-15	4.323E-15
8.00	4.045E-15	3.785E-15	3.542E-15	3.314E-15	3.101E-15	2.901E-15	2.715E-15	2.540E-15	2.376E-15	2.223E-15
8.10	2.080E-15	1.946E-15	1.821E-15	1.703E-15	1.593E-15	1.491E-15	1.395E-15	1.305E-15	1.220E-15	1.142E-15
8.20	1.068E-15	9.991E-16	9.346E-16	8.742E-16	8.177E-16	7.649E-16	7.155E-16	6.692E-16	6.280E-16	5.855E-16
8.30	5.477E-16	5.122E-16	4.791E-16	4.481E-16	4.191E-16	3.920E-16	3.666E-16	3.429E-16	3.207E-16	2.999E-16
8.40	2.806E-16	2.624E-16	2.454E-16	2.295E-16	2.146E-16	2.007E-16	1.877E-16	1.755E-16	1.642E-16	1.535E-16
8.50	1.436E-16	1.342E-16	1.255E-16	1.174E-16	1.098E-16	1.027E-16	9.601E-17	8.978E-17	8.395E-17	7.851E-17
8.60	7.341E-17	6.865E-17	6.419E-17	6.003E-17	5.613E-17	5.249E-17	4.908E-17	4.589E-17	4.291E-17	4.013E-17
8.70	3.752E-17	3.508E-17	3.281E-17	3.068E-17	2.868E-17	2.682E-17	2.508E-17	2.345E-17	2.193E-17	2.050E-17
8.80	1.971E-17	1.792E-17	1.676E-17	1.567E-17	1.465E-17	1.370E-17	1.281E-17	1.198E-17	1.120E-17	1.047E-17
8.90	9.792E-18	9.155E-18	8.560E-18	8.004E-18	7.484E-18	6.998E-18	6.543E-18	6.118E-18	5.702E-18	5.349E-18
9.00	5.001E-18	4.676E-18	4.372E-18	4.088E-18	3.823E-18	3.574E-18	3.342E-18	3.124E-18	2.922E-18	2.732E-18
9.10	2.555E-18	2.389E-18	2.234E-18	2.089E-18	1.953E-18	1.826E-18	1.707E-18	1.597E-18	1.493E-18	1.396E-18
9.20	1.305E-18	1.221E-18	1.141E-18	1.067E-18	9.979E-19	9.332E-19	8.726E-19	8.160E-19	7.630E-19	7.136E-19
9.30	6.672E-19	6.239E-19	5.834E-19	5.456E-19	6.102E-19	4.771E-19	4.482E-19	4.172E-19	3.902E-19	3.649E-19
9.40	3.412E-19	3.191E-19	2.984E-19	2.791E-19	2.610E-19	2.441E-19	2.283E-19	2.135E-19	1.998E-19	1.867E-19
9.50	1.746E-19	1.633E-19	1.527E-19	1.428E-19	1.336E-19	1.250E-19	1.169E-19	1.093E-19	1.022E-19	9.562E-12
9.60	8.943E-20	8.365E-20	7.824E-20	7.318E-20	6.845E-20	6.402E-20	5.988E-20	5.601E-20	5.240E-20	4.901E-20
9.70	4.584E-20	4.288E-20	4.011E-20	3.752E-20	3.510E-20	3.284E-20	3.072E-20	2.873E-20	2.688E-20	2.515E-20
9.80	2.352E-20	2.201E-20	2.059E-20	1.926E-20	1.802E-20	1.686E-20	1.577E-20	1.476E-20	1.381E-20	1.292E-20

Reprinted with permission from Mikel J. Harry and Regle Steward "Six Sigma Mechanical Design Tolerance" (1988), Motorola, Inc., pp. 57–58.

Appendix D

Tables for Testing Skewness and Kurtosis

(a) Table for Testing Skewness (One-Tailed Percentage Points
of the Distribution of $\sqrt{b_1} = g_1 = m_3 / m_2^{3/2}$)

Size of Sample n	Percentage Points 5%	1%	Standard Deviation	Size of Sample n	Percentage Points 5%	1%	Standard Deviation
25	0.711	1.061	0.4354	100	0.389	0.567	0.2377
30	0.661	0.982	0.4052	125	0.350	0.508	0.2139
35	0.621	0.921	0.3804	150	0.321	0.464	0.1961
40	0.587	0.869	0.3596	175	0.298	0.430	0.1820
45	0.558	0.825	0.3418	200	0.280	0.403	0.1706
50	0.533	0.787	0.3264	250	0.251	0.360	0.1531
60	0.492	0.723	0.3009	300	0.230	0.329	0.1400
70	0.459	0.673	0.2806	350	0.213	0.305	0.1298
80	0.432	0.631	0.2638	400	0.200	0.285	0.1216
90	0.409	0.596	0.2498	450	0.188	0.269	0.1147
100	0.389	0.567	0.2377	500	0.179	0.255	0.1089

Since the distribution of $\sqrt{b_1}$ is symmetrical about zero, the percentage points represent 10% and 2% two-tailed values. Reproduced from Table 34B of "Tables for Statisticians and Biometricians," Vol. 1, by permission of Dr. E. S. Pearson and the Biometrika trustees.

(b) Table for Testing Kurtosis (Percentage Points of the Distribution of $b_2 = m_4 / m_2^2$)

Size of Sample n	Percentage Points				Size of Sample n	Percentage Points			
	Upper		Lower			Upper		Lower	
	1%	5%	5%	1%		1%	5%	5%	1%
50	4.88	3.99	2.15	1.95	600	3.54	3.34	2.70	2.60
75	4.59	3.87	2.27	2.08	650	3.52	3.33	2.71	2.61
100	4.39	3.77	2.35	2.18	700	3.50	3.31	2.72	2.62
125	4.24	3.71	2.40	2.24	750	3.48	3.30	2.73	2.64
150	4.13	3.65	2.45	2.29	800	3.46	3.29	2.74	2.65
					850	3.45	3.28	2.74	2.66
200	3.98	3.57	2.51	2.37	900	3.45	3.28	2.75	2.66
250	3.87	3.52	2.55	2.42	950	3.42	3.27	2.76	2.67
300	3.79	3.47	2.59	2.46	1000	3.41	3.26	2.76	2.68
350	3.72	3.44	2.62	2.50					
400	3.67	3.41	2.64	2.52	1200	3.37	3.24	2.78	2.71
450	3.63	3.39	2.66	2.55	1400	3.34	3.22	2.80	2.72
500	3.60	3.37	2.67	2.57	1600	3.32	3.21	2.81	2.74
550	3.57	3.35	2.69	2.58	1800	3.30	3.20	2.82	2.76
600	3.54	3.34	2.70	2.60	2000	3.28	3.18	2.83	2.77

Appendix **E**

Percentage Points of the *F* Distribution

$$F_{0.25, \nu_1, \nu_2}$$

Degrees of Freedom for the Numerator (ν_1)

ν_2	1	2	3	4	5	6	7	8	9	10	12	15	20	24	30	40	60	120	∞
1	5.83	7.50	8.20	8.58	8.82	8.98	9.10	9.19	9.26	9.32	9.41	9.49	9.58	9.63	9.67	9.71	9.76	9.80	9.85
2	2.57	3.00	3.15	3.23	3.28	3.31	3.34	3.35	3.37	3.38	3.39	3.41	3.43	3.43	3.44	3.45	3.46	3.47	3.48
3	2.02	2.28	2.36	2.39	2.41	2.42	2.43	2.44	2.44	2.44	2.45	2.46	2.46	2.46	2.47	2.47	2.47	2.47	2.47
4	1.81	2.00	2.05	2.06	2.07	2.08	2.08	2.08	2.08	2.08	2.08	2.08	2.08	2.08	2.08	2.08	2.08	2.08	2.08
5	1.69	1.85	1.88	1.89	1.89	1.89	1.89	1.89	1.89	1.89	1.89	1.89	1.88	1.88	1.88	1.88	1.87	1.87	1.87
6	1.62	1.76	1.78	1.79	1.79	1.78	1.78	1.78	1.77	1.77	1.77	1.76	1.76	1.75	1.75	1.75	1.74	1.74	1.74
7	1.57	1.70	1.72	1.72	1.71	1.71	1.70	1.70	1.70	1.69	1.68	1.68	1.67	1.67	1.66	1.66	1.65	1.65	1.65
8	1.54	1.66	1.67	1.66	1.66	1.65	1.64	1.64	1.63	1.63	1.62	1.62	1.61	1.60	1.60	1.59	1.59	1.58	1.58
9	1.51	1.62	1.63	1.63	1.62	1.61	1.60	1.60	1.59	1.59	1.58	1.57	1.56	1.56	1.55	1.54	1.54	1.53	1.53
10	1.49	1.60	1.60	1.59	1.59	1.58	1.57	1.56	1.56	1.55	1.54	1.53	1.52	1.52	1.51	1.51	1.50	1.49	1.48
11	1.47	1.58	1.58	1.57	1.56	1.55	1.54	1.53	1.53	1.52	1.51	1.50	1.49	1.49	1.48	1.47	1.47	1.46	1.45
12	1.46	1.56	1.56	1.55	1.54	1.53	1.52	1.51	1.51	1.50	1.49	1.48	1.47	1.46	1.45	1.45	1.44	1.43	1.42
13	1.45	1.55	1.55	1.53	1.52	1.51	1.50	1.49	1.49	1.48	1.47	1.46	1.45	1.44	1.43	1.42	1.42	1.41	1.40
14	1.44	1.53	1.53	1.52	1.51	1.50	1.49	1.48	1.47	1.46	1.45	1.44	1.43	1.42	1.41	1.41	1.40	1.39	1.38
15	1.43	1.52	1.52	1.51	1.49	1.48	1.47	1.46	1.46	1.45	1.44	1.43	1.41	1.41	1.40	1.39	1.38	1.37	1.36
16	1.42	1.51	1.51	1.50	1.48	1.47	1.46	1.45	1.44	1.44	1.43	1.41	1.40	1.39	1.38	1.37	1.36	1.35	1.34
17	1.42	1.51	1.50	1.49	1.47	1.46	1.45	1.44	1.43	1.43	1.41	1.40	1.39	1.38	1.37	1.36	1.35	1.34	1.33
18	1.41	1.50	1.49	1.48	1.46	1.45	1.44	1.43	1.42	1.42	1.40	1.39	1.38	1.37	1.36	1.35	1.34	1.33	1.32
19	1.41	1.49	1.49	1.47	1.46	1.44	1.43	1.42	1.41	1.41	1.40	1.38	1.37	1.36	1.35	1.34	1.33	1.32	1.30
20	1.40	1.49	1.48	1.47	1.45	1.44	1.43	1.42	1.41	1.40	1.39	1.37	1.36	1.35	1.34	1.33	1.32	1.31	1.29
21	1.40	1.48	1.48	1.46	1.44	1.43	1.42	1.41	1.40	1.39	1.38	1.37	1.35	1.34	1.33	1.32	1.31	1.30	1.28
22	1.40	1.48	1.47	1.45	1.44	1.42	1.41	1.40	1.39	1.39	1.37	1.36	1.34	1.33	1.32	1.31	1.30	1.29	1.28
23	1.39	1.47	1.47	1.45	1.43	1.42	1.41	1.40	1.39	1.38	1.37	1.35	1.34	1.33	1.32	1.31	1.30	1.28	1.27
24	1.39	1.47	1.46	1.44	1.43	1.41	1.40	1.39	1.38	1.38	1.36	1.35	1.33	1.32	1.31	1.30	1.29	1.28	1.26
25	1.39	1.47	1.46	1.44	1.42	1.41	1.40	1.39	1.38	1.37	1.36	1.34	1.33	1.32	1.31	1.29	1.28	1.27	1.25
26	1.38	1.46	1.45	1.44	1.42	1.41	1.39	1.38	1.37	1.37	1.35	1.34	1.32	1.31	1.30	1.29	1.28	1.26	1.25
27	1.38	1.46	1.45	1.43	1.42	1.40	1.39	1.38	1.37	1.36	1.35	1.33	1.32	1.31	1.30	1.28	1.27	1.26	1.24
28	1.38	1.46	1.45	1.43	1.41	1.40	1.39	1.38	1.37	1.36	1.34	1.33	1.31	1.30	1.29	1.28	1.27	1.26	1.24
29	1.38	1.45	1.45	1.43	1.41	1.40	1.38	1.37	1.36	1.35	1.34	1.32	1.31	1.30	1.29	1.27	1.26	1.25	1.23
30	1.38	1.45	1.44	1.42	1.41	1.39	1.38	1.37	1.36	1.35	1.34	1.32	1.30	1.29	1.28	1.27	1.26	1.24	1.23
40	1.36	1.44	1.42	1.40	1.39	1.37	1.36	1.35	1.34	1.33	1.31	1.30	1.28	1.26	1.25	1.24	1.22	1.21	1.19
60	1.35	1.42	1.41	1.38	1.37	1.35	1.33	1.32	1.31	1.30	1.29	1.27	1.25	1.24	1.22	1.21	1.19	1.17	1.15
120	1.34	1.40	1.39	1.37	1.35	1.33	1.31	1.30	1.29	1.28	1.26	1.24	1.22	1.21	1.19	1.18	1.16	1.13	1.10
∞	1.32	1.39	1.37	1.35	1.33	1.31	1.29	1.28	1.27	1.25	1.24	1.22	1.19	1.18	1.16	1.14	1.12	1.08	1.00

DEGREES OF FREEDOM FOR THE DENOMINATOR (ν_2)

$$F_{0.10,\nu_1,\nu_2}$$

Degrees of Freedom for the Numerator (ν_1)

ν_2	1	2	3	4	5	6	7	8	9	10	12	15	20	24	30	40	60	120	∞
1	39.86	49.50	53.59	55.83	57.24	58.20	58.91	59.44	59.86	60.19	60.71	61.22	61.74	62.00	62.26	62.53	62.79	63.06	63.33
2	8.53	9.00	9.16	9.24	9.29	9.33	9.35	9.37	9.38	9.39	9.41	9.42	9.44	9.45	9.46	9.47	9.47	9.48	9.49
3	5.54	5.46	5.39	5.34	5.31	5.28	5.27	5.25	5.24	5.23	5.22	5.20	5.18	5.18	5.17	5.16	5.15	5.14	5.13
4	4.54	4.32	4.19	4.11	4.05	4.01	3.98	3.95	3.94	3.92	3.90	3.87	3.84	3.83	3.82	3.80	3.79	3.78	3.76
5	4.06	3.78	3.62	3.52	3.45	3.40	3.37	3.34	3.32	3.30	3.27	3.24	3.21	3.19	3.17	3.16	3.14	3.12	3.10
6	3.78	3.46	3.29	3.18	3.11	3.05	3.01	2.98	2.96	2.94	2.90	2.87	2.84	2.82	2.80	2.78	2.76	2.74	2.72
7	3.59	3.26	3.07	2.96	2.88	2.83	2.78	2.75	2.72	2.70	2.67	2.63	2.59	2.58	2.56	2.54	2.51	2.49	2.47
8	3.46	3.11	2.92	2.81	2.73	2.67	2.62	2.59	2.56	2.54	2.50	2.46	2.42	2.40	2.38	2.36	2.34	2.32	2.29
9	3.36	3.01	2.81	2.69	2.61	2.55	2.51	2.47	2.44	2.42	2.38	2.34	2.30	2.28	2.25	2.23	2.21	2.18	2.16
10	3.29	2.92	2.73	2.61	2.52	2.46	2.41	2.38	2.35	2.32	2.28	2.24	2.20	2.18	2.16	2.13	2.11	2.08	2.06
11	3.23	2.86	2.66	2.54	2.45	2.39	2.34	2.30	2.27	2.25	2.21	2.17	2.12	2.10	2.08	2.05	2.03	2.00	1.97
12	3.18	2.81	2.61	2.48	2.39	2.33	2.28	2.24	2.21	2.19	2.15	2.10	2.06	2.04	2.01	1.99	1.96	1.93	1.90
13	3.14	2.76	2.56	2.43	2.35	2.28	2.23	2.20	2.16	2.14	2.10	2.05	2.01	1.98	1.96	1.93	1.90	1.88	1.85
14	3.10	2.73	2.52	2.39	2.31	2.24	2.19	2.15	2.12	2.10	2.05	2.01	1.96	1.94	1.91	1.89	1.86	1.83	1.80
15	3.07	2.70	2.49	2.36	2.27	2.21	2.16	2.12	2.09	2.06	2.02	1.97	1.92	1.90	1.87	1.85	1.82	1.79	1.76
16	3.05	2.67	2.46	2.33	2.24	2.18	2.13	2.09	2.06	2.03	1.99	1.94	1.89	1.87	1.84	1.81	1.78	1.75	1.72
17	3.03	2.64	2.44	2.31	2.22	2.15	2.10	2.06	2.03	2.00	1.96	1.91	1.86	1.84	1.81	1.78	1.75	1.72	1.69
18	3.01	2.62	2.42	2.29	2.20	2.13	2.08	2.04	2.00	1.98	1.93	1.89	1.84	1.81	1.78	1.75	1.72	1.69	1.66
19	2.99	2.61	2.40	2.27	2.18	2.11	2.06	2.02	1.98	1.96	1.91	1.86	1.81	1.79	1.76	1.73	1.70	1.67	1.63
20	2.97	2.59	2.38	2.25	2.16	2.09	2.04	2.00	1.96	1.94	1.89	1.84	1.79	1.77	1.74	1.71	1.68	1.64	1.61
21	2.96	2.57	2.36	2.23	2.14	2.08	2.02	1.98	1.95	1.92	1.87	1.83	1.78	1.75	1.72	1.69	1.66	1.62	1.59
22	2.95	2.56	2.35	2.22	2.13	2.06	2.01	1.97	1.93	1.90	1.86	1.81	1.76	1.73	1.70	1.67	1.64	1.60	1.57
23	2.94	2.55	2.34	2.21	2.11	2.05	1.99	1.95	1.92	1.89	1.84	1.80	1.74	1.72	1.69	1.66	1.62	1.59	1.55
24	2.93	2.54	2.33	2.19	2.10	2.04	1.98	1.94	1.91	1.88	1.83	1.78	1.73	1.70	1.67	1.64	1.61	1.57	1.53
25	2.92	2.53	2.32	2.18	2.09	2.02	1.97	1.93	1.89	1.87	1.82	1.77	1.72	1.69	1.66	1.63	1.59	1.56	1.52
26	2.91	2.52	2.31	2.17	2.08	2.01	1.96	1.92	1.88	1.86	1.81	1.76	1.71	1.68	1.65	1.61	1.58	1.54	1.50
27	2.90	2.51	2.30	2.17	2.07	2.00	1.95	1.91	1.87	1.85	1.80	1.75	1.70	1.67	1.64	1.60	1.57	1.53	1.49
28	2.89	2.50	2.29	2.16	2.06	2.00	1.94	1.90	1.87	1.84	1.79	1.74	1.69	1.66	1.63	1.59	1.56	1.52	1.48
29	2.89	2.50	2.28	2.15	2.06	1.99	1.93	1.89	1.86	1.83	1.78	1.73	1.68	1.65	1.62	1.58	1.55	1.51	1.47
30	2.88	2.49	2.28	2.14	2.03	1.98	1.93	1.88	1.85	1.82	1.77	1.72	1.67	1.64	1.61	1.57	1.54	1.50	1.46
40	2.84	2.44	2.23	2.09	2.00	1.93	1.87	1.83	1.79	1.76	1.71	1.66	1.61	1.57	1.54	1.51	1.47	1.42	1.38
60	2.79	2.39	2.18	2.04	1.95	1.87	1.82	1.77	1.74	1.71	1.66	1.60	1.54	1.51	1.48	1.44	1.40	1.35	1.29
120	2.75	2.35	2.13	1.99	1.90	1.82	1.77	1.72	1.68	1.65	1.60	1.55	1.48	1.45	1.41	1.37	1.32	1.26	1.19
∞	2.71	2.30	2.08	1.94	1.85	1.77	1.72	1.67	1.63	1.60	1.55	1.49	1.42	1.38	1.34	1.30	1.24	1.17	1.00

DEGREES OF FREEDOM FOR THE DENOMINATOR (ν_2)

$$F_{0.05, \nu_1, \nu_2}$$

Degrees of Freedom for the Numerator (ν_1)

ν_2	1	2	3	4	5	6	7	8	9	10	12	15	20	24	30	40	60	120	∞
1	161.4	199.5	215.7	224.6	230.2	234.0	236.8	238.9	240.5	241.9	243.9	245.9	248.0	249.1	250.1	251.1	252.2	253.3	254.3
2	18.51	19.00	19.16	19.25	19.30	19.33	19.35	19.37	19.38	19.40	19.41	19.43	19.45	19.45	19.46	19.47	19.48	19.49	19.50
3	10.13	9.55	9.28	9.12	9.01	8.94	8.89	8.85	8.81	8.79	8.74	8.70	8.66	8.64	8.62	8.59	8.57	8.55	8.53
4	7.71	6.94	6.59	6.39	6.26	6.16	6.09	6.04	6.00	5.96	5.91	5.86	5.80	5.77	5.75	5.72	5.69	5.66	5.63
5	6.61	5.79	5.41	5.19	5.05	4.95	4.88	4.82	4.77	4.74	4.68	4.62	4.56	4.53	4.50	4.46	4.43	4.40	4.36
6	5.99	5.14	4.76	4.53	4.39	4.28	4.21	4.15	4.10	4.06	4.00	3.94	3.87	3.84	3.81	3.77	3.74	3.70	3.67
7	5.59	4.74	4.35	4.12	3.97	3.87	3.79	3.73	3.68	3.64	3.57	3.51	3.44	3.41	3.38	3.34	3.30	3.27	3.23
8	5.32	4.46	4.07	3.84	3.69	3.58	3.50	3.44	3.39	3.35	3.28	3.22	3.15	3.12	3.08	3.04	3.01	2.97	2.93
9	5.12	4.26	3.86	3.63	3.48	3.37	3.29	3.23	3.18	3.14	3.07	3.01	2.94	2.90	2.86	2.83	2.79	2.75	2.71
10	4.96	4.10	3.71	3.48	3.33	3.22	3.14	3.07	3.02	2.98	2.91	2.85	2.77	2.74	2.70	2.66	2.62	2.58	2.54
11	4.84	3.98	3.59	3.36	3.20	3.09	3.01	2.95	2.90	2.85	2.79	2.72	2.65	2.61	2.57	2.53	2.49	2.45	2.40
12	4.75	3.89	3.49	3.26	3.11	3.00	2.91	2.85	2.80	2.75	2.69	2.62	2.54	2.51	2.47	2.43	2.38	2.34	2.30
13	4.67	3.81	3.41	3.18	3.03	2.92	2.83	2.77	2.71	2.67	2.60	2.53	2.46	2.42	2.38	2.34	2.30	2.25	2.21
14	4.60	3.74	3.34	3.11	2.96	2.85	2.76	2.70	2.65	2.60	2.53	2.46	2.39	2.35	2.31	2.27	2.22	2.18	2.13
15	4.54	3.68	3.29	3.06	2.90	2.79	2.71	2.64	2.59	2.54	2.48	2.40	2.33	2.29	2.25	2.20	2.16	2.11	2.07
16	4.49	3.63	3.24	3.01	2.85	2.74	2.66	2.59	2.54	2.49	2.42	2.35	2.28	2.24	2.19	2.15	2.11	2.06	2.01
17	4.45	3.59	3.20	2.96	2.81	2.70	2.61	2.55	2.49	2.45	2.38	2.31	2.23	2.19	2.15	2.10	2.06	2.01	1.96
18	4.41	3.55	3.16	2.93	2.77	2.66	2.58	2.51	2.46	2.41	2.34	2.27	2.19	2.15	2.11	2.06	2.02	1.97	1.92
19	4.38	3.52	3.13	2.90	2.74	2.63	2.54	2.48	2.42	2.38	2.31	2.23	2.16	2.11	2.07	2.03	1.98	1.93	1.88
20	4.35	3.49	3.10	2.87	2.71	2.60	2.51	2.45	2.39	2.35	2.28	2.20	2.12	2.08	2.04	1.99	1.95	1.90	1.84
21	4.32	3.47	3.07	2.84	2.68	2.57	2.49	2.42	2.37	2.32	2.25	2.18	2.10	2.05	2.01	1.96	1.92	1.87	1.81
22	4.30	3.44	3.05	2.82	2.66	2.55	2.46	2.40	2.34	2.30	2.23	2.15	2.07	2.03	1.98	1.94	1.89	1.84	1.78
23	4.28	3.42	3.03	2.80	2.64	2.53	2.44	2.37	2.32	2.27	2.20	2.13	2.05	2.01	1.96	1.91	1.86	1.81	1.76
24	4.26	3.40	3.01	2.78	2.62	2.51	2.42	2.36	2.30	2.25	2.18	2.11	2.03	1.98	1.94	1.89	1.84	1.79	1.73
25	4.24	3.39	2.99	2.76	2.60	2.49	2.40	2.34	2.28	2.24	2.16	2.09	2.01	1.96	1.92	1.87	1.82	1.77	1.71
26	4.23	3.37	2.98	2.74	2.59	2.47	2.39	2.32	2.27	2.22	2.15	2.07	1.99	1.95	1.90	1.85	1.80	1.75	1.69
27	4.21	3.35	2.96	2.73	2.57	2.46	2.37	2.31	2.25	2.20	2.13	2.06	1.97	1.93	1.88	1.84	1.79	1.73	1.67
28	4.20	3.34	2.95	2.71	2.56	2.45	2.36	2.29	2.24	2.19	2.12	2.04	1.96	1.91	1.87	1.82	1.77	1.71	1.65
29	4.18	3.33	2.93	2.70	2.55	2.43	2.35	2.28	2.22	2.18	2.10	2.03	1.94	1.90	1.85	1.81	1.75	1.70	1.64
30	4.17	3.32	2.92	2.69	2.53	2.42	2.33	2.27	2.21	2.16	2.09	2.01	1.93	1.89	1.84	1.79	1.74	1.68	1.62
40	4.08	3.23	2.84	2.61	2.45	2.34	2.25	2.18	2.12	2.08	2.00	1.92	1.84	1.79	1.74	1.69	1.64	1.58	1.51
60	4.00	3.15	2.76	2.53	2.37	2.25	2.17	2.10	2.04	1.99	1.92	1.84	1.75	1.70	1.65	1.59	1.53	1.47	1.39
120	3.92	3.07	2.68	2.45	2.29	2.17	2.09	2.02	1.96	1.91	1.83	1.75	1.66	1.61	1.55	1.50	1.43	1.35	1.25
∞	3.84	3.00	2.60	2.37	2.21	2.10	2.01	1.94	1.88	1.83	1.75	1.67	1.57	1.52	1.46	1.39	1.32	1.22	1.00

DEGREES OF FREEDOM FOR THE DENOMINATOR (ν_2)

$$F_{0.025,\nu_1,\nu_2}$$

Degrees of Freedom for the Numerator (ν_1)

ν_2	1	2	3	4	5	6	7	8	9	10	12	15	20	24	30	40	60	120	∞
1	647.8	799.5	864.2	899.6	921.8	937.1	948.2	956.7	963.3	968.6	976.7	984.9	993.1	997.2	1001	1006	1010	1014	1018
2	38.51	39.00	39.17	39.25	39.30	39.33	39.36	39.37	39.39	39.40	39.41	39.43	39.45	39.46	39.46	39.47	39.48	39.49	39.50
3	17.44	16.04	15.44	15.10	14.88	14.73	14.62	14.54	14.47	14.42	14.34	14.25	14.17	14.12	14.08	14.04	13.99	13.95	13.90
4	12.22	10.65	9.90	9.60	9.36	9.20	9.07	8.98	8.90	8.84	8.75	8.66	8.56	8.51	8.46	8.41	8.36	8.31	8.26
5	10.01	8.43	7.76	7.39	7.15	6.98	6.85	6.76	6.68	6.62	6.52	6.43	6.33	6.28	6.23	6.18	6.12	6.07	6.02
6	8.81	7.26	6.60	6.23	5.99	5.82	5.70	5.60	5.52	5.46	5.37	5.27	5.17	5.12	5.07	5.01	4.96	4.90	4.85
7	8.07	6.54	5.89	5.52	5.29	5.12	4.99	4.90	4.82	4.76	4.67	4.57	4.47	4.42	4.36	4.31	4.25	4.20	4.14
8	7.57	6.06	5.42	5.05	4.82	4.65	4.53	4.43	4.36	4.30	4.20	4.10	4.00	3.95	3.89	3.84	3.78	3.73	3.67
9	7.21	5.71	5.08	4.72	4.48	4.32	4.20	4.10	4.03	3.96	3.87	3.77	3.67	3.61	3.56	3.51	3.45	3.39	3.33
10	6.94	5.46	4.83	4.47	4.24	4.07	3.95	3.85	3.78	3.72	3.62	3.52	3.42	3.37	3.31	3.26	3.20	3.14	3.08
11	6.72	5.26	4.63	4.28	4.04	3.88	3.76	3.66	3.59	3.53	3.43	3.33	3.23	3.17	3.12	3.06	3.00	2.94	2.88
12	6.55	5.10	4.47	4.12	3.89	3.73	3.61	3.51	3.44	3.37	3.28	3.18	3.07	3.02	2.96	2.91	2.85	2.79	2.72
13	6.41	4.97	4.35	4.00	3.77	3.60	3.48	3.39	3.31	3.25	3.15	3.05	2.95	2.89	2.84	2.78	2.72	2.66	2.60
14	6.30	4.86	4.24	3.89	3.66	3.50	3.38	3.29	3.21	3.15	3.05	2.95	2.84	2.79	2.73	2.67	2.61	2.55	2.49
15	6.20	4.77	4.15	3.80	3.58	3.41	3.29	3.20	3.12	3.06	2.96	2.86	2.76	2.70	2.64	2.59	2.52	2.46	2.40
16	6.12	4.69	4.08	3.73	3.50	3.34	3.22	3.12	3.05	2.99	2.89	2.79	2.68	2.63	2.57	2.51	2.45	2.38	2.32
17	6.04	4.62	4.01	3.66	3.44	3.28	3.16	3.06	2.98	2.92	2.82	2.72	2.62	2.56	2.50	2.44	2.38	2.32	2.25
18	5.98	4.56	3.95	3.61	3.38	3.22	3.10	3.01	2.93	2.87	2.77	2.67	2.56	2.50	2.44	2.38	2.32	2.26	2.19
19	5.92	4.51	3.90	3.56	3.33	3.17	3.05	2.96	2.88	2.82	2.72	2.62	2.51	2.45	2.39	2.33	2.27	2.20	2.13
20	5.87	4.46	3.86	3.51	3.29	3.13	3.01	2.91	2.84	2.77	2.68	2.57	2.46	2.41	2.35	2.29	2.22	2.16	2.09
21	5.83	4.42	3.82	3.48	3.25	3.09	2.97	2.87	2.80	2.73	2.64	2.53	2.42	2.37	2.31	2.25	2.18	2.11	2.04
22	5.79	4.38	3.78	3.44	3.22	3.05	2.93	2.84	2.76	2.70	2.60	2.50	2.39	2.33	2.27	2.21	2.14	2.08	2.00
23	5.75	4.35	3.75	3.41	3.18	3.02	2.90	2.81	2.73	2.67	2.57	2.47	2.36	2.30	2.24	2.18	2.11	2.04	1.97
24	5.72	4.32	3.72	3.38	3.15	2.99	2.87	2.78	2.70	2.64	2.54	2.44	2.33	2.27	2.21	2.15	2.08	2.01	1.94
25	5.69	4.29	3.69	3.35	3.13	2.97	2.85	2.75	2.68	2.61	2.51	2.41	2.30	2.24	2.18	2.12	2.05	1.98	1.91
26	5.66	4.27	3.67	3.33	3.10	2.94	2.82	2.73	2.65	2.59	2.49	2.39	2.28	2.22	2.16	2.09	2.03	1.95	1.88
27	5.63	4.24	3.65	3.31	3.08	2.92	2.80	2.71	2.63	2.57	2.47	2.36	2.25	2.19	2.13	2.07	2.00	1.93	1.85
28	5.61	4.22	3.63	3.29	3.06	2.90	2.78	2.69	2.61	2.55	2.45	2.34	2.23	2.17	2.11	2.05	1.98	1.91	1.83
29	5.59	4.20	3.61	3.27	3.04	2.88	2.76	2.67	2.59	2.53	2.43	2.32	2.21	2.15	2.09	2.03	1.96	1.89	1.81
30	5.57	4.18	3.59	3.25	3.03	2.87	2.75	2.65	2.57	2.51	2.41	2.31	2.20	2.14	2.07	2.01	1.94	1.87	1.79
40	5.42	4.05	3.46	3.13	2.90	2.74	2.62	2.53	2.45	2.39	2.29	2.18	2.07	2.01	1.94	1.88	1.80	1.72	1.64
60	5.29	3.93	3.34	3.01	2.79	2.63	2.51	2.41	2.33	2.27	2.17	2.06	1.94	1.88	1.82	1.74	1.67	1.58	1.48
120	5.15	3.80	3.23	2.89	2.67	2.52	2.39	2.30	2.22	2.16	2.05	1.94	1.82	1.76	1.69	1.61	1.53	1.43	1.31
∞	5.02	3.69	3.12	2.79	2.57	2.41	2.29	2.19	2.11	2.05	1.94	1.83	1.71	1.64	1.57	1.48	1.39	1.27	1.00

DEGREES OF FREEDOM FOR THE DENOMINATOR (ν_2)

$$F_{0.01,\nu_1,\nu_2}$$

ν_2	Degrees of Freedom for the Numerator (ν_1)																		
	1	2	3	4	5	6	7	8	9	10	12	15	20	24	30	40	60	120	∞
1	4052	4999.5	5403	5625	5764	5859	5928	5982	6022	6056	6106	6157	6209	6235	6261	6287	6313	6339	6366
2	98.50	99.00	99.17	99.25	99.30	99.33	99.36	99.37	99.39	99.40	99.42	99.43	99.45	99.46	99.47	99.47	99.48	99.49	99.50
3	34.12	30.82	29.46	28.71	28.24	27.91	27.67	27.49	27.35	27.23	27.05	26.87	26.69	26.60	26.50	26.41	26.32	26.22	26.13
4	21.20	18.00	16.69	15.98	15.52	15.21	14.98	14.80	14.66	14.55	14.37	14.20	14.02	13.93	13.84	13.75	13.65	13.56	13.46
5	16.26	13.27	12.06	11.39	10.97	10.67	10.46	10.29	10.16	10.05	9.89	9.72	9.55	9.47	9.38	9.29	9.20	9.11	9.02
6	13.75	10.92	9.78	9.15	8.75	8.47	8.26	8.10	7.98	7.87	7.72	7.56	7.40	7.31	7.23	7.14	7.06	6.97	6.88
7	12.25	9.55	8.45	7.85	7.46	7.19	6.99	6.84	6.72	6.62	6.47	6.31	6.16	6.07	5.99	5.91	5.82	5.74	5.65
8	11.26	8.65	7.59	7.01	6.63	6.37	6.18	6.03	5.91	5.81	5.67	5.52	5.36	5.28	5.20	5.12	5.03	4.95	4.86
9	10.56	8.02	6.99	6.42	6.06	5.80	5.61	5.47	5.35	5.26	5.11	4.96	4.81	4.73	4.65	4.57	4.48	4.40	4.31
10	10.04	7.56	6.55	5.99	5.64	5.39	5.20	5.06	4.94	4.85	4.71	4.56	4.41	4.33	4.25	4.17	4.08	4.00	3.91
11	9.65	7.21	6.22	5.67	5.32	5.07	4.89	4.74	4.63	4.54	4.40	4.25	4.10	4.02	3.94	3.86	3.78	3.69	3.60
12	9.33	6.93	5.95	5.41	5.06	4.82	4.64	4.50	4.39	4.30	4.16	4.01	3.86	3.78	3.70	3.62	3.54	3.45	3.36
13	9.07	6.70	5.74	5.21	4.86	4.62	4.44	4.30	4.19	4.10	3.96	3.82	3.66	3.59	3.51	3.43	3.34	3.25	3.17
14	8.86	6.51	5.56	5.04	4.69	4.46	4.28	4.14	4.03	3.94	3.80	3.66	3.51	3.43	3.35	3.27	3.18	3.09	3.00
15	8.68	6.36	5.42	4.89	4.56	4.32	4.14	4.00	3.89	3.80	3.67	3.52	3.37	3.29	3.21	3.13	3.05	2.96	2.87
16	8.53	6.23	5.29	4.77	4.44	4.20	4.03	3.89	3.78	3.69	3.55	3.41	3.26	3.18	3.10	3.02	2.93	2.84	2.75
17	8.40	6.11	5.18	4.67	4.34	4.10	3.93	3.79	3.68	3.59	3.46	3.31	3.16	3.08	3.00	2.92	2.83	2.75	2.65
18	8.29	6.01	5.09	4.58	4.25	4.01	3.84	3.71	3.60	3.51	3.37	3.23	3.08	3.00	2.92	2.84	2.75	2.66	2.57
19	8.18	5.93	5.01	4.50	4.17	3.94	3.77	3.63	3.52	3.43	3.30	3.15	3.00	2.92	2.84	2.76	2.67	2.58	2.49
20	8.10	5.85	4.94	4.43	4.10	3.87	3.70	3.56	3.46	3.37	3.23	3.09	2.94	2.86	2.78	2.69	2.61	2.52	2.42
21	8.02	5.78	4.87	4.37	4.04	3.81	3.64	3.51	3.40	3.31	3.17	3.03	2.88	2.80	2.72	2.64	2.55	2.46	2.36
22	7.95	5.72	4.82	4.31	3.99	3.76	3.59	3.45	3.35	3.26	3.12	2.98	2.83	2.75	2.67	2.58	2.50	2.40	2.31
23	7.88	5.66	4.76	4.26	3.94	3.71	3.54	3.41	3.30	3.21	3.07	2.93	2.78	2.70	2.62	2.54	2.45	2.35	2.26
24	7.82	5.61	4.72	4.22	3.90	3.67	3.50	3.36	3.26	3.17	3.03	2.89	2.74	2.66	2.58	2.49	2.40	2.31	2.21
25	7.77	5.57	4.68	4.18	3.85	3.63	3.46	3.32	3.22	3.13	2.99	2.85	2.70	2.62	2.54	2.45	2.36	2.27	2.17
26	7.72	5.53	4.64	4.14	3.82	3.59	3.42	3.29	3.18	3.09	2.96	2.81	2.66	2.58	2.50	2.42	2.33	2.23	2.13
27	7.68	5.49	4.60	4.11	3.78	3.56	3.39	3.26	3.15	3.06	2.93	2.78	2.63	2.55	2.47	2.38	2.29	2.20	2.10
28	7.64	5.45	4.57	4.07	3.75	3.53	3.36	3.23	3.12	3.03	2.90	2.75	2.60	2.52	2.44	2.35	2.26	2.17	2.06
29	7.60	5.42	4.54	4.04	3.73	3.50	3.33	3.20	3.09	3.00	2.87	2.73	2.57	2.49	2.41	2.33	2.23	2.14	2.03
30	7.56	5.39	4.51	4.02	3.70	3.47	3.30	3.17	3.07	2.98	2.84	2.70	2.55	2.47	2.39	2.30	2.21	2.11	2.01
40	7.31	5.18	4.31	3.83	3.51	3.29	3.12	2.99	2.89	2.80	2.66	2.52	2.37	2.29	2.20	2.11	2.02	1.92	1.80
60	7.08	4.98	4.13	3.65	3.34	3.12	2.95	2.82	2.72	2.63	2.50	2.35	2.20	2.12	2.03	1.94	1.84	1.73	1.60
120	6.85	4.79	3.95	3.48	3.17	2.96	2.79	2.66	2.56	2.47	2.34	2.19	2.03	1.95	1.86	1.76	1.66	1.53	1.38
∞	6.63	4.61	3.78	3.32	3.02	2.80	2.64	2.51	2.41	2.32	2.18	2.04	1.88	1.79	1.70	1.59	1.47	1.32	1.00

DEGREES OF FREEDOM FOR THE DENOMINATOR (ν_2)

Source: Adapted with permission from "Biometrika Tables for Statisticians," Vol. 1, 3rd ed., by E. S. Pearson and H. O. Hartley, Cambridge University Press, Cambridge, 1966.

Appendix F

Orthogonal Arrays

L4 Array

Trial Number	Column Number		
	1	2	3
1	1	1	1
2	1	2	2
3	2	1	2
4	2	2	1

L8 Array

Trial Number	Column Number						
	1	2	3	4	5	6	7
1	1	1	1	1	1	1	1
2	1	1	1	2	2	2	2
3	1	2	2	1	1	2	2
4	1	2	2	2	2	1	1
5	2	1	2	1	2	1	2
6	2	1	2	2	1	2	1
7	2	2	1	1	2	2	1
8	2	2	1	2	1	1	2

L12 Array

Trial Number	Column Number										
	1	2	3	4	5	6	7	8	9	10	11
1	1	1	1	1	1	1	1	1	1	1	1
2	1	1	1	1	1	2	2	2	2	2	2
3	1	1	2	2	2	1	1	1	2	2	2
4	1	2	1	2	2	1	2	2	1	1	2
5	1	2	2	1	2	2	1	2	1	2	1
6	1	2	2	2	1	2	2	1	2	1	1
7	2	1	2	2	1	1	2	2	1	2	1
8	2	1	2	1	2	2	2	1	1	1	2
9	2	1	1	2	2	2	1	2	2	1	1
10	2	2	2	1	1	1	1	2	2	1	2
11	2	2	1	2	1	2	1	1	1	2	2
12	2	2	1	1	2	1	2	1	2	2	1

Caution: An interaction of two columns is confounded with the remaining columns; assign only main effects to this array.

L16 Array

Trial Number	Column Number														
	1	2	3	4	5	6	7	8	9	10	11	12	13	14	15
1	1	1	1	1	1	1	1	1	1	1	1	1	1	1	1
2	1	1	1	1	1	1	1	2	2	2	2	2	2	2	2
3	1	1	1	2	2	2	2	1	1	1	1	2	2	2	2
4	1	1	1	2	2	2	2	2	2	2	2	1	1	1	1
5	1	2	2	1	1	2	2	1	1	2	2	1	1	2	2
6	1	2	2	1	1	2	2	2	2	1	1	2	2	1	1
7	1	2	2	2	2	1	1	1	1	2	2	2	2	1	1
8	1	2	2	2	2	1	1	2	2	1	1	1	1	2	2
9	2	1	2	1	2	1	2	1	2	1	2	1	2	1	2
10	2	1	2	1	2	1	2	2	1	2	1	2	1	2	1
11	2	1	2	2	1	2	1	1	2	1	2	2	1	2	1
12	2	1	2	2	1	2	1	2	1	2	1	1	2	1	2
13	2	2	1	1	2	2	1	1	2	2	1	1	2	2	1
14	2	2	1	1	2	2	1	2	1	1	2	2	1	1	2
15	2	2	1	2	1	1	2	1	2	2	1	2	1	1	2
16	2	2	1	2	1	1	2	2	1	1	2	1	2	2	1

L9 Array

Trial Number	Column Number			
	1	2	3	4
1	1	1	1	1
2	1	2	2	2
3	1	3	3	3
4	2	1	2	3
5	2	2	3	1
6	2	3	1	2
7	3	1	3	2
8	3	2	1	3
9	3	3	2	1

L18 Array

Trial Number	Column Number							
	1	2	3	4	5	6	7	8
1	1	1	1	1	1	1	1	1
2	1	1	2	2	2	2	2	2
3	1	1	3	3	3	3	3	3
4	1	2	1	1	2	2	3	3
5	1	2	2	2	3	3	1	1
6	1	2	3	3	1	1	2	2
7	1	3	1	2	1	3	2	3
8	1	3	2	3	2	1	3	1
9	1	3	3	1	3	2	1	2
10	2	1	1	3	3	2	2	1
11	2	1	2	1	1	3	3	2
12	2	1	3	2	2	1	1	3
13	2	2	1	2	3	1	3	2
14	2	2	2	3	1	2	1	3
15	2	2	3	1	2	3	2	1
16	2	3	1	3	2	3	1	2
17	2	3	2	1	3	1	2	3
18	2	3	3	2	1	2	3	1

L27 Array

Trial Number	Column Number												
	1	2	3	4	5	6	7	8	9	10	11	12	13
1	1	1	1	1	1	1	1	1	1	1	1	1	1
2	1	1	1	1	2	2	2	2	2	2	2	2	2
3	1	1	1	1	3	3	3	3	3	3	3	3	3
4	1	2	2	2	1	1	1	2	2	2	3	3	3
5	1	2	2	2	2	2	2	3	3	3	1	1	1
6	1	2	2	2	3	3	3	1	1	1	2	2	2
7	1	3	3	3	1	1	1	3	3	3	2	2	2
8	1	3	3	3	2	2	2	1	1	1	3	3	3
9	1	3	3	3	3	3	3	2	2	2	1	1	1
10	2	1	2	3	1	2	3	1	2	3	1	2	3
11	2	1	2	3	2	3	1	2	3	1	2	3	1
12	2	1	2	3	3	1	2	3	1	2	3	1	2
13	2	2	3	1	1	2	3	2	3	1	3	1	2
14	2	2	3	1	2	3	1	3	1	2	1	2	3
15	2	2	3	1	3	1	2	1	2	3	2	3	1
16	2	3	1	2	1	2	3	3	1	2	2	3	1
17	2	3	1	2	2	3	1	1	2	3	3	1	2
18	2	3	1	2	3	1	2	2	3	1	1	2	3
19	3	1	3	2	1	3	2	1	3	2	1	3	2
20	3	1	3	2	2	1	3	2	1	3	2	1	3
21	3	1	3	2	3	2	1	3	2	1	3	2	1
22	3	2	1	3	1	3	2	2	1	3	3	2	1
23	3	2	1	3	2	1	3	3	2	1	1	3	2
24	3	2	1	3	3	2	1	1	3	2	2	1	3
25	3	3	2	1	1	3	2	3	2	1	2	1	3
26	3	3	2	1	2	1	3	1	3	2	3	2	1
27	3	3	2	1	3	2	1	2	1	3	1	3	2

Reprinted with permission from G. Taguchi and S. Konishi, "Orthogonal Arrays and Linear Graphs" (1987), American Supplier Institute, Inc., pp. 1–3, 36–37.

Appendix G

Omega Transformation Table

p (%)	dB	p (%)	dB	p (%)	dB	p (%)	dB	p (%)	dB	p (%)	dB
0.0	∞	3.0	− 15.096	6.0	− 11.949	9.0	− 10.047	12.0	− 8.652	15.0	− 7.532
0.1	− 29.995	3.1	− 14.949	6.1	− 11.872	9.1	− 9.994	12.1	− 8.611	15.1	− 7.498
0.2	− 26.980	3.2	− 14.806	6.2	− 11.797	9.2	− 9.942	12.2	− 8.570	15.2	− 7.465
0.3	− 25.215	3.3	− 14.668	6.3	− 11.723	9.3	− 9.890	12.3	− 8.530	15.3	− 7.431
0.4	− 23.961	3.4	− 14.534	6.4	− 11.650	9.4	− 9.839	12.4	− 8.490	15.4	− 7.397
0.5	− 22.988	3.5	− 14.404	6.5	− 11.578	9.5	− 9.788	12.5	− 8.450	15.5	− 7.364
0.6	− 22.191	3.6	− 14.227	6.6	− 11.507	9.6	− 9.738	12.6	− 8.410	15.6	− 7.331
0.7	− 21.518	3.7	− 14.153	6.7	− 11.437	9.7	− 9.688	12.7	− 8.371	15.7	− 7.298
0.8	− 20.933	3.8	− 14.033	6.8	− 11.368	9.8	− 9.639	12.8	− 8.332	15.8	− 7.266
0.9	− 20.417	3.9	− 13.916	6.9	− 11.300	9.9	− 9.590	12.9	− 8.293	15.9	− 7.233
1.0	− 19.955	4.0	− 13.801	7.0	− 11.233	10.0	− 9.541	13.0	− 8.255	16.0	− 7.201
1.1	− 19.537	4.1	− 13.689	7.1	− 11.167	10.1	− 9.493	13.1	− 8.216	16.1	− 7.168
1.2	− 19.155	4.2	− 13.580	7.2	− 11.101	10.2	− 9.446	13.2	− 8.178	16.2	− 7.136
1.3	− 18.803	4.3	− 13.473	7.3	− 11.037	10.3	− 9.399	13.3	− 8.141	16.3	− 7.104
1.4	− 18.476	4.4	− 13.369	7.4	− 10.973	10.4	− 9.352	13.4	− 8.103	16.4	− 7.073
1.5	− 18.172	4.5	− 13.267	7.5	− 10.910	10.5	− 9.305	13.5	− 8.066	16.5	− 7.041
1.6	− 17.888	4.6	− 13.167	7.6	− 10.848	10.6	− 9.259	13.6	− 8.029	16.6	− 7.010
1.7	− 17.620	4.7	− 13.069	7.7	− 10.786	10.7	− 9.214	13.7	− 7.992	16.7	− 6.978
1.8	− 17.367	4.8	− 12.973	7.8	− 10.725	10.8	− 9.168	13.8	− 7.955	16.8	− 6.947
1.9	− 17.128	4.9	− 12.879	7.9	− 10.665	10.9	− 9.124	13.9	− 7.919	16.9	− 6.916
2.0	− 16.901	5.0	− 12.787	8.0	− 10.606	11.0	− 9.079	14.0	− 7.883	17.0	− 6.885
2.1	− 16.685	5.1	− 12.696	8.1	− 10.547	11.1	− 9.035	14.1	− 7.847	17.1	− 6.855
2.2	− 16.478	5.2	− 12.607	8.2	− 10.489	11.2	− 8.991	14.2	− 7.811	17.2	− 6.824
2.3	− 16.281	5.3	− 12.520	8.3	− 10.432	11.3	− 8.947	14.3	− 7.775	17.3	− 6.794
2.4	− 16.091	5.4	− 12.434	8.4	− 10.373	11.4	− 8.904	14.4	− 7.740	17.4	− 6.763
2.5	− 15.910	5.5	− 12.350	8.5	− 10.319	11.5	− 8.861	14.5	− 7.705	17.5	− 6.733
2.6	− 15.735	5.6	− 12.267	8.6	− 10.263	11.6	− 8.819	14.6	− 7.670	17.6	− 6.703
2.7	− 15.566	5.7	− 12.185	8.7	− 10.209	11.7	− 8.777	14.7	− 7.635	17.7	− 6.673
2.8	− 15.404	5.8	− 12.105	8.8	− 10.154	11.8	− 8.735	14.8	− 7.601	17.8	− 6.644
2.9	− 15.247	5.9	− 12.026	8.9	− 10.100	11.9	− 8.693	14.9	− 7.566	17.9	− 6.614

p (%)	dB	p (%)	dB	p (%)	dB	p (%)	dB	p (%)	dB	p (%)	dB
18.0	− 6.584	23.0	− 5.427	28.0	− 4.101	78.0	5.497	83.0	6.886	88.0	8.653
18.1	− 6.555	23.1	− 5.222	28.1	− 4.079	78.1	5.522	83.1	6.917	88.1	8.694
18.2	− 6.526	23.2	− 5.198	28.2	− 4.058	78.2	5.548	83.2	6.948	88.2	8.786
18.3	− 6.497	23.3	− 5.173	28.3	− 4.036	78.3	5.573	83.3	6.979	88.3	8.778
18.4	− 6.468	23.4	− 5.149	28.4	− 4.015	78.4	5.599	83.4	7.011	88.4	8.820
18.5	− 6.439	23.5	− 5.125	28.5	− 3.994	78.5	5.624	83.5	7.042	88.5	8.862
18.6	− 6.410	23.6	− 5.101	28.6	− 3.972	78.6	5.650	83.6	7.074	88.6	8.905
18.7	− 6.381	23.7	− 5.077	28.7	− 3.951	78.7	5.676	83.7	7.105	88.7	8.948
18.8	− 6.353	23.8	− 5.053	28.8	− 3.930	78.8	5.702	83.8	7.137	88.8	8.992
18.9	− 6.325	23.9	− 5.029	28.9	− 3.909	78.9	5.728	83.9	7.169	88.9	9.036
19.0	− 6.296	24.0	− 5.005	29.0	− 3.888	79.0	5.754	84.0	7.202	89.0	9.080
19.1	− 6.268	24.1	− 4.981	29.1	− 3.867	79.1	5.780	84.1	7.234	89.1	9.125
19.2	− 6.240	24.2	− 4.958	29.2	− 3.846	79.2	5.807	84.2	7.267	89.2	9.169
19.3	− 6.212	24.3	− 4.934	29.3	− 3.825	79.3	5.833	84.3	7.299	89.3	9.215
19.4	− 6.184	24.4	− 4.910	29.4	− 3.804	79.4	5.860	84.4	7.332	89.4	9.260
19.5	− 6.157	24.5	− 4.887	29.5	− 3.783	79.5	5.886	84.5	7.365	89.5	9.306
19.6	− 6.129	24.6	− 4.863	29.6	− 3.762	79.6	5.913	84.6	7.398	89.6	9.353
19.7	− 6.101	24.7	− 4.840	29.7	− 3.741	79.7	5.940	84.7	7.432	89.7	9.400
19.8	− 6.074	24.8	− 4.817	29.8	− 3.720	79.8	5.967	84.8	7.466	89.8	9.447
19.9	− 6.047	24.9	− 4.793	29.9	− 3.699	79.9	5.994	84.9	7.499	89.9	9.494
20.0	− 6.020	25.0	− 4.770	75.0	4.771	80.0	6.021	85.0	7.533	90.0	9.542
20.1	− 5.993	25.1	− 4.747	75.1	4.794	80.1	6.048	85.1	7.567	90.1	9.591
20.2	− 5.966	25.2	− 4.724	75.2	4.818	80.2	6.075	85.2	7.602	90.2	9.640
20.3	− 5.939	25.3	− 4.701	75.3	4.841	80.3	6.102	85.3	7.636	90.3	9.689
20.4	− 5.912	25.4	− 4.678	75.4	4.864	80.4	6.130	85.4	7.671	90.4	9.739
20.5	− 5.885	25.5	− 4.655	75.5	4.888	80.5	6.158	85.5	7.706	90.5	9.789
20.6	− 5.859	25.6	− 4.632	75.6	4.911	80.6	6.185	85.6	7.741	90.6	9.840
20.7	− 5.832	25.7	− 4.610	75.7	4.935	80.7	6.213	85.7	7.776	90.7	9.891
20.8	− 5.806	25.8	− 4.587	75.8	4.959	80.8	6.241	85.8	7.812	90.8	9.943
20.9	− 5.779	25.9	− 4.564	75.9	4.982	80.9	6.269	85.9	7.848	90.9	9.995
21.0	− 5.753	26.0	− 4.542	76.0	5.006	81.0	6.297	86.0	7.884	91.0	10.048
21.1	− 5.727	26.1	− 4.519	76.1	5.030	81.1	6.326	86.1	7.920	91.1	10.111
21.2	− 5.701	26.2	− 4.497	76.2	5.054	81.2	6.354	86.2	7.956	91.2	10.155
21.3	− 5.675	26.3	− 4.474	76.3	5.078	81.3	6.382	86.3	7.993	91.3	10.210
21.4	− 5.649	26.4	− 4.452	76.4	5.102	81.4	6.411	86.4	8.080	91.4	10.264
21.5	− 5.623	26.5	− 4.429	76.5	5.126	81.5	6.440	86.5	8.067	91.5	10.320
21.6	− 5.598	26.6	− 4.407	76.6	5.150	81.6	6.469	86.6	8.104	91.6	10.376
21.7	− 5.572	26.7	− 4.385	76.7	5.174	81.7	6.498	86.7	8.142	91.7	10.433
21.8	− 5.547	26.8	− 4.363	76.8	5.199	81.8	6.527	86.8	8.179	91.8	10.490
21.9	− 5.521	26.9	− 4.341	76.9	5.223	81.9	6.556	86.9	8.217	91.9	10.548
22.0	− 5.496	27.0	− 4.319	77.0	5.248	82.0	6.585	87.0	8.256	92.0	10.607
22.1	− 5.470	27.1	− 4.297	77.1	5.272	82.1	6.615	87.1	8.294	92.1	10.666
22.2	− 5.445	27.2	− 4.275	77.2	5.297	82.2	6.654	87.2	8.333	92.2	10.726
22.3	− 5.420	27.3	− 4.253	77.3	5.322	82.3	6.674	87.3	8.372	92.3	10.787
22.4	− 5.395	27.4	− 4.231	77.4	5.346	82.4	6.704	87.4	8.411	92.4	10.840
22.5	− 5.370	27.5	− 4.209	77.5	5.371	82.5	6.734	87.5	8.451	92.5	10.911
22.6	− 5.345	27.6	− 4.187	77.6	5.396	82.6	6.764	87.6	8.491	92.6	10.974
22.7	− 5.321	27.7	− 4.166	77.7	5.421	82.7	6.795	87.7	8.531	92.7	11.038
22.8	− 5.296	27.8	− 4.144	77.8	5.446	82.8	6.825	87.8	8.571	92.8	11.102
22.9	− 5.271	27.9	− 4.122	77.9	5.471	82.9	6.856	87.9	8.612	92.9	11.168

p (%)	dB	p (%)	dB	p (%)	dB	p (%)	dB	p (%)	dB	p (%)	dB
93.0	11.234	94.2	12.106	95.4	13.168	96.6	14.535	97.8	16.479	99.0	19.956
93.1	11.301	94.3	12.186	95.5	13.268	96.7	14.669	97.9	16.686	99.1	20.418
93.2	11.369	94.4	12.268	95.6	13.370	96.8	14.807	98.0	16.902	99.2	20.934
93.3	11.438	94.5	12.351	95.7	13.474	96.9	14.950	98.1	17.129	99.3	21.519
93.4	11.508	94.6	12.435	95.8	13.581	97.0	15.097	98.2	17.368	99.4	22.192
93.5	11.579	94.7	12.521	95.9	13.690	97.1	15.248	98.3	17.621	99.5	22.989
93.6	11.651	94.8	12.608	96.0	13.802	97.2	15.405	98.4	17.889	99.6	23.962
93.7	11.724	94.9	12.697	96.1	13.917	97.3	15.567	98.5	18.173	99.7	25.216
93.8	11.798	95.0	12.783	96.2	14.034	97.4	15.736	98.6	18.447	99.8	26.981
93.9	11.873	95.1	12.880	96.3	14.154	97.5	15.911	98.7	18.804	99.9	29.996
94.0	11.950	95.2	12.974	96.4	14.278	97.6	16.092	98.8	19.156	100.0	∞
94.1	12.027	95.3	13.070	96.5	14.405	97.7	16.282	98.9	19.538		

$$\Omega \ (\text{dB}) = 10 \log \left[\frac{p/100}{1 - p/100} \right].$$

Note: The table only contains values for p (%) from 0 to 29.9 and from 75 to 100. For the complete table, see Genichi Taguchi and Yu-In Wu, "Off-Line Quality Control" (1979), Central Japan Quality Control Association, Nagaya, pp. 99–102.

Appendix H

Percentage Points of the *t* Distribution

ν \ α	0.40	0.25	0.10	0.05	0.025	0.01	0.005	0.0025	0.001	0.0005
1	0.325	1.000	3.078	6.314	12.706	31.821	63.657	127.32	318.31	636.62
2	0.289	0.816	1.886	2.920	4.303	6.965	9.925	14.089	23.326	31.598
3	0.277	0.765	1.638	2.353	3.182	4.541	5.841	7.453	10.213	12.924
4	0.271	0.741	1.533	2.132	2.776	3.747	4.604	5.598	7.173	8.610
5	0.267	0.727	1.476	2.015	2.571	3.365	4.032	4.773	5.893	6.869
6	0.265	0.718	1.440	1.943	2.447	3.143	3.707	4.317	5.208	5.959
7	0.263	0.711	1.415	1.895	2.365	2.998	3.499	4.029	4.785	5.408
8	0.262	0.706	1.397	1.860	2.306	2.896	3.355	3.833	4.501	5.041
9	0.261	0.703	1.383	1.833	2.262	2.821	3.250	3.690	4.297	4.781
10	0.260	0.700	1.372	1.812	2.228	2.764	3.169	3.581	4.144	4.587
11	0.260	0.697	1.363	1.796	2.201	2.718	3.106	3.497	4.025	4.437
12	0.259	0.695	1.356	1.782	2.179	2.681	3.055	3.428	3.930	4.318
13	0.259	0.694	1.350	1.771	2.160	2.650	3.012	3.372	2.852	4.221
14	0.258	0.692	1.345	1.761	2.145	2.624	2.977	3.326	2.787	4.140
15	0.258	0.691	1.341	1.753	2.131	2.602	2.947	3.286	3.733	4.073
16	0.258	0.690	1.337	1.746	2.120	2.583	2.921	3.252	3.686	4.015
17	0.257	0.689	1.333	1.740	2.110	2.567	2.898	3.222	3.646	3.965
18	0.257	0.688	1.330	1.734	2.101	2.552	2.878	3.197	3.610	3.922
19	0.257	0.688	1.328	1.729	2.093	2.539	2.861	3.174	3.579	3.883
20	0.257	0.687	1.325	1.725	2.086	2.528	2.845	3.153	3.552	3.850
21	0.257	0.686	1.323	1.721	2.080	2.518	2.831	3.135	3.527	3.819
22	0.256	0.686	1.321	1.717	2.074	2.508	2.819	3.119	3.505	3.792
23	0.256	0.685	1.319	1.714	2.069	2.500	2.807	3.104	3.485	3.767
24	0.256	0.685	1.318	1.711	2.064	2.492	2.797	3.091	3.467	3.745

ν \ α	0.40	0.25	0.10	0.05	0.025	0.01	0.005	0.0025	0.001	0.0005
25	0.256	0.684	1.316	1.708	2.060	2.485	2.787	3.078	3.450	3.725
26	0.256	0.684	1.315	1.706	2.056	2.479	2.779	3.067	3.435	3.707
27	0.256	0.684	1.314	1.703	2.052	2.473	2.771	3.057	3.421	3.690
28	0.256	0.683	1.313	1.701	2.048	2.467	2.763	3.047	3.408	3.674
29	0.256	0.683	1.311	1.699	2.045	2.462	2.756	3.038	3.396	3.659
30	0.256	0.683	1.310	1.697	2.042	2.457	2.750	3.030	3.385	3.646
40	0.255	0.681	1.303	1.684	2.021	2.423	2.704	2.971	3.307	3.551
60	0.254	0.679	1.296	1.671	2.000	2.390	2.660	2.915	3.232	3.460
120	0.254	0.677	1.289	1.658	1.980	2.358	2.617	2.860	3.160	3.373
∞	0.253	0.674	1.282	1.645	1.960	2.326	2.576	2.807	3.090	3.291

Source: This table is adapted from "Biometrika Tables for Statisticians", Vol. 1, 3rd ed., 1966, by permission of the Biometrika trustees.

Appendix I

Percentage Points of the χ² Distribution*

ν \ α	0.995	0.990	0.975	0.950	0.900	0.500	0.100	0.050	0.025	0.010	0.005
1	0.00+	0.00+	0.00+	0.00+	0.02	0.45	2.71	3.84	5.02	6.63	7.88
2	0.01	0.02	0.05	0.10	0.21	1.39	4.61	5.99	7.38	9.21	10.60
3	0.07	0.11	0.22	0.35	0.58	2.37	6.25	7.81	9.35	11.34	12.84
4	0.21	0.30	0.48	0.71	1.06	3.36	7.78	9.49	11.14	13.28	14.86
5	0.41	0.55	0.83	1.15	1.61	4.35	9.24	11.07	12.83	15.09	16.75
6	0.68	0.87	1.24	1.64	2.20	5.35	10.65	12.59	14.45	16.81	18.55
7	0.99	1.24	1.69	2.17	2.83	6.35	12.02	14.07	16.01	18.48	20.28
8	1.34	1.65	2.18	2.73	3.49	7.34	13.36	15.51	17.53	20.09	21.96
9	1.73	2.09	2.70	3.33	4.17	8.34	14.68	16.92	19.02	21.67	23.59
10	2.16	2.56	3.25	3.94	4.87	9.34	15.99	18.31	20.48	23.21	25.19
11	2.60	3.05	3.82	4.57	5.58	10.34	17.28	19.68	21.92	24.72	26.76
12	3.07	3.57	4.40	5.23	6.30	11.34	18.55	21.03	23.34	26.22	28.30
13	3.57	4.11	5.01	5.89	7.04	12.34	19.81	22.36	24.74	27.69	29.82
14	4.07	4.66	5.63	6.57	7.79	13.34	21.06	23.68	26.12	29.14	31.32
15	4.60	5.23	6.27	7.26	8.55	14.34	22.31	25.00	27.49	30.58	32.80
16	5.14	5.81	6.91	7.96	9.31	15.34	23.54	26.30	28.85	32.00	34.27
17	5.70	6.41	7.56	8.67	10.09	16.34	24.77	27.59	30.19	33.41	35.72
18	6.26	7.01	8.23	9.39	10.87	17.34	25.99	28.87	31.53	34.81	37.16
19	6.84	7.63	8.91	10.12	11.65	18.34	27.20	30.14	32.85	36.19	38.58
20	7.43	8.26	9.59	10.85	12.44	19.34	28.41	31.41	34.17	37.57	40.00
21	8.03	8.90	10.28	11.59	13.24	20.34	29.62	32.67	35.48	38.93	41.40
22	8.64	9.54	10.98	12.34	14.04	21.34	30.81	33.92	36.78	40.29	42.80
23	9.26	10.20	11.69	13.09	14.85	22.34	32.01	35.17	38.08	41.64	44.18
24	9.89	10.86	12.40	13.85	15.66	23.34	33.20	36.42	39.36	42.98	45.56
25	10.52	11.52	13.12	14.61	16.47	24.34	34.28	37.65	40.65	44.31	46.93

ν \ α	0.995	0.990	0.975	0.950	0.900	0.500	0.100	0.050	0.025	0.010	0.005
26	11.16	12.20	13.84	15.38	17.29	25.34	35.56	38.89	41.92	45.64	48.29
27	11.81	12.88	14.57	16.15	18.11	26.34	36.74	40.11	43.19	46.96	49.65
28	12.46	13.57	15.31	16.93	18.94	27.34	37.92	41.34	44.46	48.28	50.99
29	13.12	14.26	16.05	17.71	19.77	28.34	39.09	42.56	45.72	49.59	52.34
30	13.79	14.95	16.79	18.49	20.60	29.34	40.26	43.77	46.98	50.89	53.67
40	20.71	22.16	24.43	26.51	29.05	39.34	51.81	55.76	59.34	63.69	66.77
50	27.99	29.71	32.36	34.76	37.69	49.33	63.17	67.50	71.42	76.15	79.49
60	35.53	37.48	40.48	43.19	46.46	59.33	74.40	79.08	83.30	88.38	91.95
70	43.28	45.44	48.76	51.74	55.33	69.33	85.53	90.53	95.02	100.42	104.22
80	51.17	53.54	57.15	60.39	64.28	79.33	96.58	101.88	106.63	112.33	116.32
90	59.20	61.75	65.65	69.13	73.29	89.33	107.57	113.14	118.14	124.12	128.30
100	67.33	70.06	74.22	77.93	82.36	99.33	118.50	124.34	129.56	135.81	140.17

*ν = degrees of freedom.

Appendix J

Cumulative Poisson Distribution*

				$\alpha = \lambda t$				
x	0.01	0.05	0.10	0.20	0.30	0.40	0.50	0.60
0	0.990	0.951	0.904	0.818	0.740	0.670	0.606	0.548
1	0.999	0.998	0.995	0.982	0.963	0.938	0.909	0.878
2		0.999	0.999	0.998	0.996	0.992	0.985	0.976
3				0.999	0.999	0.999	0.998	0.996
4						0.999	0.999	0.999
5							0.999	0.999

				$\alpha = \lambda t$				
x	0.70	0.80	0.90	1.00	1.10	1.20	1.30	1.40
0	0.496	0.449	0.406	0.367	0.332	0.301	0.272	0.246
1	0.844	0.808	0.772	0.735	0.699	0.662	0.626	0.591
2	0.965	0.952	0.937	0.919	0.900	0.879	0.857	0.833
3	0.994	0.990	0.986	0.981	0.974	0.966	0.956	0.946
4	0.999	0.998	0.997	0.996	0.994	0.992	0.989	0.985
5	0.999	0.999	0.999	0.999	0.999	0.998	0.997	0.996
6		0.999	0.999	0.999	0.999	0.999	0.999	0.999
7				0.999	0.999	0.999	0.999	0.999
8							0.999	0.999

				$\alpha = \lambda t$				
x	1.50	1.60	1.70	1.80	1.90	2.00	2.10	2.20
0	0.223	0.201	0.182	0.165	0.149	0.135	0.122	0.110
1	0.557	0.524	0.493	0.462	0.433	0.406	0.379	0.354
2	0.808	0.783	0.757	0.730	0.703	0.676	0.649	0.622
3	0.934	0.921	0.906	0.891	0.874	0.857	0.838	0.819
4	0.981	0.976	0.970	0.963	0.955	0.947	0.937	0.927

$$\alpha = \lambda t$$

x	1.50	1.60	1.70	1.80	1.90	2.00	2.10	2.20
5	0.995	0.993	0.992	0.989	0.986	0.983	0.979	0.975
6	0.999	0.998	0.998	0.997	0.996	0.995	0.994	0.992
7	0.999	0.999	0.999	0.999	0.999	0.998	0.998	0.998
8	0.999	0.999	0.999	0.999	0.999	0.999	0.999	0.999
9			0.999	0.999	0.999	0.999	0.999	0.999
10							0.999	0.999

$$\alpha = \lambda t$$

x	2.30	2.40	2.50	2.60	2.70	2.80	2.90	3.00
0	0.100	0.090	0.082	0.074	0.067	0.060	0.055	0.049
1	0.330	0.308	0.287	0.267	0.248	0.231	0.214	0.199
2	0.596	0.569	0.543	0.518	0.493	0.469	0.445	0.423
3	0.799	0.778	0.757	0.736	0.714	0.691	0.669	0.647
4	0.916	0.904	0.891	0.877	0.862	0.847	0.831	0.815
5	0.970	0.964	0.957	0.950	0.943	0.934	0.925	0.916
6	0.990	0.988	0.985	0.982	0.979	0.975	0.971	0.966
7	0.997	0.996	0.995	0.994	0.993	0.991	0.990	0.988
8	0.999	0.999	0.998	0.998	0.998	0.997	0.996	0.996
9	0.999	0.999	0.999	0.999	0.999	0.999	0.999	0.998
10	0.999	0.999	0.999	0.999	0.999	0.999	0.999	0.999
11			0.999	0.999	0.999	0.999	0.999	0.999
12							0.999	0.999

$$\alpha = \lambda t$$

x	3.50	4.00	4.50	5.00	5.50	6.00	6.50	7.00
0	0.030	0.018	0.011	0.006	0.004	0.002	0.001	0.000
1	0.135	0.091	0.061	0.040	0.026	0.017	0.011	0.007
2	0.320	0.238	0.173	0.124	0.088	0.061	0.043	0.029
3	0.536	0.433	0.342	0.265	0.201	0.151	0.111	0.081
4	0.725	0.628	0.532	0.440	0.357	0.285	0.223	0.172
5	0.857	0.785	0.702	0.615	0.528	0.445	0.369	0.300
6	0.934	0.889	0.831	0.762	0.686	0.606	0.526	0.449
7	0.973	0.948	0.913	0.866	0.809	0.743	0.672	0.598
8	0.990	0.978	0.959	0.931	0.894	0.847	0.791	0.729
9	0.996	0.991	0.982	0.968	0.946	0.916	0.877	0.830
10	0.998	0.997	0.993	0.986	0.974	0.957	0.933	0.901
11	0.999	0.999	0.997	0.994	0.989	0.979	0.966	0.946
12	0.999	0.999	0.999	0.997	0.995	0.991	0.983	0.973
13	0.999	0.999	0.999	0.999	0.998	0.996	0.992	0.987
14		0.999	0.999	0.999	0.999	0.998	0.997	0.994
15			0.999	0.999	0.999	0.999	0.998	0.997
16				0.999	0.999	0.999	0.999	0.999
17					0.999	0.999	0.999	0.999
18						0.999	0.999	0.999
19							0.999	0.999
20								0.999

$$\alpha = \lambda t$$

x	7.50	8.00	8.50	9.00	9.50	10.0	15.0	20.0
0	0.000	0.000	0.000	0.000	0.000	0.000	0.000	0.000
1	0.004	0.003	0.001	0.001	0.000	0.000	0.000	0.000
2	0.020	0.013	0.009	0.006	0.004	0.002	0.000	0.000
3	0.059	0.042	0.030	0.021	0.014	0.010	0.000	0.000
4	0.132	0.099	0.074	0.054	0.040	0.029	0.000	0.000
5	0.241	0.191	0.149	0.115	0.088	0.067	0.002	0.000
6	0.378	0.313	0.256	0.206	0.164	0.130	0.007	0.000
7	0.524	0.452	0.385	0.323	0.268	0.220	0.018	0.000
8	0.661	0.592	0.523	0.455	0.391	0.332	0.037	0.002
9	0.776	0.716	0.652	0.587	0.521	0.457	0.069	0.005
10	0.862	0.815	0.763	0.705	0.645	0.583	0.118	0.010
11	0.920	0.888	0.848	0.803	0.751	0.696	0.184	0.021
12	0.957	0.936	0.909	0.875	0.836	0.791	0.267	0.039
13	0.978	0.965	0.948	0.926	0.898	0.864	0.363	0.066
14	0.989	0.982	0.972	0.958	0.940	0.916	0.465	0.104
15	0.995	0.991	0.986	0.977	0.966	0.951	0.568	0.156
16	0.998	0.996	0.993	0.988	0.982	0.972	0.664	0.221
17	0.999	0.998	0.997	0.994	0.991	0.985	0.748	0.297
18	0.999	0.999	0.998	0.997	0.995	0.992	0.819	0.381
19	0.999	0.999	0.999	0.998	0.998	0.996	0.875	0.470
20	0.999	0.999	0.999	0.999	0.999	0.998	0.917	0.559
21	0.999	0.999	0.999	0.999	0.999	0.999	0.946	0.643
22		0.999	0.999	0.999	0.999	0.999	0.967	0.720
23			0.999	0.999	0.999	0.999	0.980	0.787
24					0.999	0.999	0.988	0.843
25						0.999	0.993	0.887
26							0.996	0.922
27							0.998	0.947
28							0.999	0.965
29							0.999	0.978
30							0.999	0.986
31							0.999	0.991
32							0.999	0.995
33							0.999	0.997
34								0.998

*Entries in the table are values of $F(x) = P(C \le x) = \sum_{c=0}^{x} e_{\alpha}^{-\alpha c}/c!$. Blank spaces below the last entry in any column may be read as 1.0; blank spaces above the first entry in any column may be read as 0.0.

Appendix K

Supporting Theory for the CCC Chart and the Process Rejection Sampling Plan

If the probability of an unacceptable item is p, then the probability of an acceptable item is $(1-p)$, of two acceptable items is $(1-p)^2$, of three acceptable items is $(1-p)^3$, and the n acceptable items is $(1-p)^n$. If a lot is accepted when a sampling of n acceptable items is randomly selected, then the probability of acceptance is

$$Pa = (1-p)^n$$

and the probability of lot rejection is

$$Pr = 1 - (1-p)^n \cong np \qquad \text{for } np < 5.$$

The probability of an unacceptable item on the nth item is

$$g(n) = (1-p)^{n-1} p$$

and is called the geometric distribution. The probability of an unacceptable item by the nth item is

$$G(n) = \sum_{1}^{n} (1-p)^{n-1} p = 1 - (1-p)^n$$

The average number of items to an unacceptable item is

$$\bar{n} = \frac{1}{p}$$

When n is very large, as expected under a zero defects program,

$$G(n) \cong 1 - e^{-n/\bar{n}}$$

and rearranging,

$$n \cong -\bar{n} \ln(1 - G(n))$$

Now the upper and lower limits can be calculated about the average $\bar{n}$. For $G(n) = 0.05$, 0.5, 0.95, $n = 0.05\bar{n}$, $0.693\bar{n}$, $3\bar{n}$, respectively. Thus, the median is about $0.7\bar{n}$ and the upper and lower 5 percentage limits are $0.05\bar{n}$ and $3\bar{n}$. If α reflects the fraction below the lower limit, then

$$\text{LCL} = \alpha\bar{n} \qquad \text{with median} = 0.7\bar{n}$$

In constructing a control chart, these two values are of primary concern. If the number of acceptable items between unacceptable items is less than $\alpha\bar{n}$, the process average is different from $\bar{n}$ at the α significance level. If five or more unacceptable items occur before a cumulative count of $0.7\bar{n}$ is reached, then the process average is significantly poorer than $\bar{n}$.

Applying the cumulative count strategy to lots rather than items, with Pr replacing p in the geometric distribution, the probability of lot rejection by the kth lot is

$$G(k) = 1 - (1 - \text{Pr})^k = 1 - \text{Pa}^k$$

$$= 1 - (1 - p)^{nk} \cong npk \qquad \text{for } np < 0.1$$

Thus, given a rejected lot, the probability of process rejection due to a second rejected lot by the kth lot is

$$\text{PR} = npk$$

Since the probability of a rejected lot is $\text{Pr} = np$, the overall probability of process rejection is

$$\text{PR}' = \text{PR}\,\text{Pr} = n^2 p^2 k$$

which becomes smaller and smaller as zero defects is approached.

Appendix **L**

Values of e^{-x}

x	e^{-x}	x	e^{-x}	x	e^{-x}	x	e^{-x}
0.0	1.000	2.5	0.082	5.0	0.0067	7.5	0.00055
0.1	0.905	2.6	0.074	5.1	0.0061	7.6	0.00050
0.2	0.819	2.7	0.067	5.2	0.0055	7.7	0.00045
0.3	0.741	2.8	0.061	5.3	0.0050	7.8	0.00041
0.4	0.670	2.9	0.055	5.4	0.0045	7.9	0.00037
0.5	0.607	3.0	0.050	5.5	0.0041	8.0	0.00034
0.6	0.549	3.1	0.045	5.6	0.0037	8.1	0.00030
0.7	0.497	3.2	0.041	5.7	0.0033	8.2	0.00028
0.8	0.449	3.3	0.037	5.8	0.0030	8.3	0.00025
0.9	0.407	3.4	0.033	5.9	0.0027	8.4	0.00023
1.0	0.368	3.5	0.030	6.0	0.0025	8.5	0.00020
1.1	0.333	3.6	0.027	6.1	0.0022	8.6	0.00018
1.2	0.301	3.7	0.025	6.2	0.0020	8.7	0.00017
1.3	0.273	3.8	0.022	6.3	0.0018	8.8	0.00015
1.4	0.247	3.9	0.020	6.4	0.0017	8.9	0.00014
1.5	0.223	4.0	0.018	6.5	0.0015	9.0	0.00012
1.6	0.202	4.1	0.017	6.6	0.0014	9.1	0.00011
1.7	0.183	4.2	0.015	6.7	0.0012	9.2	0.00010
1.8	0.165	4.3	0.014	6.8	0.0011	9.3	0.00009
1.9	0.150	4.4	0.012	6.9	0.0010	9.4	0.00008
2.0	0.135	4.5	0.011	7.0	0.0009	9.5	0.00008
2.1	0.122	4.6	0.010	7.1	0.0008	9.6	0.00007
2.2	0.111	4.7	0.009	7.2	0.0007	9.7	0.00006
2.3	0.100	4.8	0.008	7.3	0.0007	9.8	0.00006
2.4	0.091	4.9	0.007	7.4	0.0006	9.9	0.00005

Reprinted with permission from "Elementary Business Statistics: The Modern Approach" by John E. Freund and Frank J. Williams (1982), Prentice-Hall, Inc., p. 569.

Appendix M(a)

One-Sided Statistical Tolerance Interval

	Values of the Coefficient k					
	$\gamma = 0.95$			$\gamma = 0.99$		
n	$p = 0.90$	$p = 0.95$	$p = 0.99$	$p = 0.90$	$p = 0.95$	$p = 0.99$
5	3.41	4.21	5.75			
6	3.01	3.71	5.07	4.41	5.41	7.33
7	2.76	3.40	4.64	3.86	4.73	6.41
8	2.58	3.19	4.36	3.50	4.29	5.81
9	2.45	3.03	4.14	3.24	3.97	5.39
10	2.36	2.91	3.98	3.05	3.74	5.08
11	2.28	2.82	3.85	2.90	3.56	4.83
12	2.21	2.74	3.75	2.77	3.41	4.63
13	2.16	2.67	3.66	2.68	3.29	4.47
14	2.11	2.61	3.59	2.59	3.19	4.34
15	2.07	2.57	3.52	2.52	3.10	4.22
16	2.03	2.52	3.46	2.46	3.03	4.12
17	2.00	2.49	3.41	2.41	2.96	4.04
18	1.97	2.45	3.37	2.36	2.91	3.96
19	1.95	2.42	3.33	2.32	2.86	3.89
20	1.93	2.40	3.30	2.28	2.81	3.83
22	1.89	2.35	3.23	2.21	2.73	3.73
24	1.85	2.31	3.18	2.15	2.66	3.64
26	1.82	2.27	3.13	2.10	2.60	3.56
28	1.80	2.24	3.09	2.06	2.35	3.50
30	1.78	2.22	3.06	2.03	2.52	3.45
35	1.73	2.17	2.99	1.96	2.43	3.33
40	1.70	2.13	2.94	1.90	2.37	3.25
45	1.67	2.09	2.90	1.86	2.31	3.18
50	1.65	2.07	2.86	1.82	2.27	3.12

n	$\gamma=0.95$			$\gamma=0.99$		
	$p=0.90$	$p=0.95$	$p=0.99$	$p=0.90$	$p=0.95$	$p=0.99$
60	1.61	2.02	2.81	1.76	2.20	3.04
70	1.58	1.99	2.77	1.72	2.15	2.98
80	1.56	1.97	2.73	1.69	2.11	2.93
90	1.54	1.94	2.71	1.66	2.08	2.89
100	1.53	1.93	2.68	1.64	2.06	2.85
150	1.48	1.87	2.62	1.57	1.97	2.74
200	1.45	1.84	2.57	1.52	1.92	2.68
250	1.43	1.81	2.54	1.50	1.89	2.64
300	1.42	1.80	2.52	1.48	1.87	2.61
400	1.40	1.78	2.49	1.45	1.84	2.57
500	1.39	1.76	2.48	1.43	1.81	2.54
1000	1.35	1.73	2.43	1.38	1.76	2.47
∞	1.28	1.64	2.33	1.28	1.64	2.33

Reprinted with permission from ISO 3207-1975, Statistical Interpretation of Data —Determination of a Statistical Tolerance Interval. International Organization for Standardization, Geneva.

Appendix M(b)

Two-Sided Statistical Tolerance Interval

			Values of the Coefficient k			
	$\gamma = 0.95$			$\gamma = 0.99$		
n	$p = 0.90$	$p = 0.95$	$p = 0.99$	$p = 0.90$	$p = 0.95$	$p = 0.99$
5	4.28	5.08	6.63	6.61	7.86	10.26
6	3.71	4.41	5.78	5.34	6.35	8.30
7	3.37	4.01	5.25	4.61	5.49	7.19
8	3.14	3.73	4.89	4.15	4.94	6.47
9	2.97	3.53	4.63	3.82	4.55	5.97
10	2.84	3.38	4.43	3.58	4.27	5.59
11	2.74	3.26	4.28	3.40	4.05	5.31
12	2.66	3.16	4.15	3.25	3.87	5.08
13	2.59	3.08	4.04	3.13	3.73	4.89
14	2.53	3.01	3.96	3.03	3.61	4.74
15	2.48	2.95	3.88	2.95	3.51	4.61
16	2.44	2.90	3.81	2.87	3.41	4.49
17	2.40	2.86	3.75	2.81	3.35	4.39
18	2.37	2.82	3.70	2.75	3.28	4.31
19	2.34	2.78	3.66	2.70	3.22	4.23
20	2.31	2.75	3.62	2.66	3.17	4.16
22	2.26	2.70	3.54	2.58	3.08	4.04
24	2.23	2.65	3.48	2.52	3.00	3.95
26	2.19	2.61	3.43	2.47	2.94	3.87
28	2.16	2.58	3.39	2.43	2.89	3.79
30	2.14	2.55	3.35	2.39	2.84	3.73
35	2.09	2.49	3.27	2.31	2.75	3.61
40	2.05	2.45	3.21	2.25	2.68	3.52
45	2.02	2.41	3.17	2.20	2.62	3.44
50	2.00	2.38	3.13	2.16	2.58	3.39

n	$\gamma = 0.95$			$\gamma = 0.99$		
	$p = 0.90$	$p = 0.95$	$p = 0.99$	$p = 0.90$	$p = 0.95$	$p = 0.99$
60	1.95	2.33	3.07	2.10	2.51	3.29
70	1.93	2.30	3.02	2.06	2.45	3.23
80	1.91	2.27	2.99	2.03	2.41	3.17
90	1.89	2.25	2.96	2.00	2.38	3.13
100	1.87	2.23	2.93	1.98	2.36	3.10
150	1.83	2.18	2.86	1.91	2.27	2.98
200	1.80	2.14	2.82	1.87	2.22	2.92
250	1.78	2.12	2.79	1.84	2.19	2.88
300	1.77	2.11	2.77	1.82	2.17	2.85
400	1.75	2.08	2.74	1.73	2.14	2.81
500	1.74	2.07	2.72	1.78	2.12	2.78
1000	1.71	2.04	2.68	1.74	2.07	2.72
∞	1.64	1.96	2.58	1.64	1.96	2.58

Glossary

accuracy[1]   A qualitative term describing the degree of closeness with which the indications of an instrument approach the true value of the physical quantity, property, or condition that is measured. *Note:* For quantitative use, the term "uncertainty" replaces "accuracy."

analysis of variance (ANOVA)[1]   A technique by which the total variation of a set of data is subdivided into meaningful component parts each of which is associated with a specific source of variation for the purpose of testing some hypothesis on the parameters of the model or estimating variance components.

assignable cause (process)[2]   A factor that contributes to variation is feasible to detect and identify.

attribute[1]   A characteristic (e.g., go or no go) that is appraised in terms of whether it meets or does not meet a given requirement.

batch (or lot)[1]   An identified quantity of a commodity manufactured by one supplier under conditions of manufacture that are presumed to be uniform.

binomial distribution[1]   The distribution of the number of occurrences X of an event in n independent trials when the probability p of the events occurring in each of the trials is constant. This is expressed mathematically as follows:

$$P_{r(X-x)} = \frac{n!}{x!(n-x)!} p^x (1-p)^{n-x}$$

where $0 < p < 1$ and x is an integer.

calibration (analysis sense)[1]   The process used in determining the values of the errors of measuring instruments, physical standards, and, as necessary, other metrological such as influence quantities.

506

calibration (comparative sense)[1] The process of comparing or measuring an instrument of gauge of known accuracy, called the standard, with a second instrument or gauge for the purpose of obtaining, through observation and adjustment as necessary, correlation of indications between the two and thereby enabling a statement of accuracy to be made for the second measuring instrument or gauge.

central line[2] A line on a control chart representing the long-term average or a standard value of the statistical measure being plotted.

certainty[1] The degree of probability with which some quantity may be estimated.

chance causes (random causes)[2] Factors, generally numerous and individually of relatively small importance, which contribute to variation, but which are not feasible to detect or identify.

change in mean (EVOP) A measure of whether the conditions being run during the current phase yield an average result better or worse than the reference conditions. It is equal to the phase mean minus the mean at the reference conditions. It provides a measure of the direct cost incurred by obtaining information in any particular phase.

characteristic[2] A property of items in a sample or population which, when measured, counted, or otherwise observed, helps to distinguish between the items.

class boundaries[1] The extreme possible magnitudes or values that can occur in a class. *Note:* In statistics the class boundaries of 0.210 to 0.232 have the extreme possible magnitudes of 0.20950 to 0.23249 (to five decimal places), whereas in engineering usage the class boundaries of 0.210 to 0.232 have the extreme possible magnitudes of 0.21000 to 0.23200 (to five decimal places).

class interval[1] In a frequency distribution having equally wide classes, the difference between any two consecutive midvalues.

class limits[1] The values defining the upper and lower bounds of a class. *Note:* The limit belonging to the class should be specified.

component Any single identifiable part of a whole.

confidence interval[3] An interval that has a designated chance of including the universal value.

confidence level[1] The value of the probability associated with a confidence interval or a statistical tolerance interval.

confidence limit(s)[1] The end points of the confidence interval that is believed to include the population parameter with a specified degree of confidence.

conformance[2] An affirmative indication or judgment that a product or service has met the requirements of the relevant specifications, contract, or regulation; also the state of meeting the requirements.

conformity[1]   The fulfillment of specification requirements by a product or service.

contingency table[1]   A tabular form of presentation having data in rows and columns so as to display clearly the relations between various factors.

control[1]   The methods by which some desired result is ensured.

control chart[1]   A chart on which limits are drawn and on which are plotted values of any statistic(s) obtained from successive or sequential samples of products or services.

$\bar{x}$ chart ($\underline{x}$ bar) (average chart)[2]   A control chart in which the subgroup average x is used for evaluating the stability of the process level.

control charts for individual observations[2]   A control chart in which a single observation per sample is used for evaluating the stability of the process.

c chart (count chart)[2]   A control chart for evaluating the stability of the process in terms of the count of events of a given classification occurring in the sample.

u chart (count per unit chart)[2]   A control chart for evaluating the stability of the process in terms of the average count of events of a given classification per unit occurring within a sample.

cumulative count control chart (CCC chart)   A special graphical method suitable for the monitoring and control of production processes maintained at a very high quality level such as those associated with a high technology manufacturing environment. This technique uses the approach of plotting on a chart the cumulative count of conforming units instead of nonconforming units.

modified control chart (control chart with modified limit)[2]   A control chart for evaluating the process level in terms of subgroup average, $\bar{x}$, modifying the usual Shewhart control limits so as to relate to the product or service tolerance.

moving range control chart[2]   A control chart in which the range of the latest n observations is used for evaluating the stability of the variability within a process where the current observation has replaced the oldest of the previous n observations.

multivariate control chart[2]   A control chart for evaluating the stability of a process in terms of the levels of two or more variables or characteristics.

np chart (number of affected units chart)[2]   A control chart for evaluating the stability of the process in terms of the total number of units in a sample in which an event of a given classification occurs.

p chart (proportion or fraction chart)[2]   A control chart for evaluating the stability of the process in terms of the proportion, or fraction, of the total number of units in a sample in which an event of a given classification occurs.

***R* chart (range chart)**[2] A control chart in which the subgroup range *R* is used for evaluating the stability of the variability within a process.

***s* chart (sample standard deviation chart)**[2] A control chart in which the subgroup standard deviation is used for evaluating the stability of the variability within a process.

control chart factor[1] A factor, usually varying with sample size, to convert specified statistics or parameters into a central line value or control limit appropriate to the control chart.

control limits[2] Limits on a control chart that are used as criteria for signaling the need for action, or for judging whether a set of data does or does not indicate a "state of statistical control."

control system[1] The system of controls by which control of some desired result is achieved.

correction factor[3] An adjustment usually associated with the measurement of cases from an origin other than their mean.

corrective action[1] Steps taken to rectify conditions adverse to quality.

cumulative frequency[1] The total frequency of values less than or equal to a class boundary.

customer[5] The party to an exchange of ownership or control of some commodity who is the recipient of the commodity or control. In a contractual situation, the party who receives the benefit of the product or service provided under the contract.

customer feedback and corrective action[1] A program of feedback and corrective action based upon customer-supplied data and including activities that embrace the corrective action.

cycle (EVOP)[4] A single performance of a complete set of operating conditions.

cycle time[10] The elapsed time between the commencement and completion of a task.

data[1] Information stored or communicated in any form; specifically, statistics obtained from measurements and observations.

defect[6] A failure to meet a requirement imposed on a unit with respect to a single quality characteristic; also an irregularity in material, surface, finish, etc.

defective (defective unit)[2] A unit of product or service containing at least one defect or having several imperfections that in combination cause the unit to fail to satisfy intended normal or reasonably foreseeable usage requirements.

degrees of freedom[1] A whole number used for entering statistical tables of distribution. It is obtained by subtracting the number of independent parameters assessed from the sample size. For example, $s^2 = \Sigma(x - \bar{x})^2/(n-1)$ has $n-1$ degrees of freedom, since the mean is estimated from the sample.

dependent variable[1] A characteristic of a feature of an item the magnitude or value of which is directly influenced by the magnitude or value of some other characteristic.

design[1] The expression of ideas in drawn, written, or physical terms.

destructive testing[1] Tests that stress one or more of the characteristics of the product or process beyond the point of recovery.

deviation[1] Any departure from the specified range of a characteristic.

dispersion[1] The degree of scatter shown by observation of a characteristic.

distribution curve[1] The line enveloping a frequency distribution.

drift[1] The change with time of the metrological properties of a measuring instrument used in accordance with stated conditions that result in variation of the instrument's indication or output characteristic(s).

effect of a factor[7] The change in response produced by a change in the level of the factor.

error of measurement[1] The maximum possible discrepancy between the result of a measurement and the true value of the quantity measured.

errors[1] The components of uncertainty that determine the degree of departure from an exact or absolute accuracy in measurement.

estimated process average[1] The average percent defective or average number of defects per 100 items, whichever is applicable, of the product or services submitted for original inspection as assessed from a sample of the population.

evolutionary operation (EVOP)[7] An experimental procedure for collecting information to improve a process without disturbing production.

experimental design[1] The planning of experiments such that statistically valid data will be obtained and will be capable of yielding valid results by statistical analysis when factors are varied under controlled conditions.

factor (statistical sense)[1] A variable characteristic or condition that is likely to affect the yield of a single treatment or operation.

factorial experiment[1] An experiment designed to determine the presence or absence of interactions and the effects of one or more factors each of which is applied at a minimum of two levels. In a complete factorial experiment all combinations of all the levels of the factors are tested.

feedback loop[5] A systematic sequence for communicating information on process performance as an input to maintenance of process stability.

fraction defective[1] The total number of defective or nonconforming items divided by the total number of items.

frequency (statistics)[1] The number of occurrences of a given type of event or the number of members of a population falling into a specified class.

frequency distribution[1] The relation between the values of a characteristic and their frequencies or relative frequencies.

gauge (noun)[1]   A measuring device normally used to measure mechanical dimensions of a product.

histogram[1]   A graphical representation of a frequency distribution in which the interval for each class is used as the base and a rectangle whose area represents the frequency in the interval is constructed.

homogeneous[1]   Having similar characteristics uniformly distributed throughout. Sample data from a homogeneous population also are assumed to be homogeneous.

independent variable[1]   A variable characteristic value of an item whose magnitude is independent of the magnitude or value of other characteristics of the item.

inherent process variability[2]   The variability that is inherent in a process when operating in a state of statistical control.

inner noise[8]   Functionally related noise factors that affect variation within a product (wear, fade, hardening, etc.).

inspect[1]   The process of ascertaining the value of product or service characteristics or performance against a set of requirements or a standard.

inspection[2]   Activities, such as measuring, examining, testing, gauging of one or more characteristics of a product and/or service, and comparing these with specified requirements to determine conformity.

inspection 100%[2]   Inspection of all the units in a lot or batch. (*Also called* **screening inspection.**)

inspection by attributes[1]   An inspection in which the item is classified as either defective or nondefective (or conforming or nonconforming) or the number of defects or nonconformities with respect to a given requirement or set of requirements is counted.

inspection by variables[1]   Inspection in which certain quality characteristics of the item are evaluated with respect to a scale of measurements and expressed as precise points along that scale. *Note:* Inspection by variables yields the degree of conformance or nonconformance of the item.

inspection system[1]   The documented program, together with personnel, equipment, and associated facilities and services, by which inspection of some product or service is carried out.

interaction (feature)[1]   An interaction between two or more characteristics of an item, usually during production or operation, such that the value of one feature's characteristic is affected by a characteristic of another feature.

kurtosis coefficient[1]   The degree to which a distribution is flattened or peaked. A normal distribution has a kurtosis of $+3$.

levels (DOE)[8]   The settings of various factors in a factorial experiment; high and low values of pressure, temperature, etc.; may also have multiple levels of a factor.

limit[1] The maximum extent of the value of some required characteristic of a product or service which, if exceeded, will require some action to be taken.

loss function[8] A continuous cost function that measures the cost impact of the variability of a product.

lower control limit (LCL)[1] The lower of two control limits or sets of control limits as on a control chart.

main effect[8] The effect of a factor acting on its own from one level to another to change the outcome in a factorial experiment.

maintenance The combination of all technical and corresponding administrative actions carried out to retain an item at or restore it to a state in which it can perform its required function(s).

mean[1] The sum of a set of values divided by the number of such values in the set.

measurement[1] The process of determining the value of a measurement in terms of the appropriate unit of measurement; also the result of a measuring process.

measurement standard[1] A measuring instrument, material measure, or given set of elements of a measuring system that physically defines, embodies, represents, reproduces, or conserves a unit of measurement or value of a quantity. *Note:* The purpose of a standard is to transmit, by comparison, some unit of measurement or value of a quantity to other measuring devices, products, or services.

measuring system[1] The assembly of physical elements necessary to achieve the objectives of a measurement by the application of a measuring process in a given environment.

median[1] The value within a distribution above and below which an equal number of values lies. *Note:* If n values are arranged in increasing order of algebraic magnitude and numbered 1 to n, the median of the n values is the $[(n+1)/2]$th value if n is odd. If n is even, the median lies between the $(n/2)$th and the $[(n/2)+1]$th values and is not uniquely defined.

method of measurement[1] The nature of the procedure used in the measurement.

metrology[1] The field of knowledge concerned with measurement.

mode[1] The value within a distribution that has the greatest probability or frequency of occurrence.

nested experiment[7] An experiment in which the level of one factor is chosen within the levels of another factor.

noise factor[8] A factor that disturbs the function of a product or process; a factor that may be controlled during an experiment but in the customer's typical use may not be controlled by the manufacturer; a factor that the manufacturer wishes not to control.

nominal value[1]   A dimension, expressed in a product design specification or drawing, from which variations within tolerance limits are permitted.

nonconforming unit[2]   A unit of product or service containing at least one nonconformity.

nonconformity[2]   A departure of a quality characteristic from its intended level of state that occurs with a severity sufficient to cause an associated product or service not to meet a specification requirement.

normal distribution[1]   The distribution of a random continuous variable such that the variability is due to the summed effect of many random independent causes. When plotted, this distribution has a single mode from which the curve falls away symmetrically on two sides, making the mode, median, and mean all one value. This is expressed as

$$f(x) = \frac{1}{\sigma\sqrt{2\pi}} \exp\left[-\frac{1}{2} \frac{(x-\mu)^2}{\sigma^2} \right]$$

orthogonal matrix (array)[8]   A fractional matrix that assures a balanced, fair comparison of levels of any factor or interaction of factors; all columns can be evaluated independently of one another.

outer noise[8]   Environmentally related noise factors that affect variation within a product (temperature, humidity, operators, etc.).

out of control[9]   The condition describing a process from which all special causes of variation have not been eliminated. This condition is evident on a control chart by nonrandom patterns within the control limits.

p[2]   Used in the sense of a proportion or fraction. The ratio of the number of units, in which at least one event of a given classification occurs, to the total number of units sampled.

Pareto chart[9]   A simple tool for problem solving that involves ranking all potential problem areas or sources of variation according to their contribution to cost, variation or other measure. Typically a few causes account for most of the output; hence the phrase "vital few and the trivial many."

percent defective[1]   The fraction defective multiplied by 100.

percent nonconforming[1]   The fraction nonconforming multiplied by 100.

performance[1]   The carrying out of specified functional by an item, a system, a person, or an organization.

phase (EVOP)[4]   The repeated running through a cycle of operating conditions.

phase mean (EVOP)[4]   The mean response over all conditions being run in the present phase.

pilot line[1]   A production line set up to gain experience on production tooling with a new or changed process.

pilot lot[1] A small batch or lot run through production tooling and processes at the start of production of a new or changed design or manufacturing system to gain experience and data about the product and system.

Poisson distribution[1] The distribution of a discrete random variable x with mean m when its probability function is such that

$$f(x) = \frac{e^{-\lambda}(-\lambda)^{x}}{x!} \qquad \text{for } \lambda > 0$$

Note: Although the Poisson distribution can be derived from first principles and be shown to apply to random events occurring with a small probability over short intervals of time on a continuum, in statistical quality control it is often derived and used as an approximation to the binomial distribution.

pooling[8] In an analysis of variance the combining of the sum of squares and the degrees of freedom of those factors and/or error estimates that are statistically to obtain a better estimate of experimental error.

population (universe sense)[1] The set or aggregate of similar sets from which samples are selected for measurement and statistical assessment.

precision (calibration)[1] The six-sigma scatter of statistics obtained by using one measuring device and repeatedly measuring the same characteristic. The maximum extent of the data so obtained, within the six-sigma spread, is the precision of the measurement. *Note:* The bias of the instrument plus one-half of the six-sigma spread usually determines the accuracy of the measurement made by using that instrument to measure normally distributed data.

precision (statistics)[1] The scatter of the variate within the distribution of results obtained by applying the prescribed procedure several times under the same defined conditions.

pre-control[9] This semigraphical method is a sequenced acceptance sampling scheme for the continuous monitoring of a process. With this method the process performance levels are divided into acceptable, caution, and unacceptable zones. They normally do not require computations or plotting, making it easier to implement; however, this means that trend and performance information are not readily available.

probability[1] A real number in the range 0 to 1 attached to a random event to infer its likelihood of occurrence. *Note:* Probability can be related to the relative frequency of actual occurrence or to the belief that an event will occur.

probability density function[1] A function $f(x)$ of an approximately continuous variate such that, for a continuous variate, the expression $f(x)\,dx$ is equal to the probability with which the values of the variate will fall in the interval from x to $x + dx$.

probability distribution[1] A function that determines the probability that a random variable takes any given value or set of values.

process[1] One event or a succession of events wherein people, tools, materials and/or environment act in concert to perform operation(s) which cause one or more characteristics of the production material to be altered or generated.

process average (quality level sense)[1] The average value of process quality in terms of the percentage or proportion of variant units.

process capability[5] A standardized evaluation of the inherent ability of a process to perform under operating conditions; the performance of a process after significant causes of variation have been eliminated; in manufacturing, process capability usually is equated to six standard deviations of the variability.

process capability study[1] A controlled collection of statistics from a process for the purpose of statistically determining the capability of the process on specified materials under specified conditions.

process control[9] Using data gathered about a process to control the output. This may include the use of controls including SPC techniques and the establishment of a feedback loop to prevent the manufacture of nonconforming products.

process performance[9] The statistical measure of the two types of variation exhibited by a process, within subgroup and between subgroup. Performance is determined from a process study which is conducted over an extended period of time under normal operating conditions.

process quality[2] A statistical measure for the quality of product from a given process.

process spread[1] The total variability, arising from all causes, that exists in items produced by the process.

process tolerance[1] The tolerance allowed on the product from a process. *Note:* The process tolerance may be considerably tighter than the design tolerance in order to take into consideration such matters as remeasurement accuracies and the effects of further processing.

process under control[1] A process in which the various contributors to variability of the product are monitored and maintained within defined control limits.

producer's risk[1] For a given sampling plan, the probability of rejection of a batch whose defective or nonconforming proportion has a value stated by the plan.

product[5] A generic term for whatever is produced by a process, whether goods or services.

proportion or fraction[2] The ratio of the number of units, in which at least one event of a given classification occurs, to the total number of units sampled.

qualification (component sense)[1] The entire process by which products are obtained from a manufacturer or distributor and then examined or tested against written standards and documented or identified as qualified products. *Note:* The qualification process usually implies a need for continuing control of production quality.

quality[1]   The totality of features and characteristics of a product, process, or service that bears on its ability to satisfy stated or implied needs.

quality (measurement sense)[1]   The probability that the values of the characteristics of a product or service will lie within specified limits and impart to the product or service the ability to satisfy given needs.

quality audit[1]   A systematic and independent examination to determine whether quality activities and results comply with planned arrangements and whether these arrangements are effectively implemented and are suitable to achieve objectives.

quality characteristic[1]   Any aspect of an item that can be measured and that contributes in any way to the acceptability and/or functioning of the item.

quality control[1]   The operational techniques and activities that are used to satisfy quality requirements.

random (general sense)[1]   A term generally used to imply that the process under consideration is in some sense probabilistic or that a process of selection applied to a set of objects is said to be simply random if each object has an equal chance of being chosen.

random sampling[3]   As commonly used in acceptance sampling, the process of selecting sample units in such a manner that all combinations of n units under consideration have an equal chance of being selected as the sample. *Note:* Though equal probabilities of being selected are not necessary, it is essential that the probability of selection be ascertainable as sample units are normally removed from the population as they are chosen. Sampling tables commonly assume random sampling with equal probability.

range[9]   The difference between the smallest and largest values in a set of observations.

range chart[1]   The part of a quality control chart on which are plotted the values of the ranges of samples to provide a measure of the variability of the product and/or process.

rational subgroup[1]   One of the small groups within which it is believed that assignable causes are constant and into which observations can be subdivided in carrying out certain methods of statistical analysis. *Note:* In the operation of a control chart, a continuing test for significance of differences between a succession of rational subgroups and some reference value is implied.

rejects[1]   The items of product or service that are not accepted because they fail to meet the requirement criteria.

relative frequency[1]   The ratio of the number of times a particular value (or a value falling within a given class) is observed to the total number of observations.

repeatability[1]   A quantitative expression of the closeness of the agreement between the results of successive measurements of the same value of the same physical quantity, property, or condition carried out by the same method, by the same observer, with the same measuring instruments, at the same location at appropriately short intervals of time.

replication[1]   The performance of an experiment or parts of an experiment more than once. Each performance, including the first one, is called a replicate.

reproducibility[9]   The variation in the average of the measurements made by different operators using the same gauge while measuring the identical characteristic on the same parts.

response[8]   The result obtained when an experiment is run at particular conditions.

rework[1]   Any process whereby defective material is altered in an effort to make it acceptable; also the act of reprocessing.

risk, consumer's (β)[2]   For a given sampling plan, the probability of acceptance of a lot the quality of which has a designated numerical value representing a level that it is seldom desired to accept. *Note:* Usually the designated value will be the limiting quality level (LQL).

risk, producer's (α)[2]   For a given sampling plan, the probability of not accepting a lot the quality of which has a designated numerical value representing a level that it is generally desired to accept. *Note:* Usually the designated value will be the acceptable quality level (AQL).

run[1]   A quantity of product produced in a continuing sequence of operations through a production system within one (usually short) period of time.

run chart[9]   A graph of a characteristic versus sampling sequence used to detect trends.

sample (acceptance sampling sense)[1]   One or more units of product, or a quantity of material, drawn from a specific lot or process for purposes of inspection to provide information that may be used as a basis for making a decision concerning acceptance of that lot or process.

sample size[2]   The number of units in a sample or the number of observations in a sample.

sampling[1]   An arrangement for taking samples; usually it is qualified by a description of the type of sampling.

sampling frequency[1]   In a sequential or continuous sampling plan, the ratio of the number of units of product randomly selected for inspection at an inspection station to the number of units of product moving past the inspection station.

sampling interval[2]   In systematic sampling, the fixed interval of time, output, running hours, etc. between samples.

sampling plan[1]   A plan according to which one or more samples are taken from a batch or population.

sampling procedure[1]   The method by which a sampling plan works.

scatter[1]   The extremities of the location of sample points in a sample space or a distribution.

scatter diagram[9]   A plot of the variables, one against the other, that displays their relationship.

scrap[1]   Nonconforming material, components, or equipment that is not usable and cannot be economically reworked or repaired.

signal-to-noise ratio[8]   A measure of the amount of observed variation present relative to the observed average of the data.

significance test[1]   A statistical procedure to assess whether some quantity that is subject to a random variation differs from a postulated figure by an amount greater than that attributable to random variation alone. Compare with **level of significance.**

skewed distribution[1]   A distribution the curve of which is asymmetric. Positive skewness means greater scatter to the right of the mean; negative skewness means greater scatter to the left.

specification[1]   The document that describes in detail the requirements with which the product, process, or service has to conform.

stability[1]   The tendency of a measuring instrument or gauge to be free of change, i.e., to retain its accuracy for a long time.

standard[1]   Any reference document, item, or quantity of material against which unknown characteristics are assigned value.

standard deviation[9]   A measure of the dispersion of a set of values about its average value. When standard deviation is denoted by "s," it represents the sample standard deviation. When denoted by "σ" it represents the population or universe standard deviation.

standard error[3]   The standard deviation of a sampling distribution.

standardization[1]   The reduction of the number of characteristics or features of a system or the reduction of the number of ways these may vary or interact.

statistic[2]   A quantity calculated from a sample of observations, most often to form an estimate of some population parameter.

statistical control[9]   The condition describing a process from which all assignable causes of variation have been eliminated and only chance causes remain, evidenced on a control chart by the absence of points beyond the control limits and by the absence of nonrandom patterns or trends within the control limits.

statistical process control[9]   The use of statistical tools such as histograms, control charts, and other variation analysis techniques to analyze a process or its output so as to take appropriate action to achieve and maintain a state of statistical control.

statistical tolerance limits[1]   A set of limits calculated from the results of sample observations and between which, under given assumptions, a stated fraction of the population will lie with a given probability.

stratification[1] The physical or conceptual division of a population into separate parts called strata.

subgroup[1] A section of a set of elements, individuals, or observations having one or more characteristics in common.

supplier[5] The party that provides inputs to a process.

system[1] A group of items or quantity of material having dependent and independent elements which act in concert to achieve a group function or group functions.

tally[1] A mark or score as in recording results by the conventional five-barred gate method, i.e., by marking every fifth score across the preceding four scores.

test[1] A critical trial or examination of one or more properties or characteristics of a material, product, service, process, set of observations, etc.

test procedure[1] A measurement instruction describing the method by which one or more quality characteristics is to be assessed and/or determined. Usually it involves one class of characteristics.

tolerance (specification sense)[2] The total allowable variation around a level or state (upper limit minus lower limit) or the maximum acceptable excursion of a characteristic.

tolerance limits[2] Limits that define the conformance boundaries for an individual unit of a manufacturing or service operation.

total process variability[2] The inherent process variability plus variations due to factors that have been allowed to change, such as operator errors, abnormal equipment adjustments, or wear, and material outside specifications, system errors (bias), or other assignable causes.

treatment combination[7] A given combination showing the levels of all factors to be run for that set of experimental conditions.

true process average[1] The actual mean percent defective or defects per 100 units of the population of original production material from a process for the given characteristic.

unit[2] An object on which a measurement or observation may be made.

upper control limit (UCL)[1] An upper limit to a range of values, usually on a control chart, such that values above the limit indicate a likely need for corrective action.

variable[1] A quantity that may take any one of a specified set or range of values.

variability (statistics)[1] The variation in the numerical magnitude of quantities.

variance (population)[1] A measure of dispersion of a finite population. The arithmetic mean of the squares of the deviations from the arithmetic mean of the population.

variance (sample)[1] The sum of the squares of the deviations from the sample mean divided by the degrees of freedom. The number of degrees of freedom of a random sample of size n is $n - 1$.

verification[2] The act of reviewing, inspecting, testing, checking, auditing, or otherwise establishing and documenting whether items, processes, services, or documents conform to specified requirements.

REFERENCES

1. James Robert Taylor, "Quality Control Systems" (1989), McGraw-Hill Book Company, pp. 423–492.
2. American National Standard ANSI/ASQC A1-A2 (1987), American Society for Quality Control.
3. Acheson J. Duncan, "Quality Control and Industrial Statistics" 4th ed. (1974), Richard D. Irwin, Inc., pp. 997–1003.
4. George E. P. Box and Norman R. Draper, "Evolutionary Operation" (1969), John Wiley & Sons, Inc., pp. 72–73 and p. 107.
5. J. M. Juran, " Juran on Planning for Quality" (1988), The Free Press, pp. 327–334.
6. Harvey C. Charbonneau and Gordon L. Webster, "Industry Quality Control" (1978), Prentice-Hall, Inc., p. 226.
7. Charles R. Hicks, "Fundamental Concepts in the Design of Experiments" 3rd ed. (1982), CBS College Publishing, pp. 373–377.
8. Phillip J. Ross, "Taguchi Techniques for Quality Engineering" (1988), McGraw-Hill Book Company, pp. 251–254.
9. ASQC, "Statistical Process Control Manual," Automotive Division, pp. 2-1–2-3.
10. Philip R. Thomas, "Competitiveness Through Total Cycle Time" (1990), McGraw-Hill Publishing Company, p. 155.

Index